INTRODUCTION TO LIVESTOCK AND POULTRY PRODUCTION: SCIENCE AND TECHNOLOGY

Including Aquatic, Companion, and Laboratory Animals

INTRODUCTION TO POULTRY SCIENCE AND

JASPER S. LEE

Agricultural Educator
Mississippi State, Mississippi

JIM HUTTER

Southwest Missouri State University
Springfield, Missouri

RICK RUDD

University of Florida
Gainesville, Florida

CHRIS EMBRY

Normal Community High School
Normal, Illinois

JODY POLLOK

Michigan State University
East Lansing, Michigan

LYLE E. WESTROM

University of Minnesota
Crookston, Minnesota

AUSTIN M. BULL

North Carolina A&T State University
Greensboro, North Carolina

LIVESTOCK AND PRODUCTION: TECHNOLOGY

AgriScience and Technology Series

Jasper S. Lee, Ph.D.
Series Editor

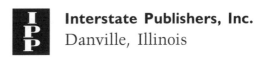

Interstate Publishers, Inc.
Danville, Illinois

INTRODUCTION TO LIVESTOCK AND POULTRY PRODUCTION: SCIENCE AND TECHNOLOGY

Library of Congress Catalog Card No. 95-77143

ISBN 0-8134-3050-X

1 2 3 4 5 6 7 8 9 10   03 02 01 99 98 97 96 95

Order from:
Interstate Publishers, Inc.
510 North Vermilion Street
P.O. Box 50
Danville, IL 61834-0050
Phone: (800) 843-4774
Fax: (217) 446-9706

Cover Photos:

Top row (from left): Quarter horse mare with colt (courtesy, American Quarter Horse Association; hybrid striped bass; and angus cow with calf.

Middle row (from left): sow nursing litter of pigs and Holstein dairy cows grazing.

Bottom row (from left): ostrich (courtesy, University of Arkansas at Pine Bluff); Suffolk sheep grazing; and hen and rooster with chicks (courtesy, Mississippi State University).

Preface

This is not a traditional book about livestock and poultry production! It is about the broad area of animal science in our daily lives.

Introduction to Livestock and Poultry Production: Science and Technology represents a new approach to education in animal science. It is about far more than traditional livestock and poultry enterprises. The book includes these, and much more. Further, a student- and teacher-friendly approach is used.

Traditional areas of animal agriculture are included. A base of animal industry with supporting science and technology is used to begin the book. The industry and science connection enhances the relationship of animal agriculture and biology. Understanding the biology helps with animal similarities and differences throughout the book. It also leads to higher abilities to solve problems in animal production.

The approach is to introduce a broad spectrum of animal science. Beef, swine, sheep, dairy, horse, and poultry production are presented. These are followed by aquatic, draft, companion, service, laboratory, and exotic animals. The content includes chapters and species that students with limited agriculture background can use and gain production skills.

The student-friendly approach is based on research with students who use books. A clear and consistent approach is used. Line art and photographs clarify and enhance important concepts. The writing level is appropriate for the intended audience. Type face and size have been selected to enhance interest and readability. The content is presented in a systematic manner.

As the first book in animal science for students, *Introduction to Livestock and Poultry Production: Science and Technology* is designed to develop a strong base for future study. The book is comfortably a part of the AgriScience and Technology (AST) Series produced by Interstate

Publishers, Inc. This series is designed to meet the needs of students and teachers as they prepare to go into the 21st century.

Explore this book for yourself. Read, look, and learn in an exciting way. You will be glad that you did!

Jasper S. Lee
Series Editor

Acknowledgements

The authors of **Introduction to Livestock and Poultry Production: Science and Technology** are indebted to many individuals who made contributions to the writing of this book. Their assistance helped the authors realize their goal of a new and innovative approach to educating students about the science and technology of animal production.

Important assistance was received from agricultural industry, associations, and government agencies. It is impossible to recognize all who made contributions with information, photographs, and in other areas. A few are acknowledged here and others are acknowledged throughout the book:

Harris Farms, Coalinga, California

Wagner Farms, Newton, Mississippi

Embrex Incorporated, Research Triangle Park, North Carolina

Gone Fishin' Exotic Pets, Starkville, Mississippi

Washington Department of Fisheries Salmon Hatchery, Wilapa, Washington

Research Farm, Virginia Tech, Blacksburg, Virginia

Maddox Dairy, Fresno, California

Aquaculture Program, South Bend High School, South Bend, Washington

FFA Chapter, South Panola High School, Batesville, Mississippi

Newton Vocational Center, Newton, Mississippi

BarLink Ranch, Madras, Oregon

Agriculture Department, Clovis High School, Clovis, California

Progressive Farmer Magazine, Birmingham, Alabama

American Breeders Service, DeForest, Wisconsin

Texas Department of Agriculture, Austin, Texas

The American Donkey and Mule Society, Denton, Texas

Agriculture Department and Bull Test Station, Hinds Community College, Raymond, Mississippi

Red Bluff Ranch, Thermopolis, Wyoming

Luck-E-G Nubians, Blanchard, Idaho

Vermont Department of Agriculture, Montpelier, Vermont

Lucky Three Ranch, Inc., Loveland, Colorado

Special acknowledgement is due individuals who gave of their time and expertise in reviewing the manuscript for technical accuracy. These individuals are: Robert C. Albin, Associate Dean, College of Agricultural Sciences and Natural Resources at Texas Tech University; Richard Joerger, Head of Agricultural Systems Technology and Education Department at Utah State University; and Tim Chamblee, Professor of Poultry Science at Mississippi State University.

The assistance of DeShannon Davis and Sharon Hunter at LEE AND ASSOCIATES is greatly appreciated. Their assistance with photographs and in manuscript work is acknowledged.

Each author has several individuals who provided considerable assistance, and these are hereby acknowledged.

The Acknowledgements would be incomplete without paying special appreciation to the staff of Interstate Publishers, Inc. These individuals worked diligently to edit and produce the best possible book. President Vernie Thomas is acknowledged for his enthusiasm for agricultural publishing and commitment to agricultural education. Marketing Vice President Dan Pentony is acknowledged for his help with photographs and support for the book. Kim Romine is acknowledged for her work in the design of the book. All others at Interstate are recognized for assistance in many ways. Thank you!

Contents

Part Four—Pleasure and Draft Animal Technology

Part Five—Service, Laboratory, and Exotic Animal Technology

Chapter 1

THE ANIMAL INDUSTRY

Animals are important to people. They have made the lives of people more enjoyable for hundreds of years. Whether for food, companionship, medicine, clothing, or work, animals continue to be important in our daily lives.

Animals and their products affect everyone. In fact, animal products are likely close by as you are reading this book! Look around the room. Is anyone wearing leather shoes or wool clothing? What is on the lunch menu for today? How many students in your class have a pet at home? All of these uses make animals important to us.

Livestock and poultry production is a large global industry. The industry ranges from the small farmer in rural China to the veterinarian in suburban America. Learning about the animal industry is exciting!

1-1. People enjoy raising animals. This sheep producer takes pride in her show lamb.

1

OBJECTIVES

To help understand the role of science and technology in the animal industry, this chapter has the following objectives:

1. Describe the animal industry
2. Explain the meaning of livestock and poultry production
3. Describe the role of science in livestock and poultry production
4. Discuss the use of animals in the United States and the world

TERMS

agriscience	animal supplies	natural selection
animal domestication	dairy cattle	nutrition
animal health	life science	pleasure animal
animal industry	livestock	poultry
animal marketing	meat animal	ration
animal production	meat animal by-product	science
animal services	mohair	technology

THE ANIMAL INDUSTRY

The *animal industry* is all of the activities involved in raising animals and meeting the needs of people for animal products. It includes many important steps that make modern production possible. The animal industry assures people of quality products. It uses procedures that provide proper care for animals.

LIVESTOCK

Livestock are animals produced on farms and ranches for food and other purposes. They are often given special care to assure good health and growth. Common livestock include cattle, swine (hogs), sheep, horses, and goats. Animals now being produced more widely include llamas, bison, and elk.

1-2. Beef cattle convert pasture into meat.

POULTRY

Poultry includes a number of fowl (birds) that are raised on farms primarily for use as food. Both meat and eggs are produced by poultry. Special kinds of feed, facilities, and care are used to keep poultry healthy and growing. Common kinds of poultry are chickens, turkeys, ducks, and geese. New species that are being farmed include quail, ostrich, and emu.

1-3. A modern poultry facility with chicks. Note the clean litter, heat source, and automatic feeders and waterers.

ANIMALS AND PRODUCTS

Most of us are well aware of common animal products. The hamburger, turkey sandwich, and fish filet are foods that we enjoy eating. Animals are used for more than food. They fill important roles as pets, in sports, and to do work.

In processing, very little of an animal is wasted. Many parts that were formerly wasted are now used for important purposes, such as making medicine from animal organs and other uses in human medicine. These materials are known as by-products. *Meat animal by-products* are products made from the parts of animals that are not used for food.

1-4. Hogs grow best in clean facilities.

Food

Animals provide many foods that people enjoy. These foods are also high in nutrients and help us live healthy lives. Without good food, people are not healthy and do not grow as they should.

Foods from animals primarily include meat, milk, and eggs. Some animals give us more than one kind of food product, such as chickens that provide meat and eggs.

We get meat from animals we raise and from wild animals. The animals that we raise for food include cattle, fish, turkeys, chickens, swine, and sheep. *Meat animals* are animals raised for their meat.

Wild animals used for food include deer, rabbit, quail, and fish. Wild animals are not included as livestock. They are known as game animals or wildlife. The meat from animals is known by different names, such as beef from cattle and pork from swine.

Milk is primarily from cattle. Cattle specially grown to produce milk are known as *dairy cattle*. Goats and a few other animals are sometimes milked.

Eggs are primarily from chickens. A few other species may produce eggs for human food, including guineas and ducks. In some instances, people enjoy fish eggs, known as caviar.

Table 1-1
Common Food Products People Get From Animals

Animal	Food Item Name*
Meat	
Cattle	
Younger than 3 months	Veal
Older animals	Beef
Swine (hogs)	Pork
Sheep	
Young (less than 1 year)	Lamb
Older (older than 1 year)	Mutton
Goat	Goat mutton
Chicken	
Young (less than 12 weeks)	Broiler
Neutered young male	Capon
Old hen	Hen
Turkey	Turkey
Fish	Fish
Milk	
Dairy cattle	Milk
Goats	Goat's milk
Eggs	
Chickens	Eggs
Fish	Caviar

*Variety meat is the general name for food made from organs and glands of different meat animals. Potted meat animals and vienna sausage are examples.

Clothing

Clothing is made from many different animal by-products. In addition, some animals are raised specifically for products to make clothing. These include mink for their fur and certain breeds of sheep raised primarily for wool.

1-5. Clothing is made from animal skins and fibers.

Clothing may be made from animal skin (hide) or hair that grows on the skin. Bones, antlers, and other animal parts may be used. The skin of animals is sometimes known as leather. Cattle, sheep, kangaroo, and snakes are a few animals whose skin is used in making belts, shoes, jackets, and other items. The hair or fleece from several different animals is used to make clothing, such as wool from sheep and fur from mink. A special quality clothing is made from **mohair**, a product of angora goats. Feathers from geese and other poultry may be used in clothing and for other purposes.

The most valuable animal product used in clothing is the pearl. Often used in jewelry, the pearl is formed by an oyster when sand or another foreign body gets inside its shell. Some oysters are grown to produce pearls, known as cultured pearls.

Pleasure

Pleasure animals are those kept as pets or to use in other ways to help people enjoy life. Examples include dogs, cats, and ornamental fish. Horses, ponies, and cattle are examples of animals kept for pets that are too big to stay inside a home. Some animals are used for sporting events, such as horse racing and goat roping. Many different kinds of pleasure

1-6. People enjoy having pets in their homes.

animals are found in North America. Some pleasure animals are known as companion animals.

Service

Some animals help people. Service animals are animals that assist people in living and work. They are used in many ways and may be given special training. Dogs may be used to lead visually impaired people, herd sheep, or guard property. Some animals,

1-7. A pet shop.

such as rats and monkeys, are carefully used in finding new medicines to cure human disease. And there are many others!

AREAS OF THE ANIMAL INDUSTRY

Getting good animal products into the forms we want involves a number of different activities. It is more than raising animals! Animal producers use supplies and services to help animals grow. Once food animals and their products have been produced, they must be made available to the consumer.

Production

Animal production is raising animals for food and other uses. How animals are produced has changed. Livestock are raised on farms or ranches, often on a large scale. Other animals may be raised in homes, aquaria, pens, and other small-scale facilities.

The nature of animal production has become larger and more specialized, but with fewer producers in North America. Fifty or more years ago, many people raised animals for food purposes. Many homes had a few chickens and a cow for milk. Nearly every farm was diversified and kept a few animals in addition to crops. Times have changed!

Today, livestock are raised on specialized farms or ranches that are often quite large. (Places where cattle, sheep, and horses are raised are often called ranches.) New approaches are used to grow them. Only one

1-8. The range area of North America is ideal for sheep. (Courtesy, National Lamb and Wool Grower, Englewood, Colorado)

kind of animal may be grown. It is unusual to find a large-scale farm that has more than one kind of livestock. Many corporations grow for a specific market. You will likely see some of these products in your local grocery store. An example is Tysons Foods, which produces poultry products.

Many small producers are still involved with animals. Service animals, such as draft horses, are often raised on small farms. Companion animals are raised by pet growers or individuals who raise them for a hobby. Animals used in laboratories are often carefully raised under close supervision so that standards are met.

Supplies and Services

Animal production requires many supplies and services. Some supplies are produced on the farm or ranch, such as hay for livestock feed. Most come from off the farm.

Animal supplies are the materials in animal production provided from sources off the farm. Supplies are made by feed mills and factories and provided by dealers or others to the farm. Common supplies include manufactured feed, animal medicines, fencing material, and equipment.

1-9. Feed mills make feed that provides a balanced ration for animals.

Animal services include the help that producers need to raise animals efficiently. Veterinarians help assure good animal health. Farriers are people skilled in putting shoes on horses. People skilled in shearing sheep (clipping off the wool) provide the service to wool producers. People in animal services usually go from one farm or ranch to another providing help.

Marketing

When animals or their products are ready for the consumer, they are marketed. *Animal marketing* is the movement of animals from the farm or ranch to the consumer. It includes preparing the animal for consumption. With pork, the swine are slaughtered and butchered into the desired pieces of meat. These cuts are delivered to your local grocery store.

1-10. Marketing involves packaging products for easy sale and use, such as eggs in a carton.

The products of animals go through a similar marketing process. Eggs and milk are major products in animal marketing.

All animal products are carefully packaged and stored to assure good quality. Meat, eggs, and milk are inspected. Products that are not wholesome are rejected and are not used for food.

SCIENCE AND TECHNOLOGY IN LIVESTOCK AND POULTRY PRODUCTION

Animals have not always been tame. Some are still wild today. Raising animals began hundreds of years ago. People found that it was easier to tend them than to hunt them in the wild. In recent years, science and technology have become very important in animal production.

DOMESTICATION

Animal domestication is taking animals from nature and raising them in a controlled environment. It involves taming wild animals and helping them adapt to being raised by humans.

Domestication is important in the history of how human societies developed. It began thousands of years ago with goats, donkeys, and a few other animals. People are striving to domesticate new species today, such as catfish and elk.

1-11. Domesticated animals are tame and friendly to humans.

Successfully producing animals requires education. The animal industry depends on people educated in many areas to produce the animals we need. Science and technology play significant roles in education about animal production.

AGRISCIENCE AND RELATED AREAS

Agriscience is the use of science in producing animals and other farm products, such as crops and forestry. Many areas of science are involved. Producers need education in basic areas of science. Many schools have classes in agriscience to help people learn about animal production.

Science is knowledge about the world we live in. It deals with facts and relationships among the facts. People who have an education in science are better animal producers. Animal science is the special application of basic science in the production of animals.

Areas of life science are very important in successful animal production. *Life science* is about living things, and includes zoology, which is the study of animals. The study includes the structure and functions of animals and their parts. It also includes studying how to keep them healthy by meeting their needs.

Another important life science that is indirectly related to animals is botany. Botany is the study of plants. Since most animals eat plants or plant products, knowing how to produce plants is essential.

Technology is the use of science in the work and lives of people. Producers of animals use technology about animals in their work. Every time a disease is treated or an animal fed, some form of technology is

1-12. Science and technology may involve using chemical processes. Note that safety mask, goggles, and gloves are being used to transfer a solution in a vent hood.

used. The technology could be in the design of equipment used, the way the treatment is given, or in other ways.

SCIENCE AND TECHNOLOGY

Science and technology are important in the production of animals. Information from many areas is used in the animal industry. The most widely used are in animal selection, animal nutrition, and animal health.

Animal Selection

Livestock and poultry production begins with animal selection. Animals and humans share many similarities. Animals are individuals, each with its own personality, body type, and rate of growth. To successfully raise animals, those that fill a need for production, service, or a product are selected.

Natural selection occurs when animals breed without control by humans. It is how nature ensures the survival of animal and plant species. Using science in animal and poultry production has resulted in many improvements.

Think for a moment about the meat we eat. Not too long ago, health conscious consumers began to demand meat products that were lower in fat. If producers had been limited to natural breeding practices in the beef industry, it would have taken many years to develop the leaner breeds of cattle we have today.

Animals are used to help in human medicine. An example is "Genie" the pig at Virginia Polytechnic Institute and State University (Virginia

1-13. Careful attention is being given in selecting this lamb.

Tech). Genie has a gene that causes her to produce human blood protein (protein C) in her milk. This protein is a blood anticoagulant and is often in short supply in hospitals. Producing this substance in an animal can help when there is a shortage of human blood products. It could reduce the cost of medical products and be the start of a new era in animal agriculture.

Animal Nutrition

A major role of animals is to convert animal feed into human food. Some animals, such as cattle and sheep, can use bulky feed materials for growth and maintenance of their bodies. Humans have no use for grass and hay as a raw material, but can consume the "hay" when it is converted to beef by a steer.

Nutrition is the kind of food an animal needs and how it is used by the animal. Much research has been done to learn the nutritional needs of animals. Several areas of science are involved in helping producers better provide for the needs of animals.

It would be difficult for a lamb to be very productive on a diet of only hay. There are many considerations that must be taken into account when preparing a *ration* for an animal. A ration is the animal's diet. It is what the animal consumes each day.

No matter what the potential of an animal, it cannot grow without a proper diet. Not so long ago, it was a common practice to feed hogs "slop." Slop consisted of a variety of ingredients, such as table scraps, garden waste, and even coal (some farmers believed that coal furnished a hog with needed minerals).

The hogs were able to maintain their bodies and grow, but at a very slow rate as compared to modern swine. One of the limiting factors for these animals was the lack of a nutritious diet. Science aided the industry by identifying the vitamin, mineral, and nutrient needs of livestock.

Raising animals efficiently is a key to success for animal producers. Nearly one-half of the farm receipts in the United States come from the sale of animals and animal products. Feed costs account for one-half to three-fourths of the expense in raising animals. It is easy to see that animal nutrition is high on the list of concerns in the livestock and poultry industry.

Science has helped the animal industry develop rations that meet the nutritional needs of animals. Research has also found ways of using feed ingredients that are more economical for producers. With the help of scientists in animal nutrition, animals can be given an affordable, nutritious diet.

1-14. Confined animals must receive adequate feed that meets their needs.

Animal Health

An animal must be healthy to live and grow. Unhealthy animals do not eat and may die. A sick animal costs money. The money spent for an animal is lost if it dies. Producers are concerned with the well-being of animals. They do not want animals to be mistreated. Successful producers take good care of animals.

Animal health is the condition in which the animal is free of disease and all parts of the animal are functioning properly. Many different diseases can attack animals if not prevented.

Veterinary medicine deals with the health and well-being of animals. Veterinarians are people trained in methods of preventing and treating disease and other animal health problems. They often have assistants and technicians who can provide valuable help.

Producers know that it is important to prevent disease. Once an animal gets a disease, it is often very difficult to successfully provide a treatment. The best approach is to keep animals healthy and avoid disease problems. Regular vaccination can be used to prevent some diseases. Having a good place for animals to live helps to keep them healthy.

1-15. Young people learn the responsibilities of raising animals with FFA and 4-H projects.

REVIEWING

MAIN IDEAS

The animal industry is a large and diverse field. Animal domestication has had a big impact on the lives of people around the world. We depend on animal products and services every day.

Science and technology are an integral part of modern animal production. Selecting and producing animals that will meet consumer needs involves using science. Providing animals with a balanced diet and adequate care to ensure animal health uses science and technology.

Animal production is the foundation of the livestock and poultry industry. Farm animal production is moving away from the small operator to commercial operations capable of raising large numbers of livestock. Others involved in animal production include pet shops, laboratory technicians, and private individuals.

The off-the-farm functions are vital to the success of the animal industry. These include the supplies and services used to produce animals as well as all of the marketing functions.

Animals are used primarily for their meat, by-products, pleasure, and companionship. Animals are used for power in a few places. Overall, animals make a huge contribution to the quality of human life.

QUESTIONS

Answer the following questions using correct spelling and complete sentences.

1. What is the animal industry?

2. What are four ways animals are used? Explain each.

3. What is animal domestication?

4. What are the three areas in the animal industry? Briefly explain each.

5. How have farms that produce animals changed in the last 50 years?

6. Briefly describe areas of science and technology in animal production.

EVALUATING

CHAPTER SELF-CHECK

Match the terms with the correct definitions. Write the letter by the term in the blank that is provided.

a. poultry
b. animal marketing
c. livestock

d. animal industry
e. animal supplies
f. agriscience

g. ration
h. meat animal by-product

1. _____ raising animals and meeting the needs of people for animal products.

2. _____ an animal's daily diet.

3. _____ materials used in animal production from off the farm.

4. _____ raising fowl on farms.

5. _____ movement of animals to consumers.

6. _____ not a major output from a meat animal, but a marketable commodity.

7. _____ use of science in producing animals and other farm products.

8. _____ animals produced on farms and ranches for food or other purposes.

EXPLORING

1. Take a field trip to a farm that produces animals. Determine the kinds of animals that are produced. Learn the major practices followed to produce the animals. What is involved in raising the animals? Prepare a written report on your findings. Give an oral report in class.

2. Investigate how animal production has changed in the last 20 or so years. Interview an animal producer who has been in business for a long time. Ask how the production practices have changed. Discuss probable reasons for farmers raising fewer livestock species and the shrinking numbers of animal producers. Prepare a written report on your findings. Give an oral report in class.

3. Investigate how animals and animal products are marketed in your local community. Tour a livestock auction, egg processing plant, pet store, or other facility. Interview the manager about the nature of the activities involved in marketing the animals. Summarize your findings in a written or oral report.

4. Assess your interests in animals. Indicate how you feel about the items below. Check the response that best represents your feelings about an activity with animals. Discuss your responses with your agriculture teacher. This should help you better understand your interest in animals. Areas that you are unsure about and don't like should be explored. You may find that you really like the activity!

	like	unsure	don't like
feed an animal	____	____	____
groom an animal	____	____	____
care for a sick animal	____	____	____
give a shot to prevent disease	____	____	____
haul an animal	____	____	____
care for a baby animal	____	____	____
train an animal	____	____	____
care for where animals live	____	____	____
exhibit an animal at a show	____	____	____
use science to study animals	____	____	____

Chapter 2

ENTREPRENEURSHIP IN THE ANIMAL INDUSTRY

Livestock and poultry production offers many opportunities for people. Some people start their own farm or ranch. Others work at jobs in the animal industry. People who like to work with animals often go into livestock and poultry production. What are your preferences?

People can be very successful in the animal industry. They can work for themselves and reap the benefits of their hard work. The animal industry is large. It accounts for nearly one-half of the money from agricultural sales. People need to get the right education and experience to be successful in the animal industry.

A good time to begin planning for a future in the animal industry is in high school. Take courses in agriculture, science, and business. Get on-the-job experience by working for other people. This will help you understand what is involved in having your own animal enterprise. You need a background that provides skills in understanding animals and the animal industry.

2-1. Cattle ranchers produce livestock for various markets.

17

OBJECTIVES

The animal industry is large and has many opportunities for entrepreneurs. The following objectives are included:

1. Explain consumers and the role of consumption in animal production

2. Describe the scope of livestock and poultry production in the United States

3. Explain entrepreneurship in livestock and poultry production and list available opportunities

4. Explain the meaning of animal well-being

TERMS

animal producer	consumer	market hog
animal well-being	consumption	part-time animal producer
beef	dairy product	per capita consumption
broiler	entrepreneur	risk

SCOPE OF LIVESTOCK AND POULTRY PRODUCTION

The animal industry is set up to meet the needs of people. It provides a supply of wholesome products in the forms that people want. And people want many products!

CONSUMERS ARE IMPORTANT

Consumers are people who eat, wear, and otherwise use goods and services. Consumers want products that satisfy needs. They use or consume products.

What consumers want is important. They often have strong preferences about what they will buy. They also want to get good value for their money. Every time a person goes to the store to buy food or other items they are consumers. They make decisions based on price, the money they have, and the desirability of the products.

Many Consumers

The world's population is nearly 5.6 billion people. It takes an extraordinary amount of food and clothing for them. The number of people will increase sharply in the future. Some forecasters say that the number will double by the year 2030. Of course, animal products are more important in some parts of the world than others.

In North America, people like animal products. Canada, Mexico, and the United States have a combined population of 444 million people. The United States has about 260 million people, with California, New York, Florida, and Texas being the states with the most population.

Consumption

Consumption is the use of animal products and other goods and services that have value. Every time you drink milk, consumption has occurred. Overall, Americans have high consumption of animal products.

2-2. Milk is a good source of nutrients.

Information is compiled on the amounts that people consume. This is often reported as the amount for one person. *Per capita consumption* is the average amount consumed per year by a person. It may be reported in weight or another measure.

A typical American eats over 180 pounds of meat per year. This includes 61.7 pounds of beef, 48.7 pounds of pork, 47.2 pounds of chicken, and over 30 pounds of other meats, like turkey, veal, lamb, and fish and shellfish. It does not include game, such as deer and rabbit.

Per capita egg consumption is 235 eggs in the United States. Some of these are served as egg foods and others are used in cooking.

Dairies have about 9.7 million cows that produce 15,554 pounds of milk (about 1,944 gallons per cow) each for a total of over 19 billion gallons of milk. That is more than 75 gallons of milk for every man, woman, and child in the United States.

These production numbers are impressive! Animal production for food is big business. Add to this the number of animals used in research and service and as companions. The total is a huge industry!

Brief summaries of five areas in the animal industry are included here: dairy, beef, poultry, hog, and sheep.

DAIRY

Nationwide, milk production earns producers about $20 billion per year. Milk is sold by the hundredweight (100 pounds or about 12.5 gallons of milk). Many dairy products are made. A *dairy product* is milk and any products in which milk is the major ingredient, such as cheese and ice cream.

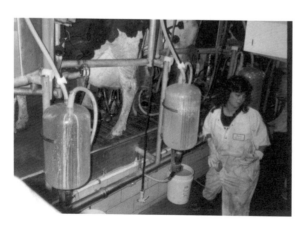

2-3. Milk production requires clean and modern facilities.

The use of dairy products has been about the same in recent years. The per capita consumption is 564.6 pounds of dairy products. In the last 25 years, dairy product consumption has varied up and down. The peak consumption was 601.2 pounds per capita in 1987, and a low of 535 pounds per capita in 1974.

For years, Wisconsin has been the number one dairy production state. In 1994, California reportedly moved ahead. Other important dairy states are New York, Pennsylvania, and Minnesota. The nature of dairy farming varies. Some places have small family-owned dairies. Other places have large corporate-owned dairies.

BEEF

Beef is the meat produced by cattle. More beef has been eaten than any other meat in the United States for many years. However, the per capita consumption has declined in recent years. In 1994, there were over 100 million beef animals in the United States. Beef cattle numbers have varied in the last 10 years.

There are over 900,000 beef producers in the United States. These producers sell nearly 54 billion pounds of beef per year. The gross income for the beef industry was over $40 billion in a recent year.

Texas has been the number one beef cattle state for many years. Texas had 14.3 million beef animals in a recent year. Nebraska ranked second, with 5.9 million beef animals. Rounding out the top five states were Kansas with 5.89 million, Oklahoma with 5.3 million, and Missouri with 4.6 million beef cattle.

2-4. Many people produce beef cattle to meet consumer demand.

POULTRY

The poultry industry is composed of two distinct segments: meat and eggs. Chickens are the primary producers of eggs. Chickens, turkeys, ducks, and a few other species are used for meat.

The poultry meat industry has expanded in the last decade. Production is now more than 31 billion pounds per year. Many chickens are marketed as broilers. A ***broiler*** is a young meat-type chicken. Most broilers grow fast and are marketed at seven weeks of age. People feel that eating poultry has certain health benefits. New technology has helped the industry expand.

Broiler farms produced 6.7 billion birds in 1993 in the United States. The value of production of the birds was more than $10.4 billion. Arkansas was the top broiler state, producing over one billion birds. Georgia produced 960 million broilers. Other states in the top five were Alabama (882 million), North Carolina (615 million), and Mississippi (528 million). Broiler production is expected to increase in the future.

Egg production has had only a slight increase in recent years. Current annual production is about 71 billion eggs. Concern about cholesterol in the human diet has resulted in some drop in egg consumption. Over 282 million hens are used to produce eggs in the United States. The total value of eggs produced was $3.8 billion dollars in a recent year.

California is the leading state in egg production, with 6.5 billion eggs in 1993. Pennsylvania is second, followed by Indiana, Ohio, and Georgia.

In addition to chickens, turkeys are produced in the United States. There were 287 million turkeys raised in 1993 (6.4 billion pounds of meat). The total value of turkeys produced was $2.5 billion.

2-5. Broiler production begins with good chicks.

2-6. Young turkeys.

SWINE

The swine industry has increased steadily over the last decade. Over 58 million hogs and pigs were on United States' farms in 1994. Though the number of breeding animals decreased in the last 10 years, the number of market animals has increased. The total number of **market hogs** in 1993 was 98 million. Market hogs are young hogs grown for slaughter. They typically weigh about 220 pounds.

2-7. The hog industry involves producing litters of healthy pigs.

Nearly 24 billion pounds of pork were marketed in the United States in 1993. The value of all pork marketed was $10.9 billion.

Iowa leads in hog production with 14.6 million head. Illinois, North Carolina, Minnesota, and Nebraska follow Iowa.

SHEEP

The United States had 9 million sheep and lambs in 1994. A total of 1.8 million sheep and 6.1 million lambs were marketed in the United States.

Texas was the number one state in sheep production with 1.7 million animals in inventory (19 percent of the total U.S. production). Other states in the top five included California, Wyoming, Colorado, and South Dakota.

2-8. Ewe with lamb. (Courtesy, National Lamb and Wool Grower, Englewood, Colorado)

ENTREPRENEURSHIP IN LIVESTOCK AND POULTRY PRODUCTION

An *entrepreneur* is a person who takes a risk by trying to meet the demands of people for a product. The entrepreneur typically owns and operates a business, including farms and ranches. A key part of being an entrepreneur is risk.

Risk is the possibility of losing what has been invested. People must have money—often a large amount—to start some kinds of animal entrepreneurship. If the business fails, they will lose their money. Often, the money is borrowed from banks and other sources.

Entrepreneurs are thought of as their own bosses. They have tremendous responsibility. The success or failure of their business depends on how they perform. Being an entrepreneur offers many advantages. You can set your own work schedule, have the opportunity to be creative, reap the rewards of your success, and get a great deal of self-satisfaction. There are, however, no guarantees of success.

There are many opportunities in animal production for entrepreneurs. The United States Department of Agriculture lists 20 career titles in the *Handbook of Agricultural Occupations* related to animal and poultry production. Many opportunities for entrepreneurship are related to raising animals. Others involve meeting the needs of producers for supplies, services, and marketing.

ANIMAL PRODUCER (FARMER OR RANCHER)

An *animal producer* raises livestock, poultry, and other animals. Animal producers may be known as farmers or ranchers. Often, they are known by the kind of animal they raise, such as trout farmer or horse rancher.

Livestock and poultry farmers engage in the breeding, feeding, and general business of raising commercial livestock and poultry. They can be classified as beef cattle farmers, dairy farmers, sheep farmers, swine farmers, small animal farmers, and fish and aquatic animal producers.

Entering the livestock and poultry production industry as a farmer requires a great deal of capital. Most types of farming require specialized facilities, equipment, and land. Many young or new entrepreneurs begin by working at home or developing partnerships with established farmers.

Some start as tenant farmers and eventually are able to save the start-up capital required.

It is best to select an area of interest and concentrate on one or two species of livestock. For example, an entrepreneur could choose to specialize in dairy production. Raising dairy cattle takes much time and effort. The producer must be familiar with approved practices for production and make good business decisions. Although it is possible to have a diver-

2-9. Being an entrepreneur in animal agriculture may involve operating hay equipment.

sified operation with many types of livestock, a farmer that concentrates on one species can operate more efficiently.

A drawback to focusing on one livestock species is that the producer is tied to one product in the market. When the market is favorable, profits are good. If the market falls, the profits also fall.

If you are considering becoming an entrepreneur, collect as much information about the area as possible. You should know what kind of facility is required, how much it will cost to get started, what the market for your product is like, and have the knowledge to successfully produce your livestock. A good place to begin is in your high school agricultural education program.

PART-TIME ANIMAL PRODUCER (FARMER OR RANCHER)

Another option for entrepreneurship lies in becoming a part-time animal producer. A *part-time animal producer* is one who raises animals only part of the time. The individual may do other work on a full-time basis.

Many farms in America today are operated on a part-time basis. Some people like to be around nature and animals. They like the thought of making a little extra money from their animals. Some families that have traditionally lived in the cities or suburbs are also starting to enter the part-time farm business. These can provide some income for the families involved, but the major source of income usually comes from the full-time job.

Part-time farms have marked differences from full-time operations. The part-time farms are usually smaller, with fewer acres and fewer numbers

2-10. Part-time producers are generally smaller and more diverse than full-time operations.

of livestock. These operations are commonly more diversified than full-time operations. Several species of livestock, fruit and vegetable production, and some crop land are the norm.

These operations are less expensive to start and can provide some additional income. Many small operations attempt to fill a marketing niche, like Angora goats or Siamese pot-bellied pigs. If you decide to enter such an enterprise, be very cautious. Be certain that there is a market for your product and that you can raise the animals. Many entrepreneurs have lost money due to bad decisions.

NON-PRODUCTION ANIMAL ENTREPRENEURSHIP

Producers rely on many supplies, services, and marketing activities to get products to consumers. People are needed to start and operate these businesses. Without these businesses, today's animal producers would not have a high level of production.

Animal Supplies

Providing supplies to animal producers can result in a good opportunity for some individuals. Animal supplies entrepreneurs setup businesses to meet the needs of animal producers for supplies. Many kinds of animal supplies businesses are possible. The kinds needed depend on the animals and products produced. Here are a few examples:

Feed mill—manufactures feed for animal producers

Feed store—sells feed to animal producers

Animal medicine store—sells animal medicines and other supplies

Animal equipment store—sells equipment used in animal production, such as feeders, waterers, milkers, gates, fencing, chutes, and pens

Animal Services

Animal producers often need specialized services to help with animal production. Opportunities often exist for people to be entrepreneurs in these areas. The kind of service depends on the animals and products produced. Here are a few examples:

Veterinarian—cares for the health of animals

Farrier—puts shoes on the feet of horses

Custom shearer—shears wool from sheep

Hauler—has specialized equipment to haul live animals, such as cattle trucks and trailers, or their products, such as refrigerated milk tanks on trucks

Construction contractor—constructs buildings, fences, and other facilities on farms and ranches for animals

2-11. Shearing sheep is an important service in harvesting wool. (Courtesy, National Lamb and Wool Grower, Englewood, Colorado)

Animal Marketing

Entrepreneurs in animal marketing may fill a wide range of roles. They link the animal producer and the consumer of the animal product. They must have good skills in working with producers and be informed on market information. Here are a few examples:

Cattle buyer—buys cattle from a producer for a processor or other user

Sale barn owner—operates a selling facility where producers bring animals to sell to buyers

Packing house operator—operates a facility that packs and/or processes animal products.

Food processor—takes raw animal products and prepares other products in forms that consumers want, such as making cheese from milk

CONSIDERING ANIMAL WELL-BEING

People who raise and care for animals want them to have a good place to live and grow. They want the animals to be well cared for and to be healthy.

ANIMAL WELL-BEING

Animal well-being is caring for animals so that their needs are met and they do not suffer. It involves understanding the animal and its needs. A good environment is needed. Food, water, and other necessities must be properly provided.

Animals that are being raised on today's farms often have better conditions than they would have if they were still wild. They are vaccinated to prevent disease. They are fed diets that meet nutritional needs. Facilities are used to protect them from injury and danger. They are moved carefully to avoid injury and mistreatment.

This book presents livestock and poultry production from the standpoint of animal well-being. The content has been carefully selected to demonstrate proven practices. Anyone who goes into animal production or works with animals in any way must consider animal well-being.

CONTROVERSY

Some producers have been criticized for how they treat animals. This is unfortunate. Animal rights and animal welfare organizations have been formed to help prevent abuse.

Animal welfare organizations and producers usually agree that proper care of animals is essential. However, they differ on the meaning of proper care. Animal welfare groups and producers advocate not causing undue pain to animals and providing them with an adequate environment, nutritious diet, and proper health care. The goal of some animal rights groups is to stop the use of animals for companionship, food, sport, fur, research, business, transportation, or other human "pleasure."

Most producers take pride in providing their animals with the best care possible. Livestock and poultry production professionals know that a healthy animal is a productive animal. Laws have been passed to ensure animal well-being.

2-12. Producers care about the well-being of their animals. This cow is being treated for a disease. Without a caring owner, she would not be treated.

REVIEWING

MAIN IDEAS

Livestock and poultry production is a large and diverse industry. Annual receipts from livestock and poultry account for nearly one-half of the income generated from agriculture.

Consumers are people who use animal products. They have preferences about what they will buy. Producers must produce animals that meet the needs of consumers. If not, the consumer will not buy the products and the price the producer receives will be low when the animals are sold.

The average American consumes over 180 pounds of meat per year. Dairies produce over 75 gallons of milk for every person in the country. United States' animal producers have about 9.7 million dairy cows, 109 million head of beef cattle, 6.7 billion broilers, 282 million laying hens, 287 million turkeys, 156 million head of hogs, and 9 million sheep and lambs on their farms and ranches. Animal and poultry production is big business!

There are many opportunities for entrepreneurs in the animal production industry. Most opportunities lie in animal farming, either in a full- or part-time operation.

Animal well-being is a goal of professionals in the livestock and poultry production industry. Herd health and productivity can only be maintained through keeping animal well-being a top priority for producers.

QUESTIONS

Answer the following questions using correct spelling and complete sentences.

1. What is a consumer? What is the role of consumption?

2. Define entrepreneur. Discuss opportunities in the livestock and poultry production industry for entrepreneurship.

3. Select one area of animal production and summarize its scope in the United States.

4. What is the role of an animal producer?

5. What is animal well-being?

EVALUATING

CHAPTER SELF-CHECK

Match the terms with the correct definitions. Write the letter by the term in the blank that is provided.

a. animal producer   c. per capita consumption   e. animal well-being
b. consumer   d. entrepreneur   f. risk

1. ____ the possibility of losing what has been invested.

2. ____ people who eat, wear, and otherwise use goods and services.

3. ____ the average amount consumed per year by a person.

4. ____ caring for animals so that their needs are met.

5. ____ one who assumes risk in a business activity.

6. ____ an individual who raises animals.

EXPLORING

1. Prepare a bar graph that shows amounts of consumption for animal products. Include beef, pork, poultry, eggs, and sheep and mutton.

2. Visit an animal producer. Determine the kinds of animals produced and the general nature of the work. Discuss the job and lifestyle of being an entrepreneur in animal agriculture.

3. Investigate animal well-being. Select a species of animal, determine what it needs to be healthy and grow, and develop a list of what a producer can do to assure its well-being. Use other references found in the library.

Chapter 3

CAREER OPPORTUNITIES IN LIVESTOCK AND POULTRY PRODUCTION

You do not have to own animals to work with them!

Some people work for others. They like the security of a good job. People who work for others do not have the same risk as entrepreneurs. Many careers are available for those who want to work with animals. Other careers are in animal supplies and services and marketing. These do not always involve contact with animals.

Many off-farm careers in livestock and poultry production support animal production. Less than 2 percent of the United States work force is in farming. However, nearly 20 percent of all workers are in careers related to agriculture. Although some careers have limited involvement with animal production, about 535,000 people work in meat and dairy production. In addition, thousands of people are in poultry and aquaculture careers. Others are in the pleasure and service animal fields.

3-1. Laboratory research is an important part of animal production.

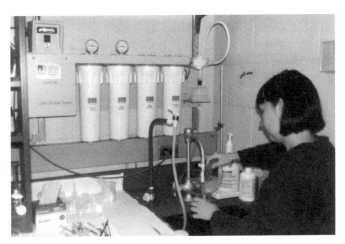

OBJECTIVES

To better understand careers in the animal industry, the following objectives are included:

1. Explain why choosing a career is important

2. List careers in animal production and describe the nature of the work

3. Describe science-oriented career areas in livestock and poultry production

4. Explain career opportunities in animal management and service

5. Describe career opportunities in education, extension, and communication

TERMS

career	gainful employment	job
employment	goal	occupation
entry level job	goal setting	work ethic

CHOOSING A CAREER IN THE ANIMAL INDUSTRY

People want to spend their lives doing useful things. They want to feel that they have contributed to society. Choosing a career is an important decision. People work thousands of hours during the 30 years of their employment. That is a long time!

Gathering information can be very helpful in making a good choice. You will want to feel good about what you are doing with your life. Being employed is an important way to make a contribution.

EMPLOYMENT

People use their time in different ways. **Employment** is the primary way people use time, and is often called work. Sometimes people serve as volunteers and are not paid. Other times they work for what they gain to help cover the costs of living.

Gainful employment is having a job that provides benefits. People are usually paid a salary or wages for their work. They may also have certain benefits, such as a ranch manager being provided free housing on the ranch. These benefits are a part of the total returns that a person gets for working.

Entrepreneurs have jobs. They are gainfully employed, though they may not be paid a salary. They receive benefits if their animal production enterprise is a success.

Work Ethic

People view work differently. **Work ethic** is how we feel about work. Do we think it is essential? Do we think it is the most important thing in our lives?

How we go about choosing a career and the one we choose is a part of our work ethic. People begin developing attitudes about work at a very young age. They learn from the people around them. Most people learn that work is a natural and integral part of life. It is how people go about having what they need and want.

3-2. Choosing a career should involve careful study of subjects related to animal production.

The way people go about work is also a part of the work ethic. People are expected to be good, dependable, honest workers. They are expected to work hard and do a job well.

Goals

People often think about what they want to do in their lives. They identify things that they want to have and do. A lot of planning and work are needed to reach goals.

A **goal** is a level of achievement that people set for themselves. Goals have a powerful influence in our lives. We all want to achieve what we set out to do. No one wants to fail. People need to be careful in setting goals to be sure that they reflect what is important to them.

Goal setting is describing what we want to accomplish. It includes how we will achieve the accomplishments. People do it all of the time. Sometimes our goals may not be very important in the long run. Goals people set about their careers are definitely important! Begin by understanding careers and the needed education. Match this information with what you want.

CAREERS, OCCUPATIONS, AND JOBS

A **career** is the general direction that a person takes with work. Most careers involve a sequence. People begin at a lower level and advance to higher levels. Moving up is climbing the career ladder or advancement.

The lowest level job is the entry level. An **entry level job** is one that requires little or no previous experience. It is the first job for many people.

Education and following instructions will help people advance. Careers may have a series of occupations and jobs.

An *occupation* is an important part of a career. Occupations involve specific work, known as duties. The same occupation has similar duties in different locations. For example, the occupation of a large animal veterinarian has similar duties regardless of where the veterinarian works.

A *job* is specific work. Jobs are carried out at certain sites and with employers. People who have jobs are employed. People can have occupations and not have jobs. For example, if the large animal veterinarian had no job, the individual still has an occupation though unemployed.

Sorting through the meaning of career, occupation, and job helps people make good decisions. The next time you hear about employment, check to see if it is an occupation or a job and how it is a part of a career.

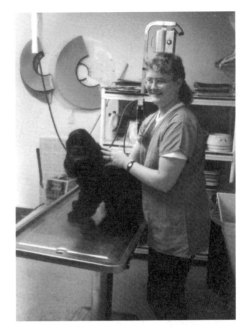

3-3. Some jobs involve caring for small animals, such as this dog.

CAREERS IN MEAT ANIMAL PRODUCTION

Owning an animal production business is one way to be involved in the production of livestock. Entrepreneurship was included in Chapter 2. However, many people want to be involved with animals, but not as entrepreneurs. They want employment with other people.

FARM OR RANCH WORKER

Many people who eventually want to own and operate their own farm start out as a farm or ranch worker. Opportunities for this type of employment will vary with the type of agriculture in your community.

Farm and ranch work varies based on the knowledge and ability of the person employed. Duties range from general farm labor in doing routine jobs, such as feeding, sanitation, and moving animals, to more complicated tasks, such as treating sick animals. Small farms may employ a person

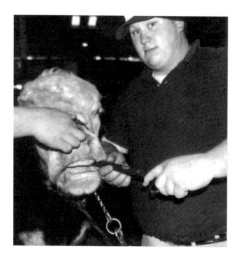

3-4. Farm workers do a variety of tasks—all useful in keeping animals healthy and productive. This shows medicine being given to a beef animal.

to do a wide range of tasks. Workers in large operations may specialize in a particular area, such as milking cows.

Many entry-level positions are as a farm or ranch worker. New employees are not expected to have the knowledge of livestock to operate the facility on their own. They are expected to have a general knowledge of livestock and work competently with animals. They need to quickly learn their job. High school agricultural education is helpful.

FARM MANAGER

Being a farm manager has responsibility. The farm manager carries out the everyday operation of the farm. It may also include long range planning and setting production goals. Farm managers are usually hired by corporate farms, institutional farms, or by individuals who own the farm but do not want to be involved in the daily operation.

Becoming a farm manager requires extensive experience or a combination of experience and education. Employers are often looking for people with education beyond high school.

Tasks a farm manager might do include: selecting and marketing animals, formulating feed, maintaining herd health, and supervising employees. Helpful experience includes high school agricultural education, being a farm worker, and postsecondary education in an agricultural college.

3-5. Farm managers may use computers to keep records.

CAREERS IN PLEASURE AND SERVICE ANIMAL INDUSTRY

Careers in the pleasure and service animal industry are just as plentiful as those in the meat animal industry. You could have a career raising animals for important research or be a pet store owner. Horses, game and fish, service animals, and laboratory animals are a few fields of specialization that may interest you.

THE HORSE INDUSTRY

The horse industry is more visible in many communities than meat animal production. Many people enjoy riding and racing horses. Stables provide a place to keep and ride a horse. In addition, there are plenty of places to enjoy horseback riding.

Increased interest in light horses for pleasure and racing has opened new jobs. These range from stable attendant to race horse trainer. As with any other career, you begin doing hand labor.

Stable hands spend most of their time doing essential work, such as cleaning the barn and feeding the horses. As you become more experienced, you may become a horse trainer, riding instructor, stable manager, or barn boss. Jobs in the race industry differ from job titles in the pleasure horse industry.

3-6. Farrier preparing a horse's foot for a shoe. (Courtesy, American Paint Horse Association)

GAME AND FISHERIES

Do you like wildlife? Do you like being out of doors? Are you concerned about fish and wildlife? Then you may be interested in a career in game and fisheries. You could be a wildlife conservation officer. Perhaps a career on a game farm or in a fish hatchery is more to your liking. Let's examine these career opportunities in the game and fish industry.

3-7. **Fish farmer examining a crawfish.**

A wildlife conservation officer works to educate the public about fish and wildlife. The duties may include conducting workshops on wildlife and the environment. The conservation officer is also responsible for the enforcement of game and fish laws. There is much competition for good jobs in this field. Conservation officers are usually hired by state and federal agencies and are considered civil service employees.

Game farm workers raise and manage different species of wildlife. Some are returned to the wild. Others are kept on game farms. Hunting clubs, conservation groups, and government agencies need wildlife to enhance populations. Although no specific qualifications exist, a high school diploma and an interest in working with game birds and animals are desirable.

Fish hatchery technicians raise and manage fish. The fish may be for stocking in public and private ponds, lakes, and waterways. People employed in this field tend eggs, maintain water quality, feed fish, provide a healthy environment for the fish, and manage the facilities. Fish hatchery workers are generally under the direction of a fisheries biologist. High school courses in mathematics, science, chemistry, biology, and aquaculture are helpful.

People in game and fish positions should have an interest in wildlife, fisheries, and conservation. Most wildlife conservation officers must pass a test and take some type of training. Experience is helpful and a college degree in a wildlife area is beneficial.

3-8. **Harvesting catfish.** (Courtesy, Delta Pride Catfish, Inc.)

SERVICE ANIMALS

Raising animals that serve humans is important. Service animals, such as leader dogs for the blind, companion animals, and draft animals are important in the animal industry.

Draft animals are used for work, primarily the power to pull machinery. Most machines today are powered by engines. Some draft animals are used as hobbies and, in a few cases, to do specialized work. In a few countries, animal power is used to grow crops.

Although there are large producers of service animals, many of them are raised by small producers or individuals.

LABORATORY ANIMALS

A laboratory animal is a special animal used to help humans solve problems. They allow research to improve human and animal life. Laboratory animals are raised and kept under the best possible conditions. Without them, progress in human medicine and related areas would nearly stop.

Many laboratory animals are delicate animals. They are raised to meet specific qualifications for research. For example, animals are bred with immune systems that are almost identical to human immune systems for AIDS research. Laboratory animal production is a complex field. It requires an in-depth knowledge of science, chemistry, and biology as well as an interest in animal research.

3-9. Keeping areas clean where laboratory animals are raised is important.

SCIENCE-ORIENTED CAREER OPPORTUNITIES

The science of animal agriculture is an expanding field. Scientists use animals and work laboratory equipment. They can be involved with producing experimental animals or increasing wildlife populations.

3-10. Animal scientists plan experiments in animal health and other areas. (Courtesy, Merch & Co., Inc.)

ANIMAL SCIENTIST

Animal scientists provide services to the livestock and poultry production industry. Positions related to animal science are: animal nutrition specialists, animal reproduction specialists, geneticists, chemists, and biotechnologists.

An animal research scientist concentrates in a specific area of interest in animal production. Research scientists work in laboratories, with experimental animals and animal production inputs, and other scientists. Most people who choose this career have an advanced college degree in the area. Personal needs are a high interest in scientific inquiry, knowledge of biology, chemistry, and microbiology.

LABORATORY ASSISTANT

Assisting with laboratory animal production or experimentation can be a rewarding career. Animal laboratory assistants help with experiments, care for laboratory animals, and raise laboratory animals. Attention to detail is required. A high school diploma with an emphasis in animal agriculture is a minimum. Most laboratory assistants have some college education. Some have a master's degree in biology, chemistry, animal science, or a related discipline.

AGRICULTURAL ENGINEER

Agricultural engineers often work in the animal industry. They work with animal facilities, waste management, environmental impact, and improving animal products. Engineers must have a bachelor's degree and the ability to assess

3-11. Engineering is important in designing a milking facility. (Courtesy, Marco Nicovich, Mississippi Cooperative Extension Service)

a problem and prepare a solution to it. Knowledge of agriculture and the animal industry is helpful.

WILDLIFE BIOLOGIST

A wildlife biologist does many of the tasks of a general biologist. They focus on wildlife animals and habitat. A wildlife biologist may study an endangered animal species or concentrate on improving the environment for many species. Their work can include genetics, studying the environmental impact of industry, or repopulating an animal where they no longer exist. Wildlife biologists help with the living conditions of wild animals and fish.

A wildlife biologist must be committed to studying animals and science. A college education is required for entry and higher degrees are needed for advancement. If you want a career in wildlife biology, you need a high interest in biology, chemistry, animal science, mathematics, and research.

CAREER OPPORTUNITIES IN ANIMAL MANAGEMENT AND SERVICE

People who provide for the needs of animals and help with marketing are important in the livestock and poultry industry. Producers often use help from animal management and service professionals. These people help to ensure herd health, meet animal needs, and market what they produce.

VETERINARIAN

Veterinarians provide care for sick and injured animals. They care for well animals to prevent illness. A veterinarian may be a specialist that only administers care to a specific species of livestock, or a generalist that cares for farm animals and pets. A few now specialize in fish health.

Becoming a veterinarian takes time and dedication. Pre-veterinary students often earn a bachelor's degree in animal science or biology. After completing the courses, they apply for admission to a school of veterinary medicine. If they are accepted, they complete several more years of edu-

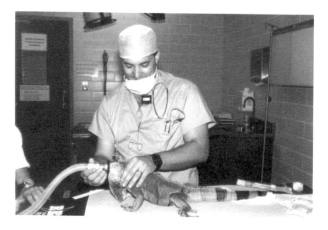

3-12. Veterinary medicine may involve treating exotic animals. (Courtesy, Texas A & M University)

cation and training. Although much is required to become a veterinarian, it is a very rewarding career.

SALES AND SERVICE

The animal industry has many professionals in supplies sales and service. These people sell feed to farmers, ensure quality products, sell retail products, service equipment, operate stock yards, and perform many other services.

Many sales positions related to livestock and poultry production require a college education. The education should be in animal science or related area. Machinery service, stockyard worker, veterinary assistant, or feedmill worker may not require a college degree. Although not a requirement, some postsecondary education is desirable. Agricultural education in high school and experience in the animal industry, will help get a job in sales and service.

3-13. Animal trainers teach animals to do certain things, such as this jumping donkey. (Courtesy, Lucky Three Ranch, Loveland, Colorado)

LIVESTOCK BUYER

Knowing what the industry and consumer demand in a meat animal is necessary to success in livestock buying and selling. A good "eye" for livestock quality on the hoof and a good business sense are essential. Buyers and sellers may work for themselves or have clients in the production or packing industry.

Training in animal selection is a requirement for this career. High school education in agriculture, biology, and mathematics is helpful. Additional training in animal science, business management, and public relations at the postsecondary level will also be beneficial.

CAREER OPPORTUNITIES IN EDUCATION, EXTENSION, AND COMMUNICATION

Careers available in education, extension, and communication emphasize the "people" side of the livestock and poultry production industry. People are involved at all levels. Teaching, sharing knowledge, and communicating information are necessary in the animal industry.

AGRISCIENCE TEACHER

Agriscience teachers work in the public schools. They teach basic agriculture and specialized classes. They work with students of all ages—kindergarten through adult. They are agriculture experts in their local communities. A bachelor's degree in agriculture and certification to teach agriculture are needed. Agriscience students are often involved in the FFA. This is a student organization that enhances agriculture instruction. It also provides personal and lead-

3-14. A Florida agriscience teacher makes a point in class.

ership skill development. Students have supervised experience programs in areas of agriculture, including animal production.

AGRICULTURAL EXTENSION AGENT

Agricultural extension agents work with local producers (usually in a particular county) to help them improve production or solve livestock and other agricultural problems. Extension agents that specialize in livestock can provide important help to producers.

A bachelor's degree in agriculture is needed to be an extension agent. Many states require a master's degree within a few years. To work as a livestock and poultry agent, a background in animal production is needed.

AGRICOMMUNICATION SPECIALIST

Agricommunication specialists develop procedures to share information about animal production. They may work with newspapers, in public relations, and with broadcast media (radio and television). An agricommunicator may host a television show, write for a livestock magazine, produce videos, or be involved in any area of communication.

3-15. Agricommunication specialists are important in many areas of animal science, such as this television editor.

A career in agricommunication will require a university degree in agriculture or communication. Sometimes people combine the study of agriculture and communication. People employed in this field need to have broad knowledge of agriculture and a people-oriented personality.

REVIEWING

MAIN IDEAS

Exciting careers are available in the livestock and poultry production industry. A wide range of interests and ability levels is found. Opportunities are in many areas. The meat animal industry provides wholesome food. The pleasure and service animal area meets important needs. Careers are also found in science-based fields, in animal management and service, and in agricommunication.

Careers in the meat animal industry include becoming an entrepreneur in animal or poultry production, working for someone on a farm or ranch, or managing a farm or ranch. The pleasure and service animal industry also offers many opportunities that include working with horses, game and fisheries, draft animals, and laboratory animals.

Science-related careers include working in animal laboratories, conducting animal research, discovering new medical treatments, and designing animal facilities. Animal management and service careers include veterinary science, providing inputs for animal production, and marketing animal products. Many people are involved in the people side of the animal industry through education, extension, and agricultural communications.

QUESTIONS

Answer the following questions using correct spelling and complete sentences.

1. What should be considered in choosing a career in the animal industry?
2. What are the career opportunities in livestock and poultry production?
3. What are the responsibilities of a farm manager?
4. What careers are in the horse industry?
5. Where might an animal research scientist work? What kinds of research would they likely conduct?
6. What are the career opportunities for people who like to work with wildlife? What would one do to prepare for these careers?
7. Discuss a science-related service career in the livestock and poultry production industry.
8. What opportunities exist for people interested in education, extension, and communication?

EVALUATING

CHAPTER SELF-CHECK

Match the terms with the correct definitions. Write the letter by the term in the blank that is provided.

a. entry-level job   c. gainful employment   e. goal
b. work ethic   d. career   f. job

1. _____ levels that people want to achieve.

2. _____ job that requires little or no previous experience.

3. _____ specific work carried out at a site.

4. _____ how people feel about work.

5. _____ general direction that a person takes with work.

6. _____ having work that provides benefits.

EXPLORING

1. Interview professionals in the livestock and poultry production industry. Discuss the job description, daily activities, and career preparation. Also, find out what the salary for beginning professionals would be. Prepare a written report that summarizes what you learn.

2. Arrange to discuss post-high school education for career opportunities with your guidance counselor and your agriculture teacher. Take the *AgriScience Interest Inventory* to help assess your interests in a career in animal production. (The Inventory is available from Interstate Publishers, Inc., P. O. Box 50, Danville, IL 61834.)

3. Construct a poster of a career that interests you in the livestock and poultry production industry. Share your poster with the class.

Chapter 4

ANIMALS AS ORGANISMS

Over one million different kinds of animals are on the Earth. Many of these animals or their products are sources of food and clothing. You probably use many of these products in your daily living!

Some animals have been domesticated. They are carefully tended on farms and ranches. Other animals remain wild, with some used for food and clothing products.

Animals are very complex creatures. By studying the functions and needs of animals, people are better able to raise them. Research is studying new areas of animal physiology, genetics, and environment. The research should help find better ways of raising animals.

The most commonly raised livestock in the United States are beef cattle, dairy cattle, swine, sheep, goats, horses, and fish. Each of these animals has the same basic organization levels. Each has tissues, organs, and organ systems. Slight variations among these animals distinguish them from each other.

4-1. Understanding the fundamental differences between animals takes hard work and study time.

47

OBJECTIVES

This chapter is about anatomy, physiology, and systems of important animals. The following objectives are included:

1. Describe the major animal groups
2. Describe the anatomy and physiology of animals
3. Explain the major organ systems of animals

TERMS

anatomy	homeostasis	organ system
circulatory system	invertebrate	physiology
classification system	Kingdom Animalia	reproductive system
connective tissue	lymphatic system	respiratory system
digestive system	muscular system	skeletal system
endocrine system	muscular tissue	taxonomy
epithelial tissue	nervous system	vertebrate
excretory system	nervous tissue	
hierarchy of classification	organism	

MAJOR ANIMAL GROUPS

Animals are organisms. An **organism** is any living thing. As organisms, animals are in the **Kingdom Animalia**. Though all animals are in this Kingdom, they also have unique differences. Some have fur, while others have scales or feathers. Some fly, while others walk or swim.

Taxonomy is the science of classifying organisms. Most classifications of animals are due to physical characteristics. A **classification system** is used to distinguish different animals from each other. The **hierarchy of classification** is used to show the relationships and differences among animals. Table 4-1 shows the hierarchy of classification.

Table 4-1
Hierarchy of Classification

Category	Description
Species	Reproductively isolated populations with physical similarities
Genus	Contains related species
Family	Contains related genera
Order	Contains related family
Class	Contains related orders
Phylum	Contains related classes
Kingdom	Contains related phyla

Moving from bottom to top—more characteristics in common.

Moving from top to bottom—common characteristics are more distinctive.

Organisms in the Kingdom Animalia share three traits. They are made up of cells, can move about on their own, and get their food from other sources. Animals take in food, transform and store it, and release it as energy.

Within the Kingdom Animalia, animals are grouped many ways. This grouping is based on the structures and functions of the body.

4-2. The Guernsey cow is an example of a vertebrate (belongs to the Phylum Chordata and the Subphylum Verbrata).

The domestic farm animals discussed in this book are all vertebrates. *Vertebrates* are animals with backbones. This is as opposed to *invertebrates* that are animals without backbones. Examples of invertebrates include shrimp, crawfish, honeybees, spiders, mites, earthworms, and snails.

Vertebrates belong to the Phylum Chordata and to the Subphylum Verbrata. They have many common characteristics. They have vertebrae (bones and cartilage), which surround the nerve cord. Bones make up the internal skeleton that provides the body's framework. They have a skull that protects the brain. Vertebrates have an axial skeleton made of the backbone and skull. Paired limbs (arms and legs) are attached to the axial skeleton. Finally, muscles provide movement by being attached to the skeleton.

Agricultural animals belong to three main classes: birds (Aves), bony fish (Osteichthyes), and mammals (Mammalia).

BIRDS

Some 9,000 species of birds have been identified. Birds belong to the class Aves. Birds can live in the air, on land, or in the water. Most birds are wild; a few have been domesticated. Chickens, turkeys, and ducks are often raised. People are eating more chicken and turkey because they are lower in fat and cholesterol.

Birds are covered with feathers. They are lightweight because of their hollow bones. Their wings help to lift them and make them better flyers. Birds have two legs and internal organ systems similar to most other animals.

4-3. Birds belong to the class Aves and are very different from fish and mammals.

FISH

Fish are aquatic animals. They live in water. There are three classes of fish, but the bony fish (Osteichthyes) are the ones commonly raised for food. Scientists have identified about 25,000 species of fish.

Many fish are caught from rivers and oceans, but aqua-farming is increasing in popularity. Catfish are the most commonly raised fish. Trout,

4-4. Turtles can be raised on farms.

tilapia, salmon, and baitfish are also raised on aquafarms. Many people raise several species of ornamental fish at home or at their business for pets or decorations.

Fish are covered with skin and/or scales. The head, tail, and trunk make up the three major parts of the body. The mouth, eyes, and gills are found on the head, while the trunk contains the internal organs. Strong muscles in the tail help the fish to move about.

Fish are physiologically similar to other animals. However, fish filter oxygen from the water as it passes over their gills. Some lay eggs that are fertilized outside the body; others give birth to live young. The body temperature is regulated by the temperature of the water.

MAMMALS

Mammals belong to the class Mammalia. Female mammals have mammary glands that produce milk for the newborn young. Male and female mammals mate to reproduce. The female will carry the developing embryo in her uterus.

Mammals have hair, a well-developed brain, a lower jawbone with teeth, and a heart with four chambers. Unlike fish, mammals internally regulate their body temperature. Examples of mammals include cattle, goats, hogs, horses, and sheep.

Appendix A lists the classifications for selected animals.

ANATOMY AND PHYSIOLOGY OF ANIMALS

Anatomy and physiology are important in knowing how organisms live and go about life processes. **Anatomy** is the study of the form, shape, and appearance of animals. **Physiology** is the study of the functions of the cells, tissues, organs, and systems of an organism. Knowing anatomy is essential to understanding physiology.

All animals have organ systems that maintain homeostasis. **Homeostasis** is the relative constancy of the internal environment. The circulatory system (an organ system) carries nutrients from the digestive system. It also carries oxygen to the cells from the respiratory system via the blood and rids metabolic waste from the cells to the excretory system. The nervous system coordinates the functions of the other systems.

Understanding life processes is important. Knowing the workings of the inner body involves anatomy and physiology of livestock. Internal and external features are closely related. Figures 4-5 through 4-9 illustrate the external parts of several animals.

MAJOR ANIMAL SYSTEMS

ORGANIZATION

Animals have several levels of organization. All organisms are made of cells. Cells form tissue. Tissues form organs and organs form organ systems. These systems make up the organism. The structure and function of an organ system depend on the structure and function of the organs, tissues, and cells within that system.

There are four types of animal tissue: epithelial, connective, muscular, and nervous.

Epithelial Tissue

Epithelial tissue covers body surfaces and lines body cavities. Its purpose as skin is to protect the body. Epithelial tissue contains cilia that are hairlike extensions. These cilia move materials. For example, cilia in the lungs move dirt and other impurities out of the body. The main organ system of the epithelial tissue is skin.

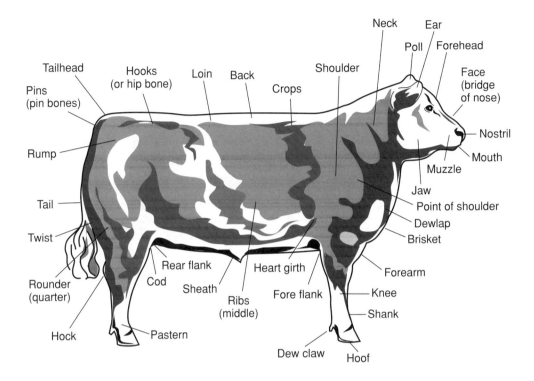

4-5. Major external parts of a steer.

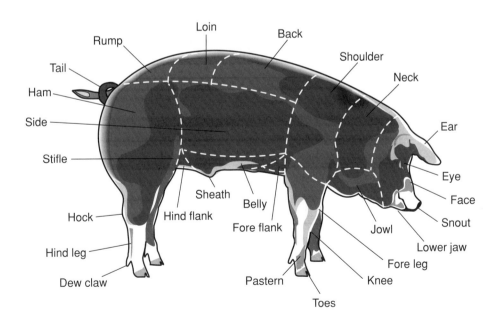

4-6. Major external parts of a hog.

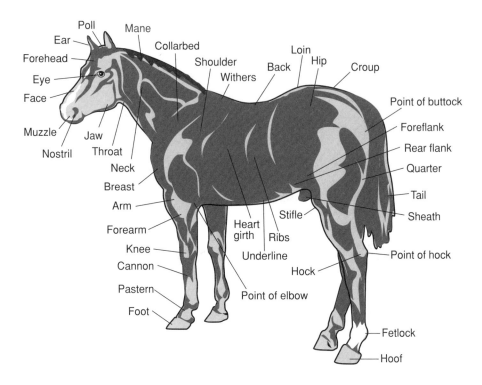

4-7. Major external parts of a horse.

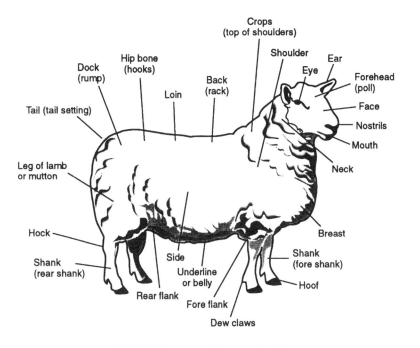

4-8. Major external parts of a sheep.

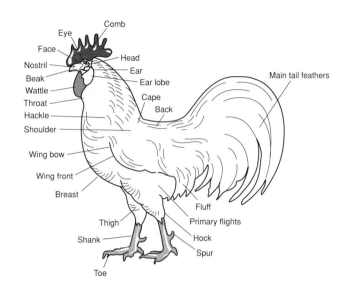

4-9. Major external
parts of a chicken.

Connective Tissue

Connective tissue holds and supports body parts. It binds body parts together providing support and protection. It also fills spaces, stores fat, and forms blood cells. Connective tissue helps to form muscular and skeletal cells.

Muscular Tissue

Muscular tissue creates movement of body parts. It is also known as contractile tissue. These tissues not only aid in movement of the entire animal, but also allow the respiratory and digestive systems to function.

Nervous Tissue

Nervous tissue responds to stimuli and transmits nerve impulses. Nervous tissue contains neurons that conduct impulses away from the cell. Environmental stimuli cause reactions to occur.

ORGAN SYSTEMS

Each kind of tissue forms organs and ***organ systems***. An organ system is made of several organs that work together to perform an activity. Each system has its own organs and specific purposes. The organ systems and chief structures in the systems are shown in Table 4-2.

Table 4-2
Organ Systems and Structures

System	Chief Structures
Circulatory	Heart
Digestive	Stomach and intestines
Integumentary	Skin
Endocrine	Ductless glands
Excretory	Kidneys and bladder
Muscular	Muscles
Nervous	Brain, spinal cord, nerves
Reproductive	Ovaries and testes
Respiratory	Lungs
Skeletal	Bones, joints

Muscular System

The *muscular system* acquires materials and energy. It creates bodily movements, maintains posture, supports, and produces heat. Figure 4-10 outlines parts of the hog's muscular system.

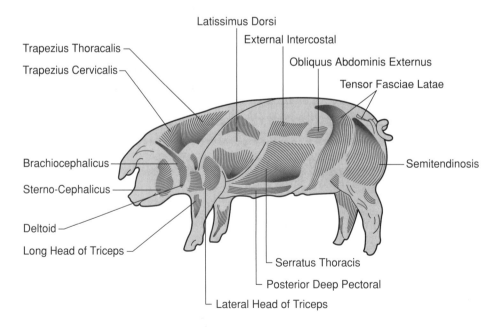

4-10. Parts of the hog's muscular system.

Skeletal System

The *skeletal system* provides a framework for the body. It also supports and protects internal organs, stores minerals, and produces blood cells. The skeletal system of a chicken and a hog are shown in Figure 4-11.

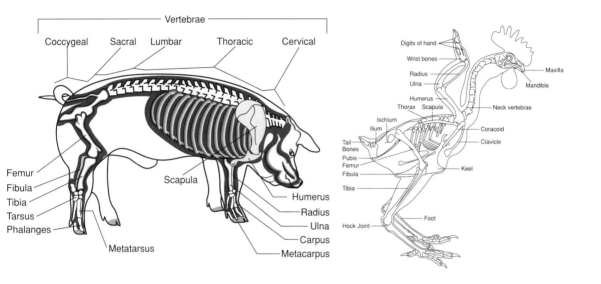

4-11. Skeletal systems of a hog and a chicken.

Digestive System

The *digestive system* breaks food into smaller parts that are used by the body. Note the differences between monogastric or nonruminant and polygastric or ruminant animals. Figure 4-12 outlines the digestive systems of a hog (monogastric or nonruminant), a cow (polygastric or ruminant), and a chicken.

Respiratory System

The *respiratory system* governs gas exchange. Oxygen is taken in by the lungs. This system also maintains blood pH by expelling carbon dioxide. Figure 4-13 shows the respiratory system of a hog.

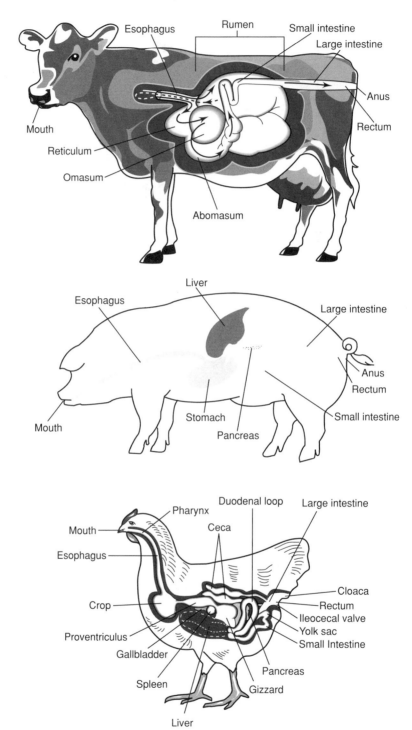

4-12. Digestive systems of a cow (polygastric or ruminant), a hog (monogastric or nonruminant), and a chicken.

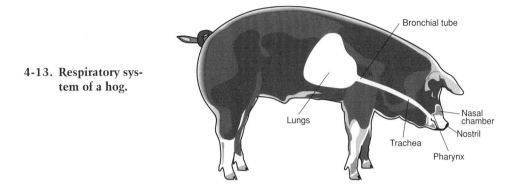

4-13. Respiratory system of a hog.

Circulatory System

Nutrients, oxygen, and metabolic wastes are moved by the ***circulatory system***. Oxygen and nutrients are transported to the cells, while wastes are removed. The circulatory system also moves hormones and protects against injury and microbes. The circulatory system of a horse is shown in Figure 4-14.

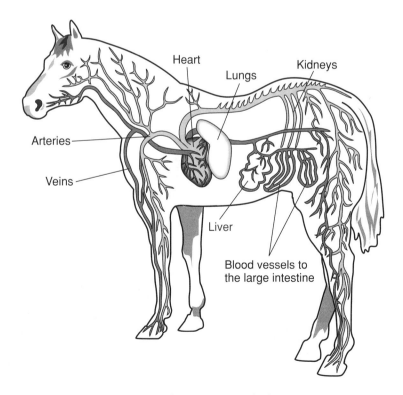

4-14. Circulatory system of a horse.

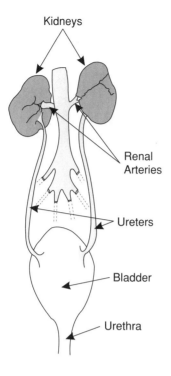

4-15. Excretory system of various animals.

Excretory System

The *excretory system* rids the body of waste. It maintains the chemical composition and volume of blood and tissue fluid. Figure 4-15 shows the excretory system of various animals.

Lymphatic System

The *lymphatic system* circulates a clear fluid known as lymph. One role of lymph is to protect the body from disease. The lymphatic system transports excess tissue fluid to the blood, moves fat to the blood, and helps give immunity against diseases.

Nervous System

The *nervous system* coordinates body activities. In coordination with the endocrine system, nerve impulses react to stimuli. The system regulates other systems and controls learning and memory. The *endocrine system* secretes hormones that regulate the body's metabolism, growth, and reproductive systems. The nervous system of a horse is shown in Figure 4-16 and the endocrine system of a cow is shown in Figure 4-17.

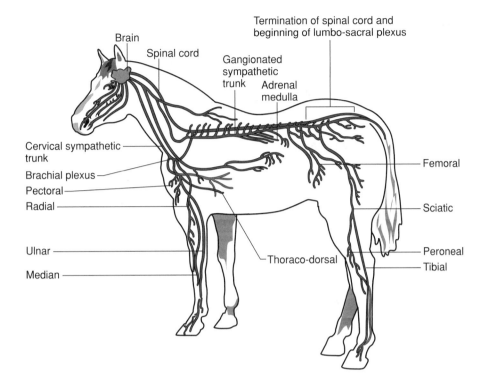

4-16. Nervous system of a horse.

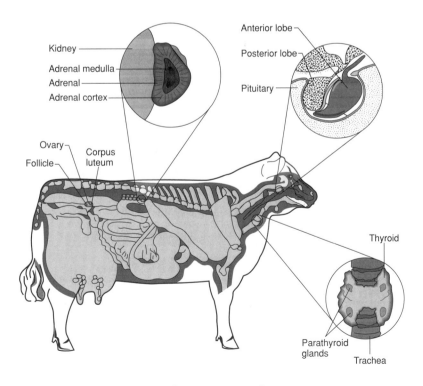

4-17. Endocrine system of a cow.

Intequmentary System

The intequmentary system is the skin. The skin protects internal body tissues from outside dangers. It keeps out foreign materials, such as bacteria and dust. It also helps regulate body temperature.

Reproductive System

The *reproductive system* produces offspring. New individuals of the same species are produced. The reproductive system is discussed in Chapter 7.

REVIEWING

MAIN IDEAS

The hierarchy of classification is a system for grouping animals. Kingdom is the broadest classification, with species the most specific classification.

The Kingdom Animalia includes all agricultural animals. Organisms in the Kingdom Animalia are made of cells, can move about on their own, and get their food from other sources. Agricultural animals belong to the phylum Chordata and the subphylum Verbrata. Agricultural animals belong to three classes: birds, fish, and mammals.

Animals have four types of tissue. Epithelial tissue covers body surfaces and lines body cavities. Connective tissue holds and supports body parts. Muscular tissue creates movement of body parts. Nervous tissue responds to stimuli and transmits nerve impulses. These tissues are organized into organ systems. Animals have the following organ systems: circulatory, digestive, integumentary, endocrine, excretory, muscular, nervous, reproductive, respiratory, and skeletal.

QUESTIONS

Answer the following questions using correct spelling and complete sentences.

1. Explain the hierarchy of classification.

2. What are the important traits of organisms in the Kingdom Animalia?

3. What are the common characteristics of vertebrates?

4. Compare and contrast the three classes of animals discussed in this chapter.

5. Distinguish between anatomy and physiology.

6. Describe the major differences between the external parts of the animals shown in this chapter.

7. Describe the organization of animals.

8. What are the four different types of tissue? Explain the function of each.

9. Explain the major functions of the ten organ systems.

10. Compare and contrast the digestive systems of a hog, a cow, and a chicken.

11. How are the nervous and endocrine systems related?

EVALUATING

CHAPTER SELF-CHECK

Match the terms with the correct definitions. Write the letter by the term in the blank that is provided.

a. vertebrate   d. invertebrate   g. taxonomy
b. anatomy   e. physiology   h. organ system
c. organism   f. homeostasis

1. ____ living things

2. ____ animal with backbone

3. ____ constancy of the internal environment of an organism

4. ____ function of cells, tissues, organs, and systems

5. ____ form, shape, and appearance of animals

6. ____ organs working together to perform an activity

7. ____ animal that does not have a backbone

8. ____ science of classifying organisms

EXPLORING

1. Dissect a pig or other farm animal. Compare the various types of tissue and organ systems and their functions. Use proper procedures in your work. Follow all safety practices.

2. Visit a local farm or ranch that raises various agricultural animals or attend a livestock show. Compare and contrast the external parts of the animals. Use a camera to make photographs of the different animals. Prepare a poster or bulletin board that summarizes your observations.

3. Make a sketch of an animal. Label the major external parts. The sketch can be made in the space below or on a separate sheet of paper.

Kind of animal _____

Chapter 5

ANIMAL NUTRITION AND FEEDING

Animals must have the right feed to live and grow properly. Feeding animals often raises questions: What does an animal need? Why are these needs important? How are the needs met?

All animals have many of the same needs for nutrients. But, animals vary in the kinds of feed they can use. Selecting the right feed and method of feeding varies with nutrient needs and species of animal.

Animals that do not get the right feed suffer from poor nutrition. Poor nutrition causes slow growth, poor reproduction, lowered production levels, and poor health. Sometimes, poor nutrition leads to the early death of an animal. Animal producers want fast growth at the lowest cost. They want to look after the well being of their animals as well as make a profit. Nutrition is a primary concern of an animal owner. Animals that are properly nourished are healthy and grow fast.

5-1. A young animal needs to be properly fed. This dairy calf receives a carefully prepared diet to meet its needs.

OBJECTIVES

This chapter is about the nutrition and feeding of animals. It has the following objectives:

1. List the major nutrient needs of animals and purposes of each
2. Describe the ways animals use nutrients
3. Describe the types of feedstuffs
4. Explain how animals are fed

TERMS

amino acids	forage	nutrient
balanced ration	free access	nutrition
carbohydrate	glucose	palatability
concentrate	growth	protein
fats	lactation ration	reproduction ration
fat soluble vitamin	macromineral	scheduled feeding
feed	maintenance	supplement
feedstuff	micromineral	vitamin
fiber	mineral	water soluble vitamin

NUTRITION AND NUTRIENTS

Nutrition is the process by which animals eat and use food. Food is used to live, grow, lactate, reproduce, and work. All animals must have the food they need for a balanced diet.

Scientists know a lot about nutrition but there is so much more to learn. Nutrition is still being explored to a great degree. For example, scientists at the University of Missouri have found that pregnant sows must have proper Vitamin A. If not, pig litters will be smaller.

5-2. As ruminants, cattle make good use of pasture grasses.

Proper animal nutrition increases feed efficiency and rate of gain. It decreases the number of days required to grow to market size. Producers want animals to grow fast. They also want them to make good use of the feed. Feeding too much or the wrong feed is a big waste!

Animals must be fed a diet that meets their needs. If their needs are not met, they will not grow, can get sick, and may die. The way animals eat and what they are fed depends upon many factors. Because of the differences in the digestive systems of animals, they must all be fed differently. For example, ruminants eat more roughages and grasses because they have four compartments to their stomach. The stages of life and the activities of the animal also affect feed intake.

NUTRIENT NEEDS OF ANIMALS

Feeds contain different substances. Each substance meets nutrient needs in varying ways. The substances are nutrients. A **nutrient** is a substance that is necessary for an organism to live and grow. They provide for the life processes of an organism.

A ration is the total amount of feed an animal is given in a 24-hour period. The ration for an animal may be fed in small amounts throughout the day or all at once. A **balanced ration** is one that contains all the nutrients that the animal needs in the correct proportions. Too much of

5-3. An elephant has a big appetite.

5-4. Feed may be packaged in convenient bags.

a nutrient causes waste and can result in toxicity. A deficiency of one nutrient can result in stunted growth and low production.

Each animal's nutritional requirements are different. Age, stage of development, environmental conditions, and genetic makeup affect the kind and amount of nutrients needed. A good understanding of nutrients helps producers meet each animal's nutritional requirements.

The nutrients required by all animals include water, carbohydrates, fats, proteins, minerals, and vitamins. Each has a unique purpose with animals.

Water

Water is necessary to sustain life. Animals can live longer without food than they could without water. Water is found in all the cells of the body. Water is a topic of many conversations these days because it is a limited natural resource. Many water supplies are polluted and methods need to be taken to preserve our current supplies.

Water makes up about 75 percent of an animal's body and as high as 90 percent in newborn animals. Dairy cows may consume up to 50 gallons of water on a hot summer day. Water even makes up 70 to 80 percent of plants.

Functions. Water has two basic functions: regulating body temperature and a major part of all biochemical processes.

Water controls temperature because it allows heat to accumulate, readily transfers heat, and loses heat through evaporation. This is because of the mobility of the blood. All biochemical reactions require water. Many molecules dissolve in water, and water transports these elements throughout the body.

5-5. **Water well, pump, and storage tank on a beef cattle feedlot in California.**

Consumption. Animals get water in several ways. Sources include drinking water, water in feed, and water produced through biochemical reactions. Water is lost through urine, feces, sweat, and vaporization from the lungs.

Consumption of water increases due to heat stress or high temperature, the ability of the animal to conserve water, and differences in activity, gestation, and lactation.

5-6. **This automatic waterer provides clean water for horses.**

Drinking water should be clean and free of dangerous chemicals. Special attention to water supplies is needed in the summer due to excessive heat. Winter time can also cause water problems. Cold temperatures cause water to freeze. Animals should have fresh water supplies made available at least once a day if not all day long.

Water provides the environment for aquatic animals. The quality of the water is important beyond just for drinking. Chapters 14 and 18 have information on water for fish.

Carbohydrates

Carbohydrates provide energy. They are a major component of plant tissues. Carbohydrates should make up about 75 percent of the animal's diet. Besides energy, carbohydrates aid in the use of proteins and fats.

Functions and Sources. Carbohydrates provide energy for growth, maintenance, work, reproduction, and lactation. There are three types of carbohydrates: sugars, starches, and fiber. Simple sugars (monosaccharides) and double sugars (disaccharides) are two types of sugars. Glucose and fructose are simple sugars. Sucrose is a double sugar and is used to make table sugar. These are found in fruit and milk.

Starches are the food supply for plants. Starch is found in bread, potatoes, rice, and other vegetables. Starches and sugars are converted to glucose. *Glucose*, a simple sugar, is the ultimate source of energy for most cells. Starch is an energy reserve found in roots, tubers, and seeds. Potatoes are full of starch.

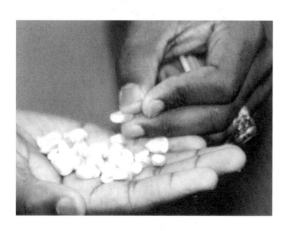

5-7. Corn is a major source of carbohydrates in animal feed.

Fiber is the material left after the food has been digested. It is made of plant cell walls and cellulose. The purpose of fiber is to help the digestive system run smoothly. Fiber absorbs water and provides bulk. It also helps to increase the bacterial population in the rumen of cattle.

As food is being digested, only monosaccharides can be absorbed from the digestive tract. Newborn animals are an exception. A newborn animal can absorb more complex molecules in its intestines than an older animal. This is why it is important to feed babies very shortly after they are born. This gives them energy and they receive antibodies from their mother's milk.

Nonruminants versus Ruminants. Nonruminants cannot break down the cellulose in fiber. Ruminants have microflora, or bacteria, in their stomachs to break down cellulose. This is why animals, such as cows and sheep, eat various forages and grasses. The fiber of young grasses and hay that is cut earlier is easiest to digest. Some monogastrics, such as horses and rabbits, have some microflora in their cecum and colon that allows them to digest some cellulose. An example of undigestible "human" food is oat bran.

5-8. Poultry are nonruminants. These growing broilers are fed a highly nutritious feed.

The major sources of carbohydrates are cereal grains, such as corn, wheat, barley, oats, hay, and rye. Abnormal carbohydrate metabolism can lead to ketosis and diabetes.

Carbohydrates are not stored in the body and must be eaten every day. A small amount of glucose is always in the blood. Unused glucose and carbohydrates are converted to fats.

Fats

Fats provide energy. They contain the highest amount of energy, with 2.25 times more energy than carbohydrates. Fats help supply energy for normal body maintenance. Fats also provide a healthy skin, keep the nervous system healthy, give food a good flavor, and carry fat soluble vitamins A, D, E, and K.

Animals do not develop a fat deficiency for energy. However, they can develop a vitamin deficiency.

Fats fed to animals include tallow, vegetable oil, tankage (processed meat and bones), cottonseed, and fish meal.

Increasing data shows that obesity in humans and animals is genetically based. This indicates the importance of genetic selection in animal breeding programs.

Humans consume fats in butter, margarine, eggs, cheese, cream, ice cream, whole milk, and oils.

Protein

Protein itself is not essential, but the amino acids that make up protein are. **Protein** is needed to build new and repair old tissue. Protein is especially important for weight gain, growth, and gestation. It is used only for energy when carbohydrates and fats are deficient.

Protein serves many functions. It is found in wool, feathers, horns, claws, beaks, DNA, RNA, skin, hair, hooves, blood plasma, enzymes, hormones, and immune antibodies. Protein is needed every day for basic functions. It is also needed to build and repair cells. Three to five percent of the body's proteins are rebuilt every day. The highest amounts of protein can be found in the muscle of animals and in the leaves of plants.

Amino Acids. Ten amino acids are known as the essential amino acids and the others are known as nonessential. **Amino acids** are the building blocks of proteins. Organisms use amino acids to build proteins.

The ten essential amino acids are used to make the other amino acids that the body needs. Animals with simple stomachs must eat a variety of food. Each food contributes only part of the essential amino acids. They can synthesize all but 10 or 11 of the amino acids.

Ruminants have microbes (bacteria and protozoa) in their rumen. The microbes synthesize the essential amino acids from nonessential amino acids. The microbes are then digested in the digestive system. This provides the ruminant with essential amino acids. Because of this, much research has been done to increase the efficiency of ruminants.

Sources of protein include soybean meal, cottonseed meal, fish meal, tankage, skim milk, and alfalfa hay. Table 5-1 lists the essential amino acids.

Table 5-1
Essential Amino Acids

arginine	methionine
histidine	phenylalanine
isoleucine	threonine
leucine	tryptophan
lysine	valine

Protein Deficiency. Protein is the most common nutrient deficiency. Most feedstuffs are low in protein and supplements are expensive. Abnormal levels of protein can create stress on the urinary system. For example, corn is deficient in lysine and tryptophan (amino acids), which causes very slow growth. By feeding soybean meal that is low in methionine and cystine (amino acids) this problem can be solved. Eating corn flakes with milk for breakfast is a similar example.

Many symptoms can show a protein deficiency. Some include: anorexia, slow growth rate, decreased feed efficiency, anemia, edema, low birth weight of young, and lower milk production. Protein consumption is like a chain. When one amino acid is deficient, the whole body suffers.

Soil deficient in minerals may produce crops that are low in the mineral. This can lead to deficiencies in feeds made from the crops. Some of these include selenium, copper, manganese, cobalt, and iodine. If pasture or hay is of poor quality, then protein supplements may be required.

Minerals

Minerals are inorganic elements found in small amounts in the body. Minerals are a major component of the skeleton and are necessary for the endocrine, circulatory, urinary, and nervous systems to function properly.

Minerals are necessary for a strong skeletal system. Minerals also make up organic compounds needed by muscles, organs, blood cells, and other soft tissue. They regulate many body functions.

Macrominerals are minerals required in large amounts. *Microminerals*, or trace minerals, are minerals required in smaller amounts. They are just as important as macrominerals. Calcium and phosphorus are macrominerals required in the highest amounts. Calcium makes up 49 percent and phosphorus 27 percent of the minerals needed by animals. Phosphorus is a key ingredient in the body's protein.

Lack of calcium causes bones and teeth to form improperly. When calcium is not consumed on a daily

5-9. Some minerals may be provided in blocks that animals can lick. This block is in a holder that keeps it clean.

basis, it is removed from the bones to be used in other body processes. This will weaken the bones, causing osteoporosis. Osteoporosis is a disease in which the bones deteriorate. This affects roughly 25 percent of older men and 50 percent of elderly women.

Iron deficiency leads to anemia and a tired feeling. Iron is needed to make hemoglobin for red blood cells. Lack of iodine will lead to fatigue, increased appetite, and a rapid pulse. Magnesium deficiency causes muscle tremors and shaking.

Sodium and potassium are needed to maintain water balance. Sodium and potassium transfer nutrients and waste through the cell membrane.

Minerals are found in alfalfa, cereal grains, bone meal, molasses, and salt.

Deaths due to mineral deficiencies are rare, but inadequate amounts can cause economic losses to the producer. Deficiencies may cause a poor rate of gain, feed inefficiency, decreased reproduction, and a decrease in production of meat, milk, eggs, and wool.

Trace mineralized salt should be made available to animals. This will give most animals all the minerals they need. Table 5-2 presents the mineral needs of animals. The functions and deficiency symptoms are also shown.

Table 5-2
Mineral Needs, Functions, and Deficiency Symptoms

Macrominerals	Function	Deficiency Symptoms
Calcium (Ca)	structural component of the skeleton; controls the excitability of nerves and muscles; required for coagulation of the blood	rickets or osteomalacia develop— the bones become soft and deformed; milk fever occurs in cows; blood clotting time increases
Phosphorus (P)	component of the skeleton providing structural support; aids in lipid transport and metabolism; component of AMP, ADP, and ATP	rickets; depressed appetite; may chew on wood or other objects
Magnesium (Mg)	needed for normal skeletal growth; needed in chemical reactions in muscles; activates enzymes during the Kreb cycle	anorexia; reduced weight gain; reduced magnesium in the blood; hyperemia of the ears and other extremities

(Continued)

Table 5-2 (Continued)

Macrominerals	Function	Deficiency Symptoms
Potassium (K)	required to move sodium; helps to maintain acid-base balance in the body; helps to uptake glucose and carbohydrates	slowed growth; unsteady walk; overall muscle weakness; Mg deficiency causes K deficiency
Sodium (Na)	maintains osmotic pressure; maintains acid-base balance; required in transmission of nerve impulses	reduced rate of growth; reduced feed efficiency; decreased milk production; weight loss in adults
Chlorine (Cl)	regulates extracellular osmotic pressure; maintains acid-base balance in the body	slowed growth; fall forward with legs extended backward when startled by a sudden noise
Sulfur (S)	required for protein synthesis; used in cartilage; in birds S is used in feathers, the lining of the gizzard, and in muscle	reduced growth; reduced weight gain
Cobalt (Co)	makes up vitamin B_{12}	loss of apetite; reduced growth; anemia; if untreated, death will result
Iodine (I)	component of thyroxin—controls the oxidation of cells	Goiter—enlargement of the thyroid gland; dry skin; brittle hair; young born without hair; reproductive problems
Zinc (Zn)	activates enzymes; used in DNA and RNA	slowed growth; anorexia; scaling and cracking of paws; poor feathering; rough hair
Iron (Fe)	found in hemoglobin in the blood; found in myoglobin in muscle	anemia—common problem among newborns; pale color; shallow breathing; rough hair; slow growth; iron deficiency affects about half of the world's human population
Copper (Cu)	enzyme activity to uptake iron; needed to maintain the central nervous system	anemia; incoordination
Manganese (Mn)	needed in bone structure	skeletal defromation—enlarged joints, lameness, shortening of legs, and bowing of the legs
Selenium (Se)	component of numerous enzymes	nutritional muscular dystrophy

Vitamins

Vitamins are organic substances needed in small quantities to perform specific functions in an animal. They do not provide energy, but are necessary in using energy. Vitamins help to regulate body functions, keep the body healthy, and develop resistance to diseases. Vitamins are required in very small quantities for these functions. The deficiency of any one can cause death.

Vitamins are either fat soluble or water soluble. *Fat soluble vitamins* are stored in the fat and are released as they are needed by the body. These include vitamins A, D, E, and K. *Water soluble vitamins* are dissolved by water and need to be consumed every day. They include vitamin C and the B vitamins.

The body needs only a small amount of vitamins. However, if the body does not get the necessary vitamins, chemical processes will not function correctly.

Ruminants do not have any problems getting the necessary nutrients. Vitamin A may be the only one that is not readily available in most feeds. Vitamin D can easily be obtained when animals are exposed to the sun.

Table 5-3 lists the functions and deficiency symptoms of these vitamins.

Table 5-3
Functions and Deficiency Symptoms of Vitamins

Fat Soluble Vitamins	Function	Deficiency Signs
Vitamin A	required in retinol for night vision; needed in epithelial cells which cover body surfaces; needed for bone growth	night blindness; dry and irritated eyes; respiratory infection; reproductive problems
Vitamin D	enhanced Ca and P levels allowing bone mineralization; prevents tetany	abnormal skeletal development—lameness, bowed, and crooked legs; slowed growth
Vitamin E	promotes health	failure of the reproductive system; changed cell permeability; muscular lesions
Vitamin K	required for blood clotting	long blook clot time; hemorrhages; in severe cases, death

(Continued)

Table 5-3 (Continued)

Fat Soluble Vitamins	Function	Deficiency Signs
Thiamin (B1)	promotes health	anorexia; beriberi in humans— numbness, weakness and stiffness in thighs, unsteady wal, edema in feet and legs, and pain along the spine
Riboflavin (B2)	functions in coenzymes	reduced growth rate; skin lesions; hair loss
Niacin	used by cells in energy metabolism	retarded growth; decreased appetite; diarrhea; vomiting; dermatitis
Pantothenic Acid	needed in energy metabolism	slowed growth; dermatitis; graying of the hair; fetal death; skin lesions
Vitamin B6	help with protein and nitrogen metabolism; involved in formation of red blood cells	convulsions; lesions around feet; face, and ears
Vitamin B12	needed in several enzyme systems	anemia*; retardation; skin pigmentation
Folacin	used in a variety of metabolic reactions	slow growth rate; anemia
Biotin	needed for several enzyme systems	scaly skin; abnormalities of the circulatory system
Choline	aids in transmission of nerve impulses	fatty liver; hemorrhaging kidney
Ascorbic Acid (Vitamin C)	prevents scurvy; causes several metabolic reactions to occur	scurvy—edema, weight loss, and diarrhea

*This can occur in monogastics fed entirely from plant material. Babies nursing from vegetarian mothers may also develop a vitamin B12 deficiency.

NUTRIENT USES AND REQUIREMENTS

Animals use different feeds and amounts to meet their changing needs. They need nutrients to maintain their bodies, for growth, reproduction, lactation, and for work. Nutrients are carried to the cells of the body via the blood. After entering the cells, nourishment and energy are provided by the nutrients.

MAINTENANCE

Maintenance is no loss or gain of the body. No matter what an animal is doing, nutrients are required. Animals require energy for internal workings of the body, heat to maintain body temperature, and small quantities of vitamins, minerals, and proteins.

A maintenance diet is usually high in carbohydrates and fats. It contains small amounts of protein, minerals, and vitamins. About one-half of the feed consumed is used for maintenance requirements. Animals on a maintenance diet are usually not growing, lactating, reproducing, or working.

GROWTH

In addition to maintenance, nutrients are needed for growth. **Growth** is an increase in the size of muscle, bones, and organs of the body. Meat production depends upon the growth of the animal. Growth requires mostly energy and smaller amounts of the other nutrients. A growth ration is high in protein, vitamins, and minerals. For example, chickens are fed a carefully prepared ration so they grow rapidly.

Young animals and breeding stock will receive a growing ration. Young animals are expected to grow rapidly and efficiently. Good animals require less feed above the maintenance level than do poor animals because they are more efficient.

5-10. Growing pigs need a complete ration high in protein.

REPRODUCTION

Nutrition is a major reason for reproductive failures. During the first trimester of pregnancy, the fetus requires large amounts of protein, minerals, and vitamins. A ***reproduction ration*** is high in nutrients needed by breeding animals. Females that do not receive a proper ration will give birth to underweight babies or abort the fetus before birth.

Males also need additional nutrients for reproduction. Lack of nutrients can lower sperm production and fertility rates. A male with an inadequate diet may be unable to breed the females.

LACTATION

Lactation is the production of milk. Females must have additional nutrients to produce large quantities of milk. A ***lactation ration*** is for females who are producing milk. Females that are pregnant and lactating require even higher levels of nutrients.

Since milk is high in protein, calcium, and phosphorus, milking animals need diets high in these nutrients. Dairy cows need a ration that has been specially prepared for them. Hens need a diet high in protein, fat, minerals, and vitamins because these are found in high amounts in eggs.

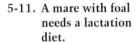

5-11. A mare with foal needs a lactation diet.

WORK

Work requires high energy feed. In the United States, work is usually limited to horses. In many other parts of the world, animals, such as oxen, require work energy. Increased carbohydrates and fats supply the needed nutrients.

TYPES OF FEEDS

Feedstuffs are ingredients in the feed for animals. Common feedstuffs include corn, soybean meal, and barley. Most feedstuffs have been carefully analyzed to determine the nutrients they contain. Some are high in certain nutrients and low in others.

Feed is what animals eat to get nutrients. Some feeds have few nutrients; others are high in nutrients. Most feeds provide more than one nutrient.

Feedstuffs can also be fed to provide flavor, color, or texture to increase palatability. *Palatability* is how well an animal likes a feed. Some feedstuffs are made more palatable by adding molasses. Many animals like the sweet taste of molasses. A feed high in nutrients is of little benefit if animals refuse to eat it!

Feeds are classified as forages, concentrates, or supplements.

FORAGES

Forages are feeds that are mostly leaves and tender stems of plants. Forages are bulky and high in fiber. They are also low in protein and energy. Forages are fed because they are relatively inexpensive.

Grasses, legumes, hay, crop residue, green chop, and corn silage are examples of forages. They contain high levels of fiber. Cattle often only eat forages. Some cattle in high production uses are provided concentrates and supplements in addition to forage.

5-12. Corn being harvested for silage. Note that the leaves, stalk, and ears are chopped together to make the silage.

5-13. Hay may be stored outside in dry climates. Rainfall is rare on this ranch in the western United States.

5-14. Round bales of hay covered to protect from damage by rain on a Virginia ranch.

Forages are harvested at the vegetative stage. As the plant matures, it loses energy because the energy is being put into the seed for reproduction. The younger the plant, the more energy and protein it will contain.

CONCENTRATES

Concentrates are feeds that are high in energy. They have more energy per pound than forages, but are usually lower in protein. Examples of

5-15. Concentrates may be fed in different forms. From the left, this shows a finely-ground meal, small pellet, and large range cube. The finely-ground meal form is for small baby animals, such as chicks or fish. The small pellet form is for calves, growing fish, dogs, and similar animals. The large range cube or pellet is for cattle on pasture that need supplemental feed.

concentrates are corn, wheat, sorghum, barley, rye, and oats. Higher producing animals need more nutrients to be productive.

Concentrates are usually not fed in a maintenance ration. If the forage is of very poor quality, concentrates may be needed.

SUPPLEMENTS

High producing animals are given supplements so they get the protein, minerals, and vitamins that they require. A **supplement** is a feed material high in a specific nutrient. Some contain several nutrients. Examples of protein supplements include meals made from soybeans, cottonseed, corn gluten, sunflower meal, rape seed, and coconut.

Protein quality is usually of less importance in ruminants than non-ruminants due to the microbes present in the rumen. Nonprotein sources of nitrogen, such as urea or ammoniated molasses, are available for ruminants to eat.

Salt and mineral blocks can be placed in a feedlot or pasture for free choice feeding.

GOOD FEEDSTUFFS

All animals require different amounts of nutrients. All feedstuffs provide differing amounts of nutrients. This makes balancing rations a difficult task to accomplish.

A good ration should possess several characteristics. It should be balanced, have variety, be succulent, be palatable, bulky, economical, and suitable.

A balanced ration will increase feed gain, decrease expense, and increase profits. A variety of feeds will make ration-balancing easier and increase palatability. A succulent ration that is juicy and fresh will increase production.

A palatable ration is agreeable to the animal's taste. If the animal does not eat the feed, maximum production will not be reached. Bulky rations aid in digestibility because of the fiber.

Economical rations should provide the essential nutrients, while maximizing profits. The ration should be suitable to the type of animal. Ruminants and nonruminants require different rations. Cattle and sheep rations consist of large amounts of roughages. Poultry and swine are mostly fed concentrates.

HOW ANIMALS ARE FED

How and when animals are fed is important in gaining production and growth. Animals need to consume the recommended amounts of each nutrient without overeating. They need feed for maintenance, growth, reproduction, lactation, and work. Feed can either be fed at scheduled times or be fed free access.

5-16. Lamb with free access to hay in a combination grain and hay feeder.

FREE ACCESS

Free access or free choice is when animals eat feed when they want it. Feed is provided in feeders and is always available. Hay is commonly fed free access to cattle. Hogs can be fed concentrates free access because they will not overeat. Cattle should not receive concentrates free access or they will overeat.

5-17. Cattle with free access to hay. The hay is protected from being trampled into the mud with a hay feeder.

Livestock on pasture are commonly provided with salt or mineral blocks free access. Of course, the pasture is free access. Livestock can graze as they wish. Water is also usually provided free access.

SCHEDULED FEEDING

Many animals are fed at certain times of day. Feeding times may be based upon the needs of the animal or due to management practices.

Scheduled feeding is providing feed at certain times of the day. It is commonly used with cattle because they will overeat. Dairy cows are

5-18. Timed feeder for scheduled feeding of fish in a tank.

usually fed at milking time. While they are in the barn, they will each be given an individual ration.

Baby fish, known as fry, are fed once every hour. Their digestive systems will not hold much food because they are small. They are unable to store much energy and thus need to be fed frequently.

Computer feeding systems are becoming more popular with cattle and hogs. Computerized chips can be placed in eartags or under the skin in the ear. A computer will read the personal identification number and give an amount of feed into a feeding trough.

RATION BALANCING

Producers want to provide animals the feed they need. This involves two areas: the nutrient needs of animals and the nutrient contents of the feedstuffs.

5-19. Feed can be no better than its ingredients. This high quality barley will be used in manufacturing feed for a feedlot.

Nutritional information about feeds is used to formulate rations. Rations are the amounts of feed that an animal is given. The amount of each essential nutrient is figured into the ration. Comparing this with the nutrient requirements of the animal yields how much roughage, concentrate, and supplement is needed by the animal. This can be done with a

5-20. Manufactured feed is often delivered in bulk form to farms.

calculator or with a computer program. Feed specialists at feed companies formulate rations for specific needs.

In ration formulation, the feedstuffs are selected to give the least cost ration. Computer programs can also compute feed rations that are economical as well as nutrient balanced. For example, more corn may be in a ration if it costs less than barley.

When balanced rations are not fed, nutritional diseases may result. Undereating and overeating can both cause disease problems.

Manufactured feeds are made to contain amounts of nutrients. Nutrients are often stated as percentages. Labels attached to feed containers provide nutrient information. Labels give the amount of protein, the ingredients, and the name and address of the manufacturer. Labels may also have a bar code. Bar codes are electronically scanned and used in tracking a feed.

7 16374 20070 9

CO-OP
FRY FISH FEED

—Manufactured By—
**ALABAMA FARMERS
COOPERATIVE, INC.
DECATUR, ALABAMA 35609**

GUARANTEED ANALYSIS
Crude Protein, not less than 49.00%
Crude Fat, not less than 9.00%
Crude Fiber, not more than 2.00%

INGREDIENTS: Fish Meal, Meat Meal, Flash Dried Blood Meal, Dried Whey, Ground Corn, Fish Oil, Vitamin A Supplement, Vitamin D-3 Supplement, Vitamin E Supplement, Riboflavin Supplement, Calcium Pantothenate, Niacin Supplement, Vitamin B-12 Supplement, Choline Chloride, Menadione Sodium Bisulfite, Thiamine Mononitrate, Ascorbic Acid, Pyridoxine Hydrochloride, Folic Acid, Ethoxyquin (A preservative), Ground Limestone, and traces of Manganous Oxide, Calcium Iodate, Copper Oxide, Cobalt Carbonate, Zinc Oxide, Iron Carbonate and Sodium Selenite.

Tag Code: 021092

5-21. Sample label showing ingredients, analysis, and manufacturer.

REVIEWING

MAIN IDEAS

Nutrition is the process by which animals eat and use food. The food is used to live, grow, reproduce, lactate, and work. Nutrients are substances necessary for the functioning of an organism. The nutrients that all animals require include water, carbohydrates, fats, proteins, minerals, and vitamins.

Water is needed to regulate body temperature. It is a major component of biochemical processes. Carbohydrates and fats provide energy. The amino acids that make up protein are needed to build and repair tissue. Proteins are the most common type of nutritional deficiency found in animals. Minerals are a major component of the skeleton and are necessary for several of the body's systems to function properly. Vitamins perform many specific functions.

Nutrients are required by animals for maintenance, growth, reproduction, lactation, and work. A maintenance diet is one in which the animal consumes enough to prevent loss or gain of the body. It is high in carbohydrates and fats. Growing animals need more protein, vitamins, and minerals. A reproduction ration includes extra protein, minerals, and vitamins for the fetus. Lactating animals require additional protein and minerals. Work utilizes high energy feed.

Forages, concentrates, and supplements are the three different types of feed. Forages are feeds made up of leaves and stalks of plants. Concentrates are feeds high in energy and low in protein. Supplements are high in protein, minerals, and vitamins.

Animals can be fed free access at all times or they can be fed at scheduled times.

QUESTIONS

Answer the following questions using correct spelling and complete sentences.

1. What is nutrition? Why is it important to understand the importance of nutrition?

2. What are nutrients? What are the six nutrients that all animals require?

3. Why is each animal's nutritional requirements different?

4. Describe the functions of water.

5. What are carbohydrates? Describe the functions of carbohydrates.

6. Compare the carbohydrate requirements of ruminants and nonruminants.

7. What are fats? What is the main function of fats?

8. What is the primary function of proteins?

9. Compare and contrast protein utilization of ruminants and nonruminants.

10. What are minerals? Why are minerals important?

11. Define vitamins. Why are vitamins required by animals?

12. Compare and contrast the five nutrient requirements of animals.

13. Compare and contrast the three types of feeds.

14. Describe the characteristics of good feedstuffs.

15. How are animals fed?

EVALUATING

CHAPTER SELF-CHECK

Match the terms with the correct definition. Write the letter by the term in the blank that is provided.

a. nutrition   e. lactation ration   i. free access
b. ration   f. feedstuff   j. scheduled feeding
c. fiber   g. concentrate
d. protein   h. supplement

1. ____ providing feed at a certain time

2. ____ nutrient used to build new tissue and repair old tissue

3. ____ process by which animals eat and use food

4. ____ material left after food has been digested

5. ____ the amount of feed an animal has in a 24-hour period

6. ____ ration for females producing milk

7. ____ ingredient used to make feed

8. ____ material fed animals for nutrients

9. ____ feed material high in a specific nutrient

10. ____ when feed is available for animals when they want to eat

EXPLORING

1. Visit a local feed mill. Determine the kinds of feeds sold and used by local animal farms. Look at how the feed is packaged. Observe where the ingredients are stored. Prepare a report on your observations.

2. Visit with a local veterinarian concerning animal nutrition and the common problems that they deal with.

3. Visit with a nutritionist at a feed company about animal feeding and rations and common problems with which they deal.

4. Tour a local farm. Compare the various rations and feedstuffs that are fed to different animals depending on their ages and uses.

5. Get a label from an animal food container (horse, cat, dog, fish, etc.) Study the label and list the following information:

Kind of animal _____

Manufacturer _____

Major ingredients _____

Percent protein _____

Percent fat _____

Percent fiber _____

What is your assessment of the quality of the food: good, poor, or unsure? Why?

Chapter 6

ANIMAL HEALTH

Health is important to people. They want to have good health and enjoy a happy life. Animal producers also want their animals to have good health. Having animals with good health is important to producers and consumers.

People want a healthy, continuous food supply. To have enough quality food, we must raise healthy animals. Health of animals is important in the production of meat, milk, eggs, and other products.

Some animal diseases are spread to humans. This makes disease control more important.

Producers need to be aware of important principles in having animals with good health. Each species and each animal within a species is unique. Sometimes the assistance of a veterinarian may be needed. Regardless, it is far better to prevent health problems than to try to treat them.

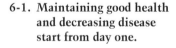

6-1. Maintaining good health and decreasing disease start from day one.

OBJECTIVES

This chapter provides guidelines to help producers in areas of animal health. It has the following objectives:

1. Explain health and how it relates to normal and abnormal behavior
2. Explain the impact of environment on animal health
3. List and explain the economic losses caused by poor animal health
4. Describe some common diseases and parasites of animals
5. Explain how good health is maintained

TERMS

anaplasmosis	endotherm	lice
anthrax	equine sleeping sickness	mastitis
antibodies	external parasite	noncontagious diseases
bacteria	foot and mouth disease	parasites
behavior	fungi	preconditioning
blackleg	grubs	protozoa
brucellosis or Bang's	health	roundworm
coccidiosis	hog cholera	shipping fever
contagious diseases	immunity	sleeping sickness
disease	internal parasite	tapeworm
disinfectant	isolation	virus
ectotherm	leptospirosis	

HEALTH AND BEHAVIOR

Health is the condition of an animal. It is a measure of how the functions of life are being performed. An animal is in good health when its life processes are functioning normally. *Behavior* is the reaction of the whole organism to certain stimuli, or the manner in which it reacts to its environment.

Disease is a disturbance in the functions or structure of an animal. Sometimes only one organ or body part may be affected. Other times, the entire animal may show disease. In short, disease is the lack of good health.

Healthy animals vary. Some animals that appear healthy may be suffering from pain or disease. Knowing each individual animal will help to easily detect and correct any ill health. A producer or animal owner will learn to diagnose good and ill health.

6-2. A healthy ewe and lamb.

GOOD HEALTH

Absence of disease is a sign of good health. Some signs of good health include: good appetite, alert and content behavior, bright eyes, shiny coat, normal feces and urine, normal vital signs, and normal reproduction.

Healthy animals will eagerly eat their food. A lingering or lazy animal may be in poor health. Healthy animals will stretch and watch the world around them. Eyes that are clear and alive are very important signs. A shiny coat is evidence of a proper diet. The presence of blood or mucus in the feces is a sure sign of a problem. Important vital signs to look for include temperature, pulse, and respiration. Table 6-1 illustrates the normal vital signs of various animals. Healthy animals will also produce healthy offspring.

Table 6-1
Normal Signs of Selected Animals

Species	Average Normal Temperature	Normal Pulse Rate	Normal Respiration Rate
	(rectal °F)	*(rate/min.)*	*(rate/min.)*
Cattle	101.5	60–70	10–30
Swine	102.6	60–80	8–15
Sheep	102.3	70–80	12–20
Goat	103.8	70–80	12–20
Horse	100.5	32–44	8–16
Chicken	106.0	200–400	15–30

Source: *Introduction to World AgriScience and Technology,* Interstate Publishers, Inc.

6-3. A swollen eye lid and discharge around the eye are signs of an eye disease.

ILL HEALTH

Animals with disease often have a behavior that is not normal. If the signs of a healthy animal are not observed, then the animal is in ill health.

Signs of ill health may include one or more of the following: lack of appetite, sunken eyes or discharge from the eyes, discharge from the mouth or nostrils, inactivity, rapid breathing, rapid pulse rate, high temperature, full hair coat, lumps or protrusions on the body, open sores, seclusion, bloody urine or feces, and loss in production levels or weight.

LOSSES DUE TO POOR HEALTH

Economic loss to producers from health problems is easy to see if animals die. The greatest losses due to poor animal health are less obvious. Diseases may lower production and threaten humans.

DEATH

Death is the most obvious loss due to poor animal health. Dead animals are of no value. The producer has lost all that was invested. Costs go up because of the expense of disposing of the animals. Chicken producers will frequently install compost pits to dispose of dead birds.

Dead animals should be buried or burned. Producers may need to get help in doing this. A veterinarian or health official may have information. Improperly disposing of animals may create bad odors and attract coyotes or wild dogs. They can also spread disease to other animals and people.

6-4. These newborn pigs have an ideal environment. The pen is designed so that the sow will not mash them. A heat lamp is used to keep them warm.

LOWER PRODUCTION

Healthy animals mean increased profits for owners. Poor health may result in lower production levels. Some easy-to-see results of poor health include: failure to reproduce or breed, slow growth rates, decreased milk production, decreased meat production and quality, and increased costs of production. All of these lead to decreased profits for producers.

HUMAN DISEASE

Diseased animals may spread diseases to humans. Examples are anthrax, brucellosis, leptospirosis, rabies, trichinosis, and tuberculosis.

Brucellosis or Bang's Disease is found in cattle. Humans may get undulant fever from cattle infected with brucellosis. Drinking raw milk from an infected cow is one way. Another is by getting milk or other fluids from an infected animal on open wounds. Dairy cattle are regularly vaccinated and checked for this disease. Pasteurization helps to make milk safe for human consumption.

Trichinosis is caused by a small worm in the flesh of some hogs. When pork is not cooked thoroughly, the disease may be transferred to humans. Very few hogs have this disease. People have stopped feeding raw garbage with pork scraps in it to hogs. Properly cooked pork and clean cooking areas have eliminated this disease. People should always thoroughly wash their hands after touching raw meat.

IMPACT OF ENVIRONMENT ON ANIMAL HEALTH

Undue stress can be caused by an animal's environment. A change in the environment or an unsuitable environment will create this stress, causing the animal to become more susceptible to diseases and parasites.

Important factors affecting the health of an animal include: temperature, light, moisture, moving, and pollution.

6-5. Our environment influences our livelihood and health just as temperature, light, moisture, moving, and pollution can affect an animal's health.

TEMPERATURE

Certain temperature ranges are best for animals to live in. If the temperature is too high or too low, health problems may occur.

Most animals raised on the farm are ***endotherms***. Endotherms maintain a constant body temperature. ***Ectotherms*** adjust their body temperature to that of their environment. Fish and other aquatic animals are examples of ectotherms.

Endothermic animals adjust to temperature extremes as best possible. During hot weather they may sweat, pant, or wade in cool water. Confined animals need proper ventilation and fans to keep cool. Animals on pasture need plenty of fresh water and shade trees.

During cold weather, animals will burn more energy to produce body heat. They must be fed more carbohydrates and fats to produce this heat. The muscles of the animal will involuntarily contract or shiver to burn energy and keep the body warm.

Younger animals have a more difficult time maintaining body temperature than older animals. They will require more care and insulation from the environment.

LIGHT

Chickens are affected by light more than most other animals. However, animals with light colored skin, such as hogs, may become burned by too much sunshine. In laying hens, light causes the pituitary gland to release hormones into the bloodstream. This causes them to grow faster and lay more eggs. Lights are often turned on in broiler and laying houses to take advantage of this. Cattle with light colors around their eyes are more likely to have eye disease. Other animals with light color skin may be more susceptible to

6-6. Young chicks have an ideal environment. Nutritious feed, water, clean bedding, and heat are provided. Lighting at night will speed up growth.

disease. For example, white cats are more likely to get skin disease than those of other colors.

MOISTURE

Moisture is due to humidity or precipitation. High humidity (moisture in the air) will not cool animals in warm weather as well as dry air. The perspiration will not evaporate as fast in a humid location.

6-7. Various bedding types absorb different amounts of moisture. Some work better than others for certain kinds of animals. This student is showing results of research comparing hay, straw, shredded newspaper, shavings, and mulch.

Diseases live and reproduce at a high rate in moisture. Animals raised in confinement are most likely to have this problem. Ventilation allows air to move about. It is needed to remove the moisture from a confined barn. Proper ventilation lowers the inside temperature and decreases the chance of a disease outbreak.

MOVING

Moving animals creates stress. The stress lowers their resistance to disease and illness. When they are corralled, hauled, or chased into pens and onto trucks, their stress level increases. Loading and hauling animals exposes them to environmental conditions, such as rain, snow, wind, and hot or cold temperatures. Trailers and loading chutes should be free of protruding nails, broken boards, wire, and other debris. Animals can be injured by these sharp objects.

Animals in good physical condition are better able to cope with stress. Producers may precondition animals to stress. This may be done by gradually acclimating them to their environment.

POLLUTION

Pollution of water sources, feed, and the environment may cause animals to develop a disease. Water and feed sources may be poisoned by pesticides or industrial waste. Storing or hauling feeds in areas that have contained chemicals may contaminate feedstuffs. Pastures sprayed with chemicals also pose a pollution problem.

COMMON DISEASES AND PARASITES

Disease is any disorder that keeps animals from carrying out life processes in a normal way. Symptoms are not the disease itself, but a sign of a problem.

CLASSIFICATION OF DISEASES

Diseases are either contagious or noncontagious. Knowing how a disease is spread will help to better control and prevent it. Diseases can be caused by many things, such as nutritional defects, morphological defects, or pathogenic organisms, such as viruses, bacteria, fungi, or protozoa.

Contagious Diseases

Contagious diseases are spread by direct or indirect contact with other animals. Pathogens, such as viruses, bacteria, fungi, protozoa, and parasites, cause contagious diseases. Pathogens are organisms that can produce disease. They will usually produce a toxin or poison that upsets the normal metabolism of the body. They are small and reproduce very rapidly.

Viruses. A **virus** is a tiny disease-producing particle too small to be seen with an ordinary microscope. Because viruses closely resemble the DNA of the animal's cells, they are hard to control. It is difficult to find a chemical that will kill the virus without killing the animal's DNA. Vaccinating animals is the most effective way of controlling viruses. Examples of viral diseases include rabies, hog cholera, foot and mouth disease, smallpox, yellow fever, and distemper.

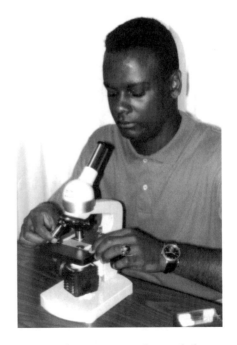

6-8. A microscope may be needed to diagnose a disease.

Bacteria. **Bacteria** are one-celled organisms sometimes called germs. Many pathogenic bacteria form spores that become resistant to control measures. Examples of dis-

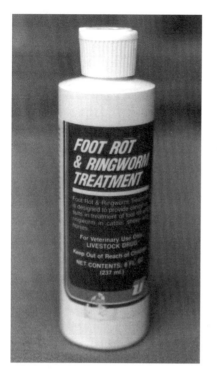

6-9. Liquid materials may be applied to the body to control fungal diseases.

eases caused by bacteria include tuberculosis, brucellosis, leptospirosis, anthrax, mastitis, and tetanus.

Fungal Diseases. *Fungi* are unicellular organisms that generally cause diseases on the outside of the body. Bacteria may be responsible for secondary infections. Examples of fungal diseases are ringworm and coccidiosis.

Protozoa. *Protozoa* are also unicellular organisms and are the simplest form of animal life. Examples of diseases caused by protozoa include malaria and anaplasmosis.

Parasites. *Parasites* are multicellular organisms that live in or on another animal. Parasites receive their nutrients from their host animal. *Internal parasites* live inside the host. They can be found in the digestive system, muscles, or other tissues of the body. They may enter the host when it is eating contaminated feed or when eating feed on a dirty floor. Most internal parasites are obtained through feces in the food. Pastures with very short grass also increase the likelihood of ingesting parasites. Examples of internal parasites include grubs, tapeworms, trichina, roundworms, and hookworms.

External parasites live on the external parts of the animal. They obtain their food from the blood and tissue of the animal. External parasites can transfer contagious diseases or other parasites to other animals. Examples of external parasites include ticks, fleas, lice, mites, and leeches.

Noncontagious Diseases

Noncontagious diseases are not spread by casual contact, but by a nutritional, physiological, or morphological problem.

Nutritional. Most nutritional diseases result from an unbalanced diet. Animals require specific amounts of fats, carbohydrates, proteins, minerals, and vitamins. Too high or too low amounts of any of these may result in a nutritional disease and decrease profitability. Mineral and protein deficiencies are quite common. (Chapter 5 discusses the nutrition of animals.)

Physiological. This type of disease results from a defect of a tissue, organ, or organ system. When an organ causes the body's metabolism to be off balance, other problems occur. Examples of physiological diseases include milk fever, heart failure, birth defects, and acetonemia.

Morphological. Poor management can result in accident or injury to animals. Possible injuries include cuts, scrapes, scratches, bruises, and broken bones. These decrease the efficiency of the animal.

Good management practices can decrease the likelihood of these problems. Pick up loose wire, old lumber, and equipment. Remove protruding nails and broken boards.

Cattle may be given magnets to collect nails and wire that they eat. Feed mills often run feed over magnets to remove any metal particles that may be in the ingredients.

6-10. Properly installing electrical wiring in animal facilities will decrease the chance of electrical shock.

SELECTED DISEASES OF ANIMALS

Some diseases affect only a select species while others affect several different animal species. Roundworms are found in hogs, cattle, sheep, horses, poultry, and fish. Blackleg is primarily found in cattle.

Several common animal diseases are briefly described below. For proper diagnosis and treatment of any disease, contact your local veterinarian.

Anaplasmosis

Anaplasmosis is a parasitic disease caused by a protozoan that attacks red blood corpuscles. Cattle are primarily affected. In chronic anaplasmosis, animals become anemic and may overcome the disease on their own. In acute forms, animals have rapid heartbeat, muscular tremors, and loss of appetite. Sick animals may become aggressive and want to fight. Death may occur in a few days. The disease is prevented by immunizing cattle. Some treatments are available, if the disease is caught early.

Anthrax

Anthrax is an acute infectious disease that affects most endothermic animals. It most frequently affects cattle during the summer when they are on pasture. Affected animals have a fever, rapid respiration, and swelling on the neck. Animals will die suddenly. Prevention of anthrax includes vaccinating animals, controlling flies, and sanitizing their environment. Penicillin may be effective if given in large doses.

Bang's or Brucellosis

Bang's or *Brucellosis* affects cattle, sheep, goats, and hogs. The reproductive tract of the female is infected, causing young to be aborted. Cattle may have to be bred several times before they become pregnant. When they do become pregnant, it may end in only a few weeks. Bang's is prevented by vaccinating heifers. Sanitation, testing cattle, and bringing only Bang's-free cattle into a herd help to control this disease. Humans can get this disease from infected animals or their milk. In humans, this is known as undulant fever. If humans survive undulant fever, they may suffer permanent crippling and disability.

Blackleg

Blackleg is an acute, highly infectious disease that usually results in death. It primarily affects cattle, although sheep and goats can also get it. Symptoms include high fever; swelling in the neck and shoulder; and muscles in the neck, shoulder, and thighs crackle when mashed. Eventually animals lose their appetite and die. Few animals recover from blackleg. Blackleg is prevented by vaccinating calves with a bacterin, which is a vaccine made from killed or inactive bacteria.

Hog Cholera

Hog cholera is a viral disease that is highly contagious. Hogs get high fevers, lose their appetite, get weak, and drink lots of water. Considerable effort has been spent eradicating hog cholera in the United States through federal government programs. There is no known treatment. Any hogs with hog cholera are destroyed.

6-11. **Properly diagnosing a disease requires study. The assistance of a veterinarian may be needed.**

Coccidiosis

Coccidiosis is a parasitic disease affecting chickens, turkeys, ducks, geese, and game birds. It is caused by a protozoan. Symptoms include bloody feces, ruffled feathers, unthrifty or sick appearance, and pale coloring. Medications known as anticoccidials are put in feed and water to prevent and treat this disease. Since it is transmitted by infected birds, wild birds should be kept away from poultry flocks and infected areas should be sanitized.

Equine Sleeping Sickness

Horses are affected by *sleeping sickness*. It is caused by a virus transmitted by insects that bite horses, mules, and wild rodents. Affected animals walk aimlessly crashing into things; later they appear sleepy, cannot swallow, grind their teeth, and possibly go blind. Some horses will recover, but they will usually die within two to four days. No effective treatment has been developed, but horses can be vaccinated to develop an immunity.

Foot and Mouth Disease

Foot and mouth disease affects animals that have cloven or divided feet. It is a highly contagious disease caused by a virus. There is no known treatment or vaccine. Quarantine is the best control at the present time. No animals or meat can be imported into the United States with this disease. Animals with the disease get watery blisters in the mouth and on the skin around the hooves. Teats of females may also blister. Animals will have a high fever. The United States does not currently have a foot and mouth disease problem, but many other countries do.

Grubs (Warbles)

Cattle **grubs** are internal parasites caused by heel flies. The fly lays its eggs in the summer around the heels of cattle. When they hatch as larva, they go into the skin and move through the body until they reach the back in a few months. In the back, the grubs cause bumps to swell in late winter and spring. The larva (about ³/₄ inch or 2 cm long) will come out of the bumps and grow into an adult fly. A good fly control program will prevent grubs. Several treatments are available and should be followed according to their directions. Grubs damage the hide by making holes in it as well as the flesh around it. Cattle grubs cause more injury to cattle than any other insect.

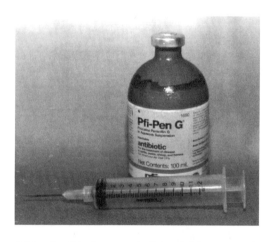

6-12. **Antibiotics may be injected into an animal using a syringe and hypodermic needle to treat some diseases.**

Leptospirosis

Leptospirosis is a bacterial disease that affects cattle, sheep, and most other farm animals, including dogs. Symptoms include high fever, poor appetite, and bloody urine. Females will abort their fetuses. Antibiotics are sometimes used to treat leptospirosis. Animals may be vaccinated against this disease.

Lice

Lice are external parasites that attack cattle, hogs, and other species. Lice are small insects that suck their host's blood. Animals may become anemic. Lice also cause animals to itch; they will often be seen scratching or rubbing against trees or posts. Lice usually appear in the winter. Back rubbers and other means can be used to administer insecticides to control lice.

Mastitis

Mastitis is the leading cause of profit loss in dairy cattle. Mastitis is a bacterial disease that affects female cattle, sheep, goats, and swine. It is an inflammation in the udder that interferes with milk production. Chronic mastitis causes the milk to be thick or lumpy. Acute mastitis is indicated by a fever and a hard, warm udder. Laboratory tests are used to detect most mastitis. Chronic mastitis that is not treated can cause the death of the animal.

Poisonous Plants

Poisonous plants are sometimes problems. Examples of poisonous plants include low larkspur, oak, tall larkspur, timber milk vetch, and water hemlock.

Most poisonous plants cannot be totally eliminated. Possible loss is reduced with good pasture management. Routine weed control will kill most poisonous plants.

Knowing the common poisonous plants in the area prevents losses. Study the symptoms of poisoned animals. Provide good pasture and feed for range animals. Treat animals that may have been poisoned promptly.

6-13. Common plants may be poisonous to animals. The College of Veterinary Medicine at Mississippi State University maintains a poisonous plant identification garden to use in training veterinarians.

Roundworms and Tapeworms

Roundworms, tapeworms, and other worms are parasites that can infect all animals. They are found in the inttestines and the stomach. Tapeworms can grow several feet long. When they become this large, they use much of the feed that the animal has eaten. Animals with low infestations of small-sized worms may not show symptoms. As the number and size of worms in the digestive tract increases, the animal loses weight, becomes anemic, and may have diarrhea. Sanitation is a good control measure.

Shipping Fever

Cattle and sheep of all ages can get *shipping fever*, though it is more of a problem for younger animals. It is an environmental disease caused by conditions animals encounter when hauled or sold. Animals will develop a high temperature, discharge from the eyes and nostrils, and may cough. Most will have difficulty breathing. It is more likely to affect thin, underfed animals that are hauled long distances. Cattle should be vaccinated three to four weeks before being moved.

MAINTAINING GOOD HEALTH

The key to personal satisfaction and the financial profitability of raising animals is to keep them healthy. Regular practices and management will increase production and decrease losses due to poor animal health. An understanding of the animal's bodily defenses will lead to healthy management practices.

BODY DEFENSES

The body's first line of defense is the epithelial tissue or skin and mucous membranes. Once the foreign substance has gotten past the epithelial tissue the body uses other mechanisms.

The digestive tract is very high in acidity that will depress bacterial growth. Tissue fluids (lymph) contain leukocytes that neutralize or engulf pathogens. The liver and other lymph organs will trap pathogens until they can be destroyed.

An inflammatory reaction usually occurs in an animal. During this, redness will occur because increased blood is flowing to the infected area; the area will feel hot; swelling or edema will occur; and increased sensitivity or pain is felt by the animal. All of these are the body's attempt at destroying pathogens. In addition, metabolic activity and immune reaction (development of antibodies) will increase.

HEALTH MANAGEMENT PRACTICES

Several methods of maintaining good animal health are: keep the animal's environment sanitary, provide proper nutrition, isolate sick ani-

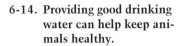
6-14. Providing good drinking water can help keep animals healthy.

mals, restrict truck and equipment traffic, restrict human access to animals, precondition animals to stress, and immunize animals.

Environmental Conditions

The environmental conditions that the animal lives in must be checked regularly. Space requirements should be met. Allow enough room for animals; do not overcrowd. Ventilation is needed to have a continuous supply of fresh and clean air. Providing a clean and dry place to lay helps to decrease the disease potential.

Sanitation

Keeping areas where animals are raised and fed clean is sanitation. Filth carries disease. Sanitation decreases the chances of animals contracting disease.

Dead animals should be removed and disposed of properly. Barns and facilities should be cleaned regularly. Waste and manure should be disposed of properly and regularly. Many farms have fenced lagoons or pits to temporarily store animal waste.

Rats, mice, and other rodents should be controlled as they can transmit diseases.

A *disinfectant* is a substance that destroys the causes of disease. Facilities, especially where young are raised, should be cleaned with a disinfectant regularly. Disinfectants include alcohol, iodine, lime, bleach, and soap. People should always follow safety precautions in using disinfectants.

6-15. Sanitation requires the proper and regular disposal of manure. Manure can be temporarily stored in a slurry store or manure pit until it can be spread or knifed into fields.

Clean facilities increase the natural beauty of the surroundings and lessen susceptibility of disease.

Proper Nutrition

Animals that have proper nutrition are more resistant to disease and stress. Providing the proper ration is important. Poorly fed, weak animals do not have the resistance needed to fend off diseases.

Isolation

Isolation is separating diseased and non-diseased animals. Diseases are often spread by animal contact. Isolating diseased animals decreases the risk of spreading a disease.

New animals brought to a farm should be isolated for three to four weeks. If the animals have a disease, the signs should be visible in that

6-16. Good quality hay can provide many nutrients for horses. (Courtesy, American Quarter Horse Association)

time. If no diseases are seen, the animals may be turned in with the herd or flock.

Restrict Truck and Equipment Traffic

Diseases can be brought to a farm by trucks, farm equipment, and other vehicles. Trucks and equipment should be disinfected between trips to other farms or feedlots. Mud, manure, sawdust, and straw can harbor disease.

Restrict Human Access

People may bring diseases with them as they travel from farm to farm. Disease can be transferred on shoes or boots. Some farms require everyone to walk through a boot tub of disinfectant solution. Other places restrict facilities to only a few individuals. Many breeding farms require visitors to wear plastic slip-on boots or only allow visitors to view the facility with video cameras.

Preconditioning

Preparing animals for stress is **preconditioning**. Stocker calves and other animals that are to be hauled are often preconditioned. Preconditioning is important with any activity that might cause stress. Castration and dehorning should be done by the time cattle are two months old, so as not to coincide with transfer to another lot or weaning. Calves should be weaned and started on feed thirty days before being sold. The amount of handling and moving time should be kept to a minimum.

Immunization

Immunity means that an animal resists disease. An animal often develops immunity when it has a disease. Artificial immunization is used to protect animals from disease.

Antibodies are immune substances produced in the body. Immunity can be developed to many different diseases, thus allowing animals to withstand exposure to disease even when coming in contact with it.

Vaccines are used to help animals develop artificial immunity to diseases. Animals should be vaccinated just like humans.

Mothers often pass immunity to newborn young. Young animals get immunity by antibody transfer from the mother through the placenta or

6-17. Preparing to inject a calf to artificially create immunity. (Courtesy, Marco Nicovich, Mississippi State University)

colostrum. Colostrum is a substance found in milk shortly after giving birth. The presence of colostrum quickly drops a few hours after birth.

Animals may be fed vaccines in their feed or given shots. Most vaccines contain dead or living organisms that cause the disease.

REVIEWING

MAIN IDEAS

Health is the condition of the body and a measure of how the functions of life are being performed. Health affects behavior of the animal. Poor health causes personal and economic losses to

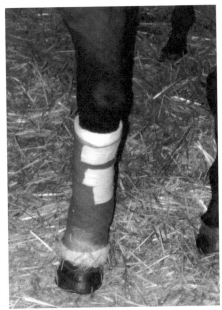

6-18. An injury to the leg of a horse has received veterinary medical treatment.

the owner or producer. Death, low production, and human contraction of disease may result.

Signs of good health include good appetite, alert and content behavior, bright eyes, shiny coat, normal feces and urine, normal vital signs, and normal reproduction.

Signs of ill health include lack of appetite, sunken eyes or discharge from the eyes, discharge from the mouth or nostrils, inactivity, rapid breathing, rapid pulse rate, high temperature, full hair coat, lumps or protrusions on the body, open sores, seclusion, bloody urine or feces, and loss in production levels or weight.

Diseases may be contagious or noncontagious. Contagious diseases may come from viruses, bacteria, fungi, protozoa, and parasites. They are spread by direct or indirect contact with other animals. Noncontagious diseases are due to poor management practices.

Maintaining good animal health is easy if the well-being of the animals is considered. Animals must be placed in a good, clean environment that is free of debris. They should not be overcrowded and must have a clean, fresh, and nutritious supply of feed and water. Dead animals must be disposed of properly. Diseased and sick animals should be isolated. A disinfectant should be used to clean stalls, feeders, and waterers. Animals may be preconditioned for stress and vaccinated to develop artificial immunity.

QUESTIONS

Answer the following questions using complete sentences and correct spelling.

1. Compare and contrast health and disease.

2. How does health affect behavior?

3. Give four signs of good health.

4. Give four signs of ill health.

5. What are some results of poor health? Explain each.

6. How does the environment affect health?

7. How can the environment be manipulated for the benefit of animals?

8. Explain the difference between contagious and noncontagious diseases.

9. List three diseases that affect cattle; horses; poultry; sheep and goats; and swine.

10. How does the body fight off disease?

11. List and explain several methods for maintaining good health.

EVALUATING

CHAPTER SELF-CHECK

Match the terms with the correct definitions. Write the letter by the term in the blank that is provided.

a. disease   d. disinfectant   g. shipping fever
b. health   e. contagious disease   h. preconditioning
c. immunity   f. isolation

1. ____ disease spread by direct or indirect contact among animals

2. ____ substance that destroys the causes of disease

3. ____ separating diseased from healthy animals

4. ____ ability to resist disease

5. ____ preparing animals for stress

6. ____ caused by conditions animals encounter when hauled

7. ____ disturbance in the functions of the body

8. ____ condition of the body and how well life functions are being carried out

EXPLORING

1. Visit a local veterinarian and tour their facilities. Ask about the most common health problems that they deal with—symptoms, causes, preventions, and treatment. Observe the treatment of an animal. Write a report on what you saw.

2. Conduct a job-shadowing project and follow a veterinarian or animal health technician for at least a day.

3. Check the vital signs of an animal. Record its pulse, respiration rate, and temperature. Compare the data to normal animals. Check the vital signs over a period and observe the fluctuations. Be sure to follow the proper procedures so that the animal is not injured.

4. Observe an animal that is in good health. List the signs that this animal exhibits. Observe an animal in ill health and compare the lists.

Chapter 7

ANIMAL REPRODUCTION

Baby animals cause excitement! People enjoy baby animals in different ways. Pet owners like a new litter of puppies or hatch of fish. Farmers and ranchers like baby animals, too, but from a different point of view. Producing new individuals is a big part of raising animals.

Animals are used for many purposes. People want a wide range of animal products. The goal of producers is to provide plenty of animals to meet the demand. If there are not enough animals, people will not have the products they want. Shortages develop and cause prices to go up.

Animal producers must raise an adequate supply of the kinds of animals that give the products people want. Animal supply is tied to reproduction. Producers must understand the processes involved in reproduction.

7-1. Cow with newborn calf that will grow to provide a useful product.

OBJECTIVES

This chapter focuses on the fundamentals of animal reproduction. The following objectives are included:

1. Explain the role of animal reproduction
2. Name and describe the functions of the major reproductive organs
3. Describe the phases of the estrus cycle
4. Explain the phases of reproductive development in the life of an animal
5. Describe the role of animal reproduction technology

TERMS

accelerated lambing	estrous synchronization	parturition
anestrous	estrus	pregnant
artificial insemination	fertilization	puberty
breed	fetus	purebred
breeding	gamete	reproduction
cloning	gestation	reproductive efficiency
conception	hormones	sexual reproduction
crossbreeding	insemination	sperm
egg	lactation	testicle
embryo	natural insemination	zygote
embryo transfer	ovary	
estrous cycle	ovulation	

ROLE OF ANIMAL REPRODUCTION

Reproduction is the process by which offspring are produced. The offspring are of the same species and have traits of their parents. For example, cattle reproduce other cattle though the offspring are not identical to their parents.

Reproduction is not essential for an organism to live. It is essential if a species is to stay in existence. Without reproduction, there would be no new animals.

SEXUAL REPRODUCTION

Sexual reproduction is the union of a sperm and an egg. A new animal begins at the time of this union. Two parents are required: male and female. The *sperm* (spermatozoa) is the sex cell produced by male animals. The *egg* or ovum is the sex cell produced by female animals.

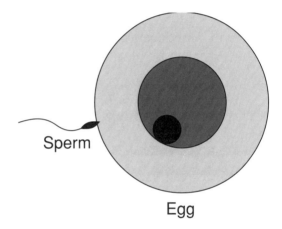

The union of sperm and egg is *fertilization*. Fertilization takes place inside the reproductive tract of female mammals, such as cattle and hogs. With other animals, fertilization is outside the reproductive tract, such as fish. In either case, time is needed for the union to grow into a new individual.

7-2. Sexual reproduction involves union of sperm and egg.

With mammals, the fertilization of an egg to create new life is *conception*. The new developing organism is formed in the reproductive tract of the female. A female that is carrying unborn young is *pregnant*.

Insemination

Insemination is placing sperm in the reproductive tract of the female. Animal producers sometimes regulate insemination. Male animals may be kept separately from the females. Other new technologies are being used in animal reproduction.

Natural insemination is the male depositing semen (fluid containing sperm) in the reproductive tract of the female. It occurs during copulation (mating). Copulation is the sexual union of a male and female animal. During copulation, the male releases semen in the reproductive tract of the female near the eggs.

Artificial insemination is collecting semen from the male and placing it in the reproductive tract of a female of the same species. Artificial insemination gives the producer control over the reproductive process.

Breeding

Breeding is helping animals reproduce so that the desired offspring result. Producers get the kinds of animals they want. Without some control over breeding, animal quality would not meet the demands in marketing.

Producers want animals that meet their needs. Animal breeds have been developed. A *breed* is a group of animals of the same species that share common traits. For example, Angus is a breed of beef cattle that are black and polled. Breed traits are inherited. They are passed from one generation to the next. Common traits offspring inherit include color, milk capacity, size, type, and presence of horns (in some species).

Purebred animals are eligible for registry in a breed association. Their parents must meet the standards for the breed. The offspring of two purebred animals of the same breed will qualify for registration as a purebred. Some animal producers raise purebred animals.

To produce animals of a certain type, purebred animals of the same species but of different breeds are mated. This is *crossbreeding*. For example, an Angus bull could be mated with a Hereford cow. It is used to improve the quality of the products produced by the offspring. In some cases, crossbred animals are mated to make offspring even more desirable. Other systems of breeding are also used.

REPRODUCTIVE EFFICIENCY

Reproductive efficiency is the timely and prolific replacement of a species. It may result in some species flourishing and others declining. Reproductive efficiency is the difference between success and failure in animal production.

The union of a sperm and an egg may not always produce a new, healthy individual. The developing animal may die before birth or hatching. With mammals, the death of a fetus is natural abortion. Some are born or hatched at low birth weights or defective in some way. Losses to abortion and other problems lower reproductive efficiency. For example, a cow that does not have a calf is not productive. She should be culled from the herd.

Research can influence a species or breed so that the cost of products is reduced. Artificial insemination is an important example. More changes are on the way.

A good example is ***accelerated lambing***. Accelerated lambing uses out-of-season lambing techniques to get three lamb crops in two years from a ewe. Most ewes have one lamb crop a year. Accelerated lambing allows sheep farmers to increase their productivity by 30 to 50 percent.

REPRODUCTIVE ORGANS AND SYSTEMS

Reproductive processes must work properly. When they do not, new animals are not produced. A new calf being born or a chick hatching depends on the proper function of reproductive organs. Anyone planning to breed animals needs to know about the reproductive organs of animals and their functions.

FEMALE REPRODUCTIVE ORGANS AND SYSTEMS

The primary reproductive organ of the female is the ***ovary***. The ovary produces female ***gametes***. Gametes are sex cells that can unite with other sex cells.

Female mammals have two ovaries. Some variation exists in other animals. Only the left ovary fully develops in chickens. Ovaries contain follicles, which are tiny structures that produce ova. A heifer has approximately 75,000 primary follicles in her ovaries at birth. A chick has about 4,000 miniature ova when hatched.

The maximum ova are present at birth or hatching. Ova are slowly used throughout a lifetime. An older cow may have only 2,500 follicles containing ova when culled from the herd.

As an ovum matures, it moves to the edge of the ovary for release. Releasing an egg is ***ovulation***. It is normally caught by the infundibulum (first segment of oviduct) and sent through the oviduct (fallopian tube).

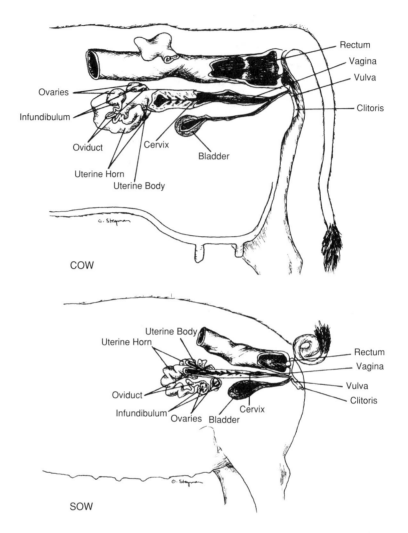

7-3. Cow and sow reproductive organs. (Courtesy, Gary Stegman, Crookston High School, Minnesota)

In mammals, the oviduct moves the ova and spermatozoa (if present) for fertilization and early cell division. The **embryo** (developing young) remains in the oviduct an additional three to five days.

Fertilization in a hen occurs in the infundibulum (sometimes called a funnel). The oviduct of the hen has four segments. A developed ovum (egg yolk) goes through these segments. The magnum secretes the albumen (thick white) of the egg. As the egg travels through the isthmus, it receives shell membranes. The uterus adds the thin white, the outer shell, and the shell pigment. The completed egg then passes through the vagina and is laid through the cloaca. The cloaca is the opening for both the digestive

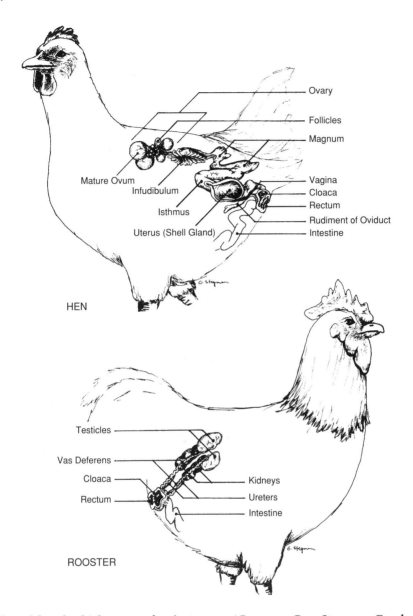

7-4. Male and female chicken reproductive organs. (Courtesy, Gary Stegman, Crookston High School, Minnesota)

and reproductive tracts. The entire time from ovulation to laying is slightly more than 24 hours.

In mammals, the embryo goes to the uterus. The uterus has two parts: uterine horn and uterine body. A placenta forms and attaches itself to the uterine wall. Cattle and swine differ from horses. A mare carries a foal in the uterine body, while a pig or calf is carried in the uterine horn.

The cervix is a thick-walled, inelastic organ (technically part of the uterus) that connects the uterus and the vagina. The cervix contains annular rings and cervical mucus that seal the uterus to keep out contaminants. During pregnancy, the mucus thickens to form a gel-like plug. Just before giving birth, ***hormones*** (chemical substances produced by the body) relax the cervix to allow the young to pass through.

The vagina is the passageway for reproduction and urine excretion. The vulva is the external opening of the reproductive and urinary systems. The clitoris is located just inside the vulva. It is a highly sensitive organ.

The reproductive organs of fish differ from mammals and hens. The ovary is a single large sac of eggs that appears as two ovaries that are joined. The eggs travel via the oviduct to the urogenital opening where they are excreted. Most fish eggs are fertilized outside the female's body.

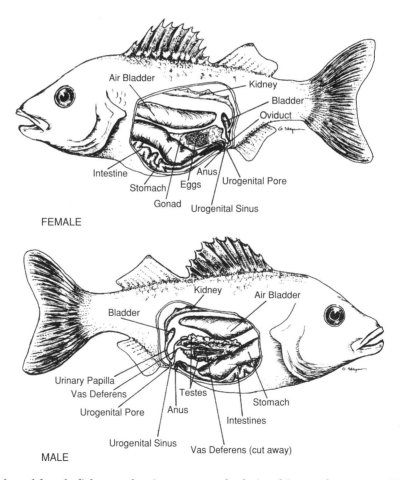

7-5. Male and female fish reproductive organs and relationship to other organs. (Courtesy, Gary Stegman, Crookston High School, Minnesota)

MALE REPRODUCTIVE SYSTEM

The primary reproductive organ of the male is the ***testicle***. Testicles produce male gametes or sperm. Two testicles are held externally in the scrotum of mammals and internally in the body cavity of male poultry and fish.

Males begin producing sperm at ***puberty*** and do so throughout life. The scrotum in mammals is a two-lobed sac that protects the testicles and helps control the temperature. Sperm production is best if testicles are cooler than the body.

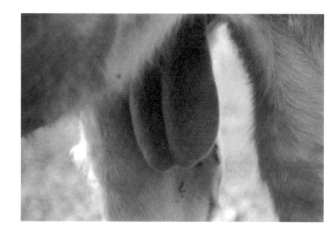

7-6. Scrotum of young bull showing properly developing testicles.

Sperm are formed in the seminiferous tubules of the testicles. Sperm pass through the epididymis before use.

The vas deferens carries the sperm and fluid (semen) from the epididymis for release. In mammals, it goes to the urethra. With poultry, sperms go to the cloaca (papillae). Sperm go to the urogenital sinus in fish. Male poultry do not have a true penis and transfer sperm to an undeveloped copulatory organ into the hen's oviduct. From the urogenital sinus of fish, sperm are excreted via the urogenital pore. Near the urethra of some mammals (bulls, stallion, and ram), sperm pool in an enlarged end of the vas deferens called an ampulla.

At ejaculation, semen passes through the penis. The penis is normally relaxed but is erect during sexual arousal. The free end of the penis (known as glans penis) serves as the sensory organ similar to the clitoris in females. It also retracts beneath the protective sheath (fold of skin) in most animals when not aroused.

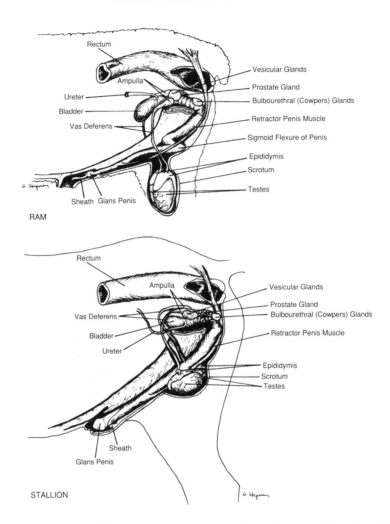

7-7. Ram and stallion reproductive organs. (Courtesy, Gary Stegman, Crookston High School, Minnesota)

PHASES OF THE ESTROUS CYCLE IN ANIMALS

The *estrous cycle* is the time between periods of estrus. The estrous cycle of female mammals has four periods: estrus, metestrus, diestrus and proestrus. These periods are cyclical (except during pregnancy) for many animals, such as cows or sows. Some animals, such as ewes, does, or mares, are seasonal breeders and may go through periods of cycling and periods of *anestrus* (absence of cycling). The estrous cycle length of common animals is shown in Table 7-1.

Table 7-1
Length of Estrus, Estrous, and Time of Ovulation

Species	Estrous Cycle (days)	Length of Estrus (heat)	Ovulation
Cow	21	12–18 hours	10–14 hours after estrus
Mare	22	6–8 days	1–2 days before estrus ends
Doe (goat)	21	30–40 hours	at the end of or just after estrus
Doe (rabbit)	Constant	Constant	8–10 hours after mating
Sow	20–21	40–72 hours	mid estrus
Ewe	17	24–36 hours	late estrus
Bitch (dog)	—	9 days	1–2 days after estrus begins
Queen (cat)	10	5 days	24 hours after mating

ESTRUS

Estrus (heat) is the period when the female is receptive to the male and will stand for mating. The length of estrus varies between species. Estrus ranges from 12 hours to six days. It is longer in some small animals.

Periods of heat or estrus are triggered by the hormone estrogen. Many changes take place, such as restlessness, mucus discharge, a swollen vulva, and standing to be ridden by other animals. Ovulation takes place during estrus for the ewe, sow, mare, and some companion animals.

METESTRUS

The period following heat or estrus is metestrus. Ovulation occurs during metestrus in the cow and doe. It is the time when LH (luteinizing hormone) triggers the corpus lutea (CL) to develop from follicular tissue that remains after release of the ova. The corpus luteum (yellow body) is important in maintaining pregnancy.

After ovulation, some capillaries break and release small amounts of blood. The blood is occasionally seen on the tail of a cow a day or so after estrus. This blood is not a sign that a cow is or is not pregnant.

DIESTRUS

Diestrus is the period in each estrous cycle in which the system assumes pregnancy. It is characterized by a fully functional corpus luteum (yellow body) that releases high levels of progesterone (the hormone that maintains pregnancy). Diestrus is often nine to 12 days in length. It is during diestrus that the uterus is prepared for pregnancy.

PROESTRUS

Proestrus begins with the regression of the corpus luteum and a drop in the hormone progesterone. FSH (follicle stimulating hormone) causes rapid follicle growth in preparation for estrus and ovulation. Late in proestrus, changes in behavior may occur as estrus approaches.

PHASES OF REPRODUCTIVE DEVELOPMENT

Reproduction involves a series of events that must be properly timed. Its success is measured in various ways, such as the number of pigs per litter or calves per 100 cows. Hatchability and liveability are important to poultry and fish operations.

PUBERTY

Puberty is when animals reach a level of sexual development that makes them capable of reproduction. Puberty in female animals is the age of the first estrus with ovulation. Puberty in males is the first ejaculate with fertile sperm.

Neither males nor females are sexually mature at puberty. The female is often too small to bear young. The male is not highly fertile nor capable of breeding regularly.

7-8. **Greatly enlarged bull sperm.** (Courtesy, Auburn University)

Both environmental and genetic factors affect the age at which puberty occurs. Weight at puberty is affected by genetic factors such as size of the parents. The primary environmental factor is anything that affects growth. Differences between breeds are the primary genetic factors within species. It is important to not breed animals until they reach the recommended weight.

FERTILIZATION

Fertilization is when a sperm penetrates an egg and pairs of genetic material are formed. The fertilized ovum is now a *zygote*.

GESTATION

Gestation is the period of pregnancy. It begins with conception and lasts until parturition. The average length of gestation varies from 114 days for a sow to 337 days for a mare. Fertilization is the beginning of gestation. A series of cell divisions without growth (cleavage) occurs while the embryo slowly migrates toward the uterus over a three to four day period.

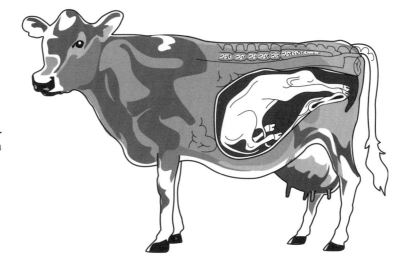

7-9. Normal presentation of calf in uterus.

The new animal is known as an embryo while the organs in the developing animal are forming, which is differentiation. Growth is rapid during this period. Differentiation is completed by day 28 in a sow and day 45 in a cow. After differentiation and until birth, the young animal is a *fetus*. The fetus gains weight and matures in preparation for birth.

PARTURITION

Parturition is the process of giving birth. Several hormone levels change and initiate this process. These initiate uterine contractions and milk production.

7-10. Ewe and newborn quadruplet lambs.

Signs of approaching parturition can be observed. Females begin to exhibit a "nesting" behavior. They become restless and attempt to separate from the herd or flock.

The first stage of parturition includes dilation of the cervix and entry of the fetus into the cervix. This usually is the longest stage; ranging from one to 12 hours, depending on the species. The second stage completes the birth of an animal through strong uterine contractions.

The birthing process and the first hour after birth are critical. Animal producers determine if the fetal position is correct. They observe the delivery for problems. They check to see that respiration has started and help the newborn nurse to get colostrum—the first milk that is high in antibodies.

The last stage of parturition is the expulsion of the afterbirth (placenta) from the uterus. This normally occurs shortly after giving birth. Sometimes the placenta is retained. This can interfere with future reproduction. The next gestation depends on a return to normal estrous cycling. The uterus must return to normal size and condition.

LACTATION

Lactation is the production of milk. Young mammals must have milk for food. The second purpose of lactation is to provide early disease resistance for the newborn. Colostrum contains important antibodies.

Lactation is a part of the reproductive cycle. Hormones that trigger the onset of lactation also play a role in parturition. Lactation requires nutrients that extend periods of anestrus following parturition. Weaning pigs or temporarily removing calves from cows can trigger estrus.

MATING BEHAVIOR

Conditions for domestic animals must be such that mating occurs. In addition to a properly cycling female, the male must be in proper condition and have the proper status (social rank) within the herd. Males need both the desire to mate (libido) and the ability to mate. Both can be enhanced with proper nutrition that keeps the males from becoming overweight and with an exercise area/breeding pen that has sound footing. Males should be kept disease and injury free so that mating is a pleasant experience.

Management of rams in a sheep flock is an example of stimulating proper estrous cycling and mating behavior. Rams should be kept away from ewes (out of sight and smelling range) until near breeding time. Reintroducing rams near breeding time will stimulate cycling in ewes. Males of any species that are very dominant can keep other males from mating. Pasture and breeding pen design help males that are lower in the social order breed their portion of females. This insures a high pregnancy rate in the total herd/flock.

INCUBATION

Incubation is the development of a new animal in a fertile poultry egg. It is outside the body of a hen. Proper incubation is needed to hatch eggs.

Four factors are important in incubating eggs: temperature, humidity, oxygen, and egg rotation. Temperature should be maintained at 99 to 103°F (37 to 39°C). The humidity should be about 60 percent during the first 18 days and 70 percent during the last three days (assuming a 21-day incubation period). Eggs should be slightly higher in the large end and be rotated two to five times daily for the first 18 days. Sufficient air exchange to prevent carbon dioxide buildup, while maintaining a 21 percent oxygen level, will enhance hatching results. Incubation times range from 21 days for chickens to 42 days for ostriches.

Table 7-2
Incubation Times for Poultry

Common Species Name	Incubation Period (days)
Chicken	21
Pheasant	24
Duck	28
Turkey	28
Goose	28–32
Ostrich	42

Early embryo development takes place in the body of the hen (107°F or 42°C) before the egg is laid. Several cell divisions take place. After laying, the egg cools and growth stops. Fertile eggs should be incubated shortly after laying. Some must be held for a while before incubation. Eggs stored below 80°F (27°C) are more likely to hatch when incubated.

Incubation continues embryo development. Organ formation (cell differentiation) is completed in the first six to eight days. Two key changes take place near the end of incubation. The beak turns toward the air cell and the yolk sac enters the body cavity. This yolk sac provides nutrition and water for the first several hours of life after hatching. The yolk is gradually used during the first ten days of life.

SPAWNING

Spawning is the releasing of eggs by a female fish and the subsequent fertilization by the male. A female may lay thousands of eggs at one spawning, depending on the species. The male fish then fertilizes the eggs by releasing sperm on the mass of eggs. The eggs of some species are protected only by a gravel covering. The males or females of some species provide the eggs with protection or incubation. Spawning is covered in more detail in Chapter 14.

ANIMAL REPRODUCTION TECHNOLOGY

The use of artificial insemination (A.I.) has increased. Today, A.I. is more common than natural service in some species, such as dairy cattle and turkeys. Synchronization and embryo transfer are also used to a lesser extent.

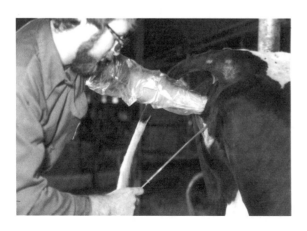

7-11. Artificial insemination of bovine.

ARTIFICIAL INSEMINATION

Artificial insemination is placing semen in the female reproductive tract by artificial techniques. A.I. has the advantage of using a quality male to breed many females. Superior traits can be extended to a large population. This has improved many species. It helps keep the cost of eggs, meat, and milk lower, while increasing producer income.

SEMEN COLLECTING

Artificial insemination requires that semen be collected from the male. A tom turkey or rooster requires manual stimulation and semen removal (sometimes called "milking of semen"). Sperm from a male fish can be obtained by applying gentle pressure on their abdomen. Bulls, rams, and stallions are usually collected with an artificial vagina (A.V.).

7-12. Bull semen being collected for artificial insemination.

Semen collection is improved if time is given for the male to become stimulated. The collected semen is evaluated, cooled slowly, and processed in preparation for freezing at temperatures of –320°F (–196°C). Semen stored in liquid nitrogen at –320°F (–196°C) can be thawed and used 30 to 40 years later.

INSEMINATION

High conception rates require much more than quality semen. The female must be cycling and in estrus (heat). The semen should be thawed and inseminated after heat detection.

Detecting estrus usually requires keen observation. Failure to detect heat is the most common cause of A.I. failure.

Estrus signs vary between species. The best indication of estrus for most species is standing heat. Standing heat is when a female stands when mounted by another animal. Other behavioral patterns change when an

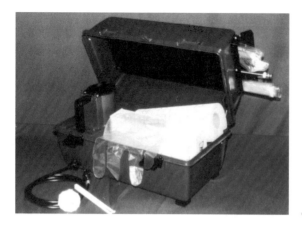

7-13. Artificial insemination kit. (Courtesy, Auburn University)

7-14. Turkeys have been selected and bred to give large amounts of meat relative to size. Some are so fleshy that they cannot perform the reproductive process.

animal is in estrus. Most become more active. Cows try to mount other animals and are generally restless. Mares urinate frequently and expose their clitoris in a process called "winking." Many animals display extra mucus and redness in the vulva. Sows may have redness in the vulva. The ewe requires a male being present before she will show signs of estrus. A vasectomized ram is often used to detect ewes in heat.

The timing and placement of semen vary depending on the species. All require that frozen semen be thawed properly (95 to 98°F or 33 to 34°C). Fresh semen should be used within an acceptable time.

Most turkeys must be artificially inseminated due to their large size. Insemination does not need to be on a daily basis. A week is usually suggested between the first two inseminations and a slightly longer interval on subsequent inseminations. This is possible in poultry because the hen has storage glands for semen.

Cows should be inseminated in the last two-thirds of estrus (heat) or the first few hours after the end of heat. Heat detection is most often done early in the morning or late at night, since, most riding activity occurs during or near the night hours. An approved practice is to breed cows detected in heat in the morning during the afternoon of the same day. Cows in heat in the evening should be bred early the next day. Insemination of cattle usually involves the rectovaginal method where a gloved arm is placed in the rectum to palpate (feel) the reproductive organs.

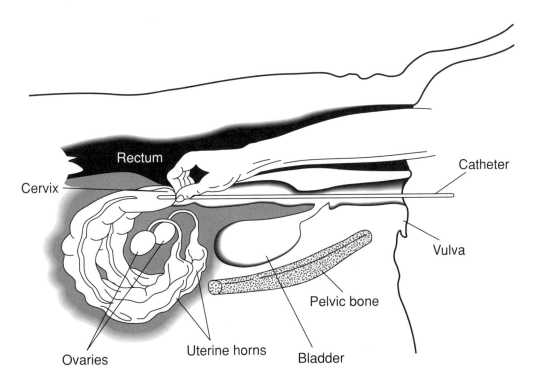

7-15. Artificial insemination of a cow with the recto-vaginal method.

Mares may need to be inseminated several times. This is because it is difficult to detect ovulation. Inseminations are made on the third, fifth, and seventh days of estrus. The procedure is different in mares. They have a delicate rectum and other anatomical differences. The arm and syringe are placed in the vagina with a finger through the cervix. This requires special emphasis on cleanliness. The semen is deposited in the body of the uterus.

Sows should be bred 24 hours after the onset of estrus. Some people prefer to breed a second time, 40 to 48 hours after the onset of estrus. An inseminating tube is inserted into the cervix. The semen is ejected with a plastic squeeze bottle syringe.

Artificial insemination of ewes is less common. Two inseminations will increase conception and multiple lambs. The size of a ewe requires smaller equipment or the use of lapriscopic artificial insemination. In lapriscopic artificial insemination, semen is placed in the uterus through the ewe's abdominal wall.

ESTROUS SYNCHRONIZATION

Estrous synchronization is bringing a group of animals into heat simultaneously. This helps schedule animal breeding and birthing.

Synchronization usually involves the use of prostaglandin, progestin, or a combination of the two. Prostaglandin causes the corpus luteum to stop production of progesterone, allowing the animal to come into estrus. Progestin has the effect of keeping progesterone levels high, holding animals in an "extended" diestrus. When the progestin source is removed, the animal quickly comes into estrus.

CIDR (Controlled Internal Drug Release) dispensers in sheep have been used in Australia and researched in the United States. CIDR's are placed in the vagina of ewes. They are removed and ewes cycle shortly afterward. This is used to breed ewes during times of the year when they do not cycle. This helps sheep producers get three lamb crops every two years.

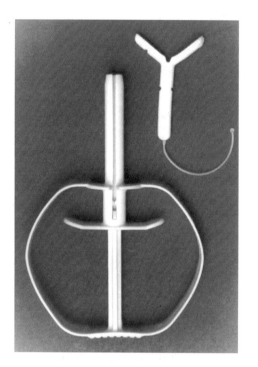

7-16. Controlled internal drug release (CIDR) dispenser for use with ewes. (Courtesy, Northwest Experiment Station, University of Minnesota, Crookston)

EMBRYO TRANSFER

Embryo transfer is moving embryos from a donor to the reproductive tract of a recipient. Donor females usually carry extraordinary genetics. Recipient animals have far less worth and are used as surrogate mothers.

Most embryo transfer work is done with beef and dairy cattle. Embryo transfer is used following superovulation. This assures a larger than normal number of eggs. Multiple breedings and careful administration of extra follicle stimulating hormone (FSH) will cause a donor to produce an average of five or six transferable embryos. Transfer of embryos can be done surgically or nonsurgically. The success rate is higher when transferring fresh embryos. Embryos can be frozen in liquid nitrogen and transferred later.

CLONING

Cloning is the production of one or more exact genetic copies of an animal. Cloning has been expanded and may be widely available soon. Cloning in animals involves letting embryos grow to the 32-cell stage before splitting into 32 identical embryos. Scientists feel that this could be repeated producing 1,024 identical copies, 13,088 identical copies, etc.

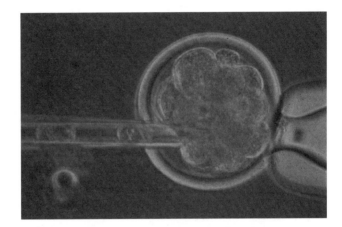

7-17. Cloning technology.

GENETIC ENGINEERING

Genetic engineering is removing, modifying, or adding genes to DNA. Genetic engineering using recombinant DNA (gene-splicing) along with other reproductive technology has the potential to greatly change animal science.

Genetic marker technology is being used to detect the presence of certain genes. This technology, along with cloning, has the potential to further increase genetic progress.

REPRODUCTIVE TECHNOLOGY ETHICS

Reproductive technology is rapidly increasing. This raises ethical questions. Technology can be good or bad depending on how it is used. People must be honest. Fear of genetic engineering abuse is not a good reason to stop the potential benefits.

People need to be educated about the use of reproductive technology. If not, poor decisions may be made.

REVIEWING

MAIN IDEAS

Reproduction is the process by which offspring are produced. Animals naturally reproduce sexually. Sperm and egg unite to begin a new individual. Producers also use artificial insemination to help assure productive animals.

Reproductive efficiency is important to animal producers. Animals must breed and regularly produce offspring. If they do not, the animals should be culled from the herd or flock.

Male and female animals have distinctive reproductive organs and systems. The testicle produces sperm and is the primary reproductive organ in the male. The ovary produces eggs and is the primary reproductive organ in the female.

Reproductive development goes through several phases. Puberty is when animals can reproduce. Fertilization is when union has occurred. Gestation is the time between conception and parturition in the female. Parturition is the birthing process.

New technology is used in reproduction. Artificial insemination is widely used. Research is developing cloning, genetic engineering, and other areas. People must consider the ethics of the newer technology.

QUESTIONS

Answer the following questions using complete sentences and correct spelling.

1. What is reproductive efficiency and why is it important?

2. Describe the structure of the female reproductive system.

3. Describe the structure of the male reproductive system.

4. Where does mammal fertilization take place? Hen fertilization? Fish fertilization?

5. What are the four phases of the estrous cycle? What happens in each phase?

6. How is puberty defined in males? In females?

7. What four factors are essential to successful incubation?

8. What reproductive technology has had the largest impact on improving animal efficiency? Why is it useful?

9. Discuss two ways that reproductive technology of the future may influence animal and poultry production.

EVALUATING

CHAPTER SELF-CHECK

Match the terms with the correct definitions. Write the letter by the term in the blank that is provided.

a. sperm   d. ovary   g. parturition
b. egg   e. reproduction h. conception
c. testicle   f. gestation

1. ____ process of producing offspring

2. ____ male sex cell

3. ____ primary female reproductive organ

4. ____ union of sperm and egg

5. ____ birthing process

6. ____ time between conception and parturition

7. ____ female sex cell

8. ____ primary male reproductive organ

EXPLORING

1. Dissect a cow's reproductive tract. Find all of the important reproductive organs. Animal slaughter houses may provide the reproductive tracts to schools for educational purposes. Be sure to follow proper safety procedures and properly dispose of the reproductive tract after the dissection.

2. Tour an artificial insemination company and observe the correct procedure used to collect semen from a bull. Observe how the semen is handled in preparation for storage and shipment.

3. Hold a debate in class on reproductive technology. Consider the ethics of the technology and other areas that are important.

4. Observe heat signs in your herd or a friend's herd or flock. Determine the kind of animal and previous breeding history. What change in behavior did you see?

5. Producers feed non-productive animals when they don't reproduce. How much does a rancher lose when cows don't calve? Assume that ZYX Ranch has 1,000 cows. How much income is lost if 5 percent of the cows don't have calves in a year? In your calculations, use a weaning weight of 500 pounds for each calf and a per pound selling price of 70 cents. (Clues: first determine the total number of calves not born and multiply by 500 pounds. This is the total weight of all lost calves. Multiply the total weight by 70 cents to get the number of dollars lost.) How would you overcome this loss if you owned ZYX Ranch?

Chapter 8

ANIMAL BIOTECHNOLOGY

Animal producers want better ways of raising animals. They also want animals that grow fast and meet the needs of people. For example, people want turkeys with a lot of white meat. New kinds of turkeys have been developed to meet the demand. Research is important in many ways in animal production.

Some scientists feel that most of our food will soon come from new kinds of plants and animals. Discoveries in biotechnology will change animal agriculture. Animal production will have many changes in the years ahead. Many of these changes will involve biotechnology.

People have different opinions about biotechnology. Some feel that it should be used. They feel that the benefits outweigh the possible problems. Other people feel that biotechnology should not be used. What do you think?

8-1. Biotechnology involves the use of many kinds of equipment, such as this inversion microscope.

OBJECTIVES

This chapter covers areas of biotechnology. The following objectives are included:

1. Explain biotechnology and how it is used

2. Distinguish between organismic biotechnology and molecular biotechnology in animals

3. Describe genetic engineering and recombinant DNA processes in animals

4. Identify issues associated with animal biotechnology

TERMS

allele	genetic code	mutation
atom	genetic engineering	organelles
biotechnology	genetics	organismic biotechnology
cell	genome	phenotype
chromosome	genotype	recessive
deoxyribonucleic acid (DNA)	heredity	recombinant DNA
dominant	heterozygous	superovulation
embryo transfer	homozygous	synthetic biology
gene	microinjection	transgenic animal
gene mapping	molecular biotechnology	
gene transfer	molecule	

8-2. Tissue culture is one example of biotechnology. Hundreds of new organisms can be "cloned" from one original.

BIOTECHNOLOGY: AGRICULTURE FOR THE FUTURE

Biotechnology is the management of biological systems for the benefit of humanity. This broad definition includes many areas of animal production. Biotechnology is often viewed as the application of science in food and fiber production. All plants and animals are included.

Biotechnology has been used for hundreds of years. Using yeast to make bread and bacteria to make cheese is biotechnology. However, today's biotechnology refers to new technologies developed since 1977. It is based on a deep and thorough understanding of life and the mechanics of living things.

SYNTHETIC BIOLOGY

Synthetic biology goes beyond biotechnology. It is the use of chemicals to create systems with some characteristics of living organisms. Now confined to laboratories, synthetic biology uses vesicles to create life-like conditions. Vesicles are tiny cell-like structures that have external membranes.

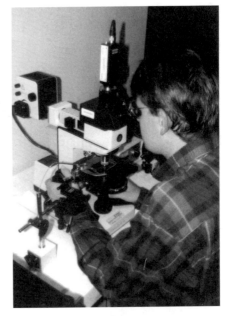

8-3. Using microscopy with attached video and computer systems to study synthetic biology.

All synthetic biology work is carried out using powerful microscopes with attached computer networks. Emory University in Atlanta, Georgia, has the only laboratory in the world doing this research.

Physical organic chemists are studying the use of synthetic biology in animal health and production. The interaction of vesicles and living organisms is a top priority. Could the creation of living organisms from nonliving substances be next?

MOLECULAR VERSUS ORGANISMIC BIOTECHNOLOGY

There are two types of biotechnology: molecular and organismic. This science combines biology, chemistry, and physics.

Molecular biotechnology involves changing the structure and parts of cells to change the organism. It begins with the *atom*, which is the smallest unit of an element. Elements are such things as hydrogen, oxygen, nitrogen, phosphorus, and sulfur. Two or more atoms make up a *molecule*, which is the smallest unit of a substance. Molecules form cells in plants and animals.

Each cell of an organism has several molecules of water within it. Water is a molecule represented by the chemical symbol of H_2O. The elements in water are hydrogen (H) and oxygen (O). Water is made up of three atoms, two hydrogen and one oxygen.

Genetics and molecular biology are important in both molecular and organismic biotechnology. Both areas are covered in more detail later in the chapter.

Organismic biotechnology deals with intact or complete organisms. The genetic makeup of the organism is not artificially changed. It is used to improve animals for the benefit of humans. This is the most widely used type of biotechnology.

8-4. A micropipet with a disposable tip is being used in this work. Proper safety and sanitation are essential.

MOLECULAR BIOTECHNOLOGY

Molecular biotechnology deals with changing the genetic make up of animals. Physical appearance may also be changed. *Genetics* deals with the laws and processes governing inheritance. The passing of traits from parents to offspring is **heredity.** Present livestock operations are much more productive than early domestic animals because of a knowledge of genetics and heredity.

Much genetic variation exists within a species. Hogs will always have piglets and cows will always have calves, but each piglet or calf is unique.

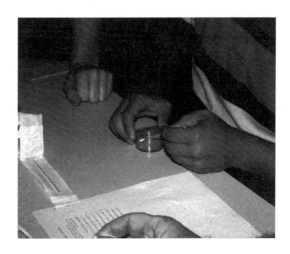

8-5. Streak planting is a simple technique that can be used in molecular biotechnology.

Cells

All living things are made up of cells. *Cells* are the basic units of life. Organisms are unicellular (one-celled) or multicellular (many cells), and all cells arise from preexisting cells.

Organelles are like tiny organs within a cell, such as the nucleus and mitochondria. These carry out specific functions. Changing these functions will change the plant or animal. A simple example is feeding the right ration, which allows the animal to grow faster. Changing the structure of the cell is a more complex way of changing the processes of an animal.

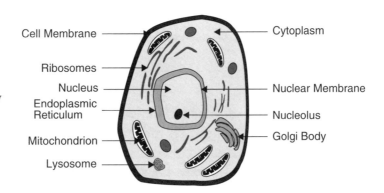

8-6. Animal cell (greatly enlarged).

Cell Membrane — Cytoplasm
Ribosomes
Nucleus — Nuclear Membrane
Endoplasmic Reticulum
Mitochondrion — Nucleolus
— Golgi Body
Lysosome

8-7. DNA extracted from *Escherichia coli* bacteria appears as white floating material in this test tube.

Chromosomes

All organisms are made of cells. A living cell consists of a cell membrane, for protection, and the cytoplasm, which is the living material of the cell. The nucleus is in the cytoplasm. The nucleus contains the ***chromosomes,*** which are threadlike parts containing the genetic material. This genetic material makes up the ***genome*** of the organism.

Chromosomes are the link between parents and offspring. When organisms reproduce sexually, the genome is the combination of the mother's and the father's traits. The offspring receive chromosomes from both parents. This gives them pairs of "like" chromosomes. All of the cells within the organism are genetically identical. Each cell contains identical numbers of chromosomes. Table 8-1 lists the number of chromosomes for different species.

Table 8-1
Number of Chromosomes for Selected Animal Species

Species	Number of Chromosomes	Species	Number of Chromosomes
Cat	38	Horse	64
Cattle	60	Human	46
Chicken	78	Mule	63
Dog	78	Sheep	54
Donkey	62	Swine	38

Chromosomes are made of genes that consist of deoxyribonucleic acid (DNA). DNA forms genes that makeup chromosomes. Chromosomes make up genetic information for cells. Cells make tissues that form organs. Organs form organ systems. These organ systems make up the organism.

Genes

Genes are the segments of chromosomes that contain the hereditary traits of organisms. Since chromosomes come in pairs, genes also come in pairs.

Alleles are the different forms of genes. Alleles may be similar or they may be different. An organism having similar alleles is considered *homozygous* for that trait. A *heterozygous* organism is one having different alleles for a particular trait.

Genotype versus Phenotype. The genetic makeup of an organism is its *genotype. Phenotype* is the organism's physical or outward appearance. In basic genetics, a homozygous allele is called "AA" or "aa" for genotype. A heterozygous allele is commonly noted as "Aa."

Breeding practices use this concept of alleles to produce various phenotypes or outcomes. Outcrossing is used to develop heterozygous offspring and inbreeding produces homozygosity. The genotype and the environment determine phenotype.

Dominant versus Recessive. Some traits are dominant while others are recessive. *Dominant* traits cover up or mask the alleles for *recessive* traits. If an animal has a dominant allele for one of its alleles, it will phenotypically show the dominant trait in complete dominance. If the animal receives two recessive alleles, then it will be homozygous recessive.

A good example is polled animals. The polled (PP) trait is dominant over the recessive (pp) trait of having horns. If a calf is homozygous dominant (PP) or heterozygous (Pp), they will be polled. If they receive an allele for the horned trait from both parents, they are homozygous recessive (pp) and will grow horns.

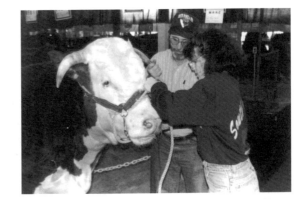

8-8. This Hereford bull received the genetic trait for horns from both parents.

Red Angus cattle are another example showing dominant and recessive alleles. Angus cattle have been black for hundreds of years. Occasionally, a red calf is born. Black is the dominant allele and red is the recessive allele. When two red calves are crossed, they will always produce a red calf. These red animals are homozygous recessive.

Incomplete Dominance. Most traits are not products of complete dominance. They are usually influenced by several alleles and by the environment. Another example is incomplete dominance. Shorthorn cattle are either red (RR), roan (Rr), or white (rr). The dominant allele is not completely masking the white color, thus the animal appears to have red and white hairs.

DNA

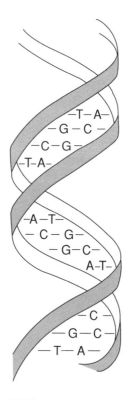

Deoxyribonucleic Acid (DNA) is a protein like nucleic acid on genes that controls inheritance. Each DNA molecule consists of two strands shaped as a double helix or spiral structure. These strands are nucleotides bonded together by pairs of nitrogen bases.

This double helix is similar to a tiny twisted ladder. The supports or backbone are made up of sugar molecules held together by phosphates. The rungs consist of nitrogen materials [cytosine (C), guanine (G), adenine (A), and thymine (T)]. The rung structure allows DNA segments to be cut out and new ones inserted.

Genetic Code

The **genetic code** is the sequence of nitrogen bases in the DNA molecule. This sequence codes for amino acids and proteins. The unique ability of DNA to replicate itself allows for the molecule to pass genetic information from one cell generation to the next.

When a chromosome divides during mitosis, the two strands unravel and separate. Each strand contains the complementary base and will immediately become a new double strand identical to

8-9. DNA structure consists of two nucleotide strands, which are bonded by the nitrogen bases cytosine, guanine, adenine, and thymine.

the original. A gene is structurally a triplet sequence of nitrogen bases (C, G, A, and T).

Genetic code determines the nature of an organism. Hogs have a different genetic code than do cattle or humans. Research is identifying the genetic code of animals.

Mutation

Mutations are changes that naturally occur in the genetic material of an organism. Mutations include changes in the chromosome number or chromosome structure. As animals grow and develop, thousands of cells are constantly dividing and millions of genes are being reproduced. This increases the risk of change resulting in a mutation.

Most mutations result in death of the cells, but occasionally mutations will result in a viable offspring that is genetically different. Mutations may be thought of as birth defects. It is likely that the first Polled Hereford resulted from a mutation.

8-10. The Polled Hereford breed originated as a result of a mutation in the genetic material of Horned Herefords.

MOLECULAR METHODS

Molecular methods have increased in the last few years. Much of the work is with unknown processes. Three areas are included here.

Genetic Engineering

Genetic engineering is a molecular form of biotechnology. The genetic information is changed to make a new product. Through this process, sections of the DNA strand are cut out and new sections are inserted.

Chromosomes contain thousands of genes making it difficult to find the right gene to remove or add. *Gene mapping* involves locating the genes on the chromosome. This is needed for successful genetic engineering. *Gene transfer* is the moving of a gene from one organism to another.

Genetic engineering is a means of complementing traditional breeding programs. It is not a replacement. The first release of a genetically engineered product was the Flvr Savr Tomato. This tomato has an altered gene to increase the time it is ripe without spoiling. This makes a higher quality final product at a lower price.

Recombinant DNA

Recombinant DNA is gene splicing. Genes are cut out of a DNA strand with a restriction enzyme that works like a scalpel. They are then inserted into circular DNA molecules in bacteria plasmids. Plasmids are circular pieces of DNA found outside the nucleus in bacteria. The plasmid is inserted into the cell that is to be altered. This allows the DNA of two different organisms to be combined. Gene splicing is a challenging process because DNA is so small!

All cells have membranes to protect them. Adding DNA to the cell's nucleus is not an easy job. Several methods may be used to insert DNA into the nucleus. Mixing DNA in a solution and microinjection are two methods used. Most cells die in the process. Many surviving cells do not have the "new" DNA present in the nucleus.

Transgenic Animals

A *transgenic animal* is an animal, which has stably incorporated a "foreign" gene into its cells. This organism can pass to its offspring this

8-11. Using a silicon graphics computer system for molecular modeling.

transgene (altered gene). All of the cells within the transgenic animal contain this transgene. ***Microinjection*** is a common method of producing transgenic animals. It is injecting DNA into a cell using a fine diameter glass needle and a microscope. Microinjecting bovine growth hormone has increased growth rates in fish, chickens, sheep, cows, hogs, and rabbits.

MOLECULAR APPLICATIONS

Molecular methods have been used to genetically alter animals. Results with animals can often be applied to humans. Medicines, nutrients, and animal quality are improved with molecular biotechnology.

Animals are used to study diseases in humans, such as Lou Gehrig's disease, sickle cell anemia, and cancer. The growth and development of animals, diseases, mutations, and the influence of male and female chromosomes can be studied in similar ways.

Insulin was once extracted from the pancreas of slaughtered cattle and hogs. This process is expensive and would sometimes cause allergic reactions. Today, high quality insulin is manufactured.

Milk composition is suited to using molecular biotechnology. Increases of beta casein decrease the time required for rennet coagulation and whey expulsion in making cheese. This reduces the amount of time needed to make cheese. Fat content in milk can be reduced to make fat-free cheeses and ice cream. If kappa casein is increased 5 percent, it is more stable and easier to ship milk to third world countries.

Livestock quality is improving. Growth rate, efficiency, and disease resistance are increasing. Reproductive performance is being enhanced. The high birth rate among Chinese hogs is being studied to increase the proliferation of American breeds.

8-12. Understanding molecular biotechnology requires a great deal of studying, skill, and careful observation.

Transgenic and other engineered animals are expensive to produce. Producing a transgenic mouse costs about $1,210. Pigs cost $250,000; sheep cost $600,000; and a cow costs $5,460,000. Research is often with mice because of relatively lower cost.

ORGANISMIC BIOTECHNOLOGY

Organismic biotechnology deals with improving organisms as they are. Most management practices are in this area of biotechnology. Several examples of organismic biotechnology are presented here.

Greater Fertility

Increasing the reproductive capacity of top animals is important to producers. Sexually reproducing more offspring has been a major limitation. Superovulating cows and embryo transfer are methods of increasing the number of young from a genetically superior individual.

Superovulation. *Superovulation* is getting a female to release more than the usual number of eggs during a single estrous cycle. Hormones are injected to assure more eggs. Superovulation is used with cattle. Cows normally produce only one egg every 21 days. By injecting the cow with a hormone such as gonadotropin, the cow can release up to 20 eggs. When in heat, the cow is inseminated and the eggs are fertilized.

The embryos are flushed out of the cow seven days after breeding. A solution made up of water and salt is used to wash the embryos out of the uterus. The embryos are no bigger than a match tip. They are either frozen or immediately placed in a recipient cow. Embryos can be frozen in straws similar to semen. After transfer, the embryos are then carried to full term by the recipient. They do not receive any genetic material from their recipient mothers, only nutrients.

Embryo Transfer. *Embryo transfer* is taking an embryo from its mother (donor) and implanting it in another female (recipient). The embryo completes developing in the recipient. Embryo transfer usually follows superovulation, but superovulation is not necessary. The valuable donor cow can continue estrus cycling and produce more eggs. Beef producers may implant embryos in dairy cows because they are larger and give more milk. These traits help the newborn calf grow fast.

Increased Production

Production can be increased by using milk and meat hormones and growth implants.

Milk Hormones. In 1994, bovine somatotropin (BST) was approved for use with animals. BST is a naturally occurring hormone that when given to cows increases their milk production. Injections cause the cow's system to become more productive.

Meat Hormones. Larger pork chops and hams are produced by injecting hogs with porcine somatotropin (PST). PST is produced in the pituitary gland of hogs. It causes them to produce more muscle cells resulting in larger and leaner cuts of valuable pork. Research is being done on mixing PST with feed.

Growth Implants. Growth implants have been used for years to promote growth. Implants are small pellets placed under the skin—usually behind the ear. They promote growth by making the animal more efficient in using feed. Implants slowly release the materials over several weeks. The materials come from hormones or molds grown on corn products.

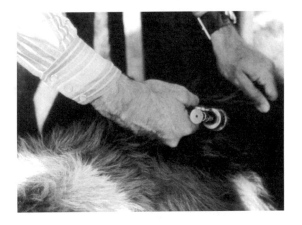

8-13. Implanting a growth enhancer into a calf. (Courtesy, Marco Nicovich, Mississippi Cooperative Extension Service)

Enhancing Animal Nutrition

Greater results have been in animal nutrition than any other area. Enhanced nutrition has increased feed efficiency, decreased time to market weight, increased milk production, and decreased overall production problems.

8-14. Steer with fistula for feed digestibility trials.

Digestibility Testing. Digestibility of feeds varies. Tests are made by getting samples from the digestive system. Cattle, which are ruminants, have been used for most of this research.

A small opening is made in the side of the cow and into the rumen (large compartment of the stomach). A rubber fistula is put in the opening. This allows taking partially digested feed samples from the stomach. Cannulas or small tubes can be placed into the duodenum (part of the small intestine) of cattle. The animal has no pain during these processes.

Fistulas make it easy to monitor feed intake and digestion. This information is used to improve feed use and animal growth.

Controlled Feeding. In a feedlot or pasture, many animals will over or under eat to meet their nutritional demands. One method of regulating this is with computer chips, which can either be put in a monitor around the animal's neck, in an ear tag, or just under the skin in their ear. A computerized feeding station will read this chip and regulate the amount

8-15. Steers fitted with sensors on chains around their necks for monitoring feed consumption in a feedlot.

8-16. As cows walk through this door, the computer identification tag registers their presence and the milking stall they enter. Once milking begins, it sends the amount of milk produced to a computer system, which is used to monitor production levels.

of feed that the animal gets. A daily feed intake sheet is used to detect health problems.

BIOTECHNOLOGY ISSUES

Ethical issues have included such areas as human disease outbreaks, destruction of entire ecosystems, and development of a new human race. These are outdated and unrealistic issues for today.

Today's biotechnology issues address threats to the welfare of people, environmental concerns, economic gains, benefits to developing countries, and personal beliefs and values.

Possible Threats to People. The risks with new, unknown products need careful study. Biotechnology has many benefits. It is not always dangerous. New products always have some risk to workers and users. Proper safety standards are set to protect people. Safety of food products and effects on humans are questions that consumers may have. Some people are reluctant to accept biotechnology because it represents change. Change brings uncertainty. To date, not one case of illness or death has resulted from recombinant DNA technology.

Possible Environmental Concerns. When developing altered organisms in molecular biotechnology, the possibility exists for upsetting ecosystems. The diversity of organisms can change and sustaining the environment becomes an issue. Controlled research reduces these concerns.

8-17. Biological processes are put to good use in decaying wastes on a hog farm in Pennsylvania.

Economic Gains. New feedstuffs and altered organisms present questions about patented products. Gaining a patent on an item allows a company to make money. On the down side, patents can slow the process of scientific knowledge.

Benefits to Developing Countries. Developing countries do not have the money or resources to develop new technologies. They may struggle to sustain a way of life. Many organizations have been set up to help educate and improve the quality of life in third world countries.

Possible Threats to Personal Beliefs. Some people are concerned about biotechnology. They often need more information to help understand issues. This could change views about creation and evolution theories. New technology could change our view of the "perfect" child. Should new plants or animals be patented? Is it right to genetically engineer an organism?

REVIEWING

MAIN IDEAS

Biotechnology is using biological systems for the benefit of humans. This uses molecular and organismic techniques.

Molecular biotechnology involves changing the structure and parts of a cell to change the organism. Understanding basic genetics is a key to successful molecular biotechnology. A cell's nucleus contains chromosomes. Chromosomes are made of genes that are in turn made up of DNA. Alleles are the different forms of genes that code for a particular trait. DNA is the ultimate controller of inheritance.

Methods of molecular biotechnology include genetic engineering, recombinant DNA, and transgenic animals. These are expensive techniques, but have great potential for helping society.

Organismic biotechnology deals with improving organisms as they are. Examples of organismic biotechnology include superovulation, embryo transfer, milk hormones, meat hormones, growth implants, digestibility testing, and controlled feeding.

Biotechnology also raises many questions. What damages could occur to individuals? How could biotechnology affect our environment? What economic gains will occur? Who will benefit from biotechnology? Will third world countries suffer because of biotechnology? Are people's personal beliefs threatened?

QUESTIONS

Answer the following questions using complete sentences and correct spelling.

1. Distinguish between the two main types of biotechnology.

2. Compare and contrast elements, atoms, and molecules.

3. What is genetics? Why is genetics important in understanding biotechnology?

4. Describe the relationship between DNA, genes, and chromosomes.

5. Distinguish between genotype and phenotype.

6. Distinguish between dominant and recessive.

7. What are mutations? How are mutations helpful?

8. What is genetic engineering?

9. What is recombinant DNA? How can recombinant DNA technology be used?

10. How is the genetic code related to recombinant DNA?

EVALUATING

CHAPTER SELF-CHECK

Match the terms with the correct definitions. Write the letter by the term in the blank that is provided.

a. biotechnology   d. gene   g. mutation
b. DNA   e. chromosome   h. molecular biotechnology
c. synthetic biology   f. transgenic animal

1. ____ changing the structure and parts of cells to change an organism

2. ____ the part of a chromosome that contains heredity material

3. ____ a protein that controls inheritance

4. ____ changes that occur naturally in the genetic material of an organism

5. ____ using biological systems for the benefit of humans

6. ____ an animal that has a foreign gene in its cells

7. ____ threadlike parts of a cell that contain genetic material

8. ____ using chemicals to create systems with the characteristics of living things

EXPLORING

1. Invite a veterinarian to serve as a resource person and give a presentation on new technologies that are affecting veterinary medicine. Ask the veterinarian to demonstrate how superovulation and embryo transfer are done. Write a report that summarizes what you have learned.

2. Conduct a debate in class addressing the question "Is animal biotechnology good for society?"

3. Take a field trip to a research or biotechnology facility to study the work that is underway. Write a report on your observations.

4. Survey students in your class to determine their opinions and perceptions about biotechnology. Carefully write questions and collect data. (This would make a science fair project.)

Chapter 9

BEEF PRODUCTION

Beef! Many people enjoy beef. Steak and other beef foods are favorites in North America. Social gatherings often feature beef on the menu. The backyard barbecue with ribs, hamburger, and loin provides good eating!

As food, people like the excellent taste and ease of cooking beef. It helps meet important nutritional needs of people. Beef is a major source of protein in the human diet. In recent years, the fat in beef has been reduced. This has made it more appealing to health-conscious consumers.

Raising beef animals is big business. Marketing and preparing the products form a large meat industry. Many people are needed to work in the cattle industry.

Beef animals also provide a lot of fun. Raising a calf is a favorite project for many youth. Feeding, grooming, and caring for animals is rewarding. Youth learn much about success in life from activities involving their beef animals.

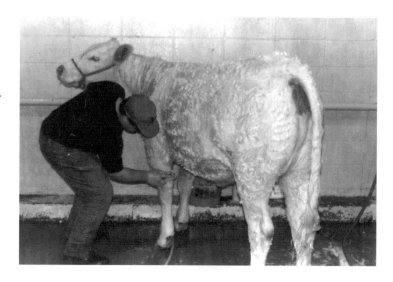

9-1. Young people enjoy taking their animals to livestock shows. This Charolais heifer is being washed to get ready for a show.

OBJECTIVES

This chapter gives basic information about raising beef animals. It has the following objectives:

1. Explain the importance of beef production
2. Describe beef cattle as organisms
3. Explain how to select beef animals, including breeds
4. List the advantages and disadvantages of beef production
5. Describe the types of beef production systems
6. Explain important management practices in beef production
7. Describe facility and equipment needs with beef cattle

TERMS

anthelmintic	conformation	forager
backgrounding	cow-calf system	grade cattle
Bos indicus	creep feeding	marbled
Bos taurus	cutability	polled
brand	dehorn	steer
castrate	dual-purpose breed	
concentrate	finishing system	

IMPORTANCE OF BEEF CATTLE PRODUCTION

Beef cattle are raised for beef. The animals are selected to yield a high amount of good meat. This is contrasted with dairy cattle, which are used for milk production.

Beef production is a large part of agriculture in North America. Providing quality beef requires good animals and efficient marketing. Products must reach the consumer in wholesome condition.

ORIGIN OF BEEF CATTLE

The origin of cattle is not fully known. It is assumed that they were first domesticated in Europe during the Stone Age. Today's cattle are believed to have originated from two species: ***Bos taurus*** and ***Bos indicus***.

Bos taurus are cattle common to the temperate regions, such as the midwestern United States. They are often known as the European breeds. Examples include Angus and Hereford. Bos indicus cattle are those cattle that are descendants of the Zebu cattle with humps on their necks. They also have large, droopy ears and loose skin. Bos indicus cattle are common to warmer tropical areas, such as the southern part of the United States.

The beef cattle in the United States were started by Christopher Columbus. Beef animals were on his second voyage to America. Early settlers from the British Isles and other European countries brought cattle with them as they settled new land in America. As the settlers moved west, so did the cattle. Abundant pasture was found in the great plains areas.

THE BEEF CATTLE INDUSTRY

Today, the cattle business is one of the largest segments in American agriculture. Sales of cattle amount to more than 20 percent of all farm markets. Nearly 100 million head of beef cattle are on the farms and ranches in the United States.

There is also a global market for beef. The United States exports approximately 423,878 metric tons to Japan, Canada, Republic of Korea, Mexico, and Hong Kong. Japan is by far the largest foreign customer of the United States' beef industry, including imports of variety meats and hides.

9-2. Cooking-out is a favorite activity with many people.

Consumption

The consumption of beef is part of a balanced diet designed for good nutrition and health. Beef continues to be one of America's favorite foods. The per capita consumption of beef is 95 pounds per year in the United States.

Beef helps consumers meet daily dietary requirements in protein, vitamin B-12, iron, zinc, and other essential nutrients. In recent years, beef has received criticism because of the amount of fat and cholesterol it can add to the diet. New research, improved genetics, and better management techniques are reducing these concerns. Moreover, the key to a healthy diet is in the balance, variety, and moderation of consumption.

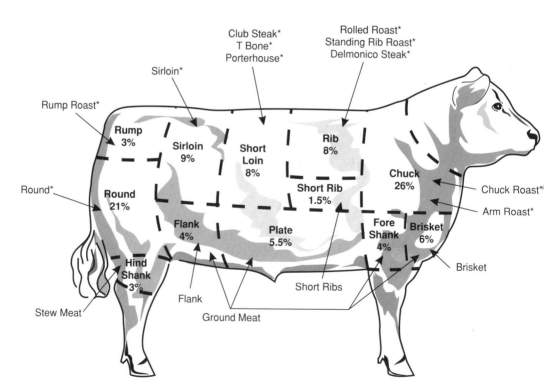

9-3. Where the beef cuts are located on an animal. (Asterisk * indicates the most desirable cuts that sell for a higher price)

Consumer Demands

Consumers want good beef. They will pay top prices for certain cuts, such as steak and loin. Consumers pay less for organs, bony pieces, and the parts made into hamburger and similar products.

Producers need to be aware of where the more valuable cuts are located on an animal. They also need to produce the type of animal that has larger amounts of meat in these cuts.

BEEF CATTLE SELECTION

Selecting a breed of beef cattle is an important decision. A lot of money and other resources are involved. The strengths and weaknesses of each breed should be considered. This will allow producers to meet their goals.

Some cattle producers decide to raise cattle that are not purebreds (entitled to registration). They may have crosses between the breeds to get the desired traits. For example, bulls of one breed may be bred to cows of another. The animals that result are crosses. Other producers raise grade animals. *Grade cattle* are not registered. Some have a purebred background but most are mixed breeds. They are selected for production potential.

Most meat production in the United States is from grade beef animals. Many purebred producers raise animals to improve the quality of the breed and to breed to grade animals.

SELECTION GUIDELINES

Remember these pointers in selecting a breed:

- There is no one breed that is the best for all traits. Producers need to have priorities on the traits that are most important to them.

- All breeds of beef cattle will have variation within their genetics. A decision has to be made by the producer on the genetic traits (i.e., maternal, growth, etc.) that are most important.

- Selection of superior animals and sound breeding practices are as important as the breed. The cattle will only be as good as their selection and management.

- Select a breed that has the desired conformation as a beef animal. *Conformation* is the type, shape, and form of an animal. It is related

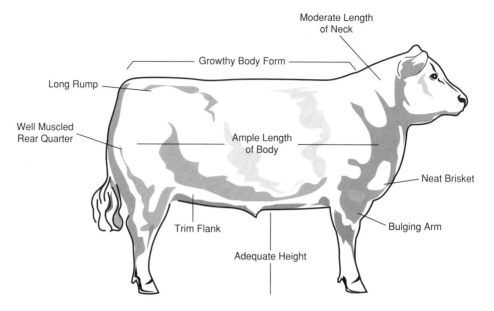

9-4. Conformation of a desirable beef animal.

to muscle (meat) in the valuable cuts and the ability of the animal to be productive. Over the years, the desired conformation has changed to moderate-framed animals that can provide a lot of meat.

- The breed needs to be selected based on its production capabilities in the specific environment where it will be raised. The producer needs to consider climate, feed, and forage resources. Some breeding animals are performance tested. This involves careful growth and feeding evaluations of young animals, particularly bulls.

- Markets for the breed should be evaluated. Can the offspring be readily sold in the area in which they are produced?

- Breeding stock needs to be available at a price that makes production economically feasible. Buy the best animals at the lowest price.

- The producer should consider personal preference and select a breed with which they are comfortable and familiar. This is probably the most commonly considered criterion in the selection process.

- Buy animals from established producers with good reputation.

BEEF BREEDS

The breeds of beef cattle common to the United States are thought to have been developed in Europe, Great Britain, or America.

Angus

The first Angus cattle were imported from Scotland by George Grant around 1873. Angus are the most popular purebred cattle. There were 193,401 purebreds registered in the United States with the American Angus Association in 1993.

9-5. A modern Angus bull. (Courtesy, American Angus Association, St. Joseph, Missouri)

The breed is black in color and has a smooth hair coat and black skin. A few Angus cattle carry a recessive gene for the color red, which are described below as Red Angus. Nearly all purebred Angus carry the dominant gene for being ***polled*** or naturally without horns.

Angus are cattle with moderate frames. They produce a high quality carcass that is well ***marbled*** or contains intramuscular fat. Marbling increases the quality of the carcass.

Red Angus

Red Angus herds developed from black Angus in the United States about 1945. Black Angus carrying the recessive red gene for color were mated to produce red offspring.

Red Angus are similar to Black Angus in many respects, except for color. However, red does not absorb as much heat as black, so red Angus are somewhat more heat tolerant.

Charolais

The Charolais is an old breed developed in central France. The first Charolais were introduced in the United States in 1936 by the King Ranch in Texas.

Charolais cattle are white or off-white in color with pink skin. They are a large framed, heavy muscled breed. Charolais may be either polled or horned. This breed is commonly used in crossbreeding programs to increase frame size and muscle.

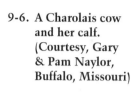

9-6. A Charolais cow and her calf. (Courtesy, Gary & Pam Naylor, Buffalo, Missouri)

Chianina

The Chianina breed originated in the Chiana Valley of Italy. It is, perhaps, one of the oldest breeds in the world. Chianina were introduced to the United States in 1971, when semen was first imported.

Chianina are large framed and are commonly used in crossbreeding programs. They have good maternal traits and generally improve the growth rate of the offspring. The color of Chianina cattle varies because of the influx of crossbreeding in the breed. Original Chianina cattle were white with a black tail switch and black skin pigment.

9-7. Young Gelbvieh bull.

Gelbvieh

The Gelbvieh breed was developed by crossing four yellow breeds of cattle. These four breeds were Glan-Donnersburg, Yello Franconian, Limburg, and Lahn. Together they form the Gelbvieh breed, which originated in 1920.

Gelbvieh cattle vary in color from cream to reddish yellow. In recent years, black Gelbvieh lines have been developed. These cattle are medium framed, produce an acceptable carcass, and have a high milking ability, when compared to other beef breeds.

Hereford

Hereford cattle originated in Hereford County, England. The first Hereford cattle were imported by Henry Clay to Kentucky. These cattle were mated to other native cattle and the first purebred herd was believed to have been established in 1840 in New York.

9-8. Hereford bull being trained for showing.

Hereford cattle have white faces and red bodies. Additionally, they may have a white feather or stripe on the top of their necks, which is commonly called a "feather neck." They are predominately white on the belly, legs, and switch. Herefords are horned, usually very docile and easy to handle, and are moderate in frame size.

Hereford cattle are extremely good *foragers,* which means they use many grasses and plants for food consumption. Furthermore, this breed is very hardy. The foraging ability and hardiness make this breed popular to the grazing regions of the western United States. It is common to hear Hereford cattle referred to as "whiteface."

Longhorn

Longhorn cattle were brought to the United States from Mexico by Spanish explorers in the 1800s. The fertility of longhorn cattle is usually

9-9. Longhorn. (Courtesy, Texas Department of Agriculture)

better than many other breeds. They are commonly used in crossbreeding because of their light birth weights.

Longhorn cattle do not grow as fast as some other breeds, but are very durable and adapt well to rough range conditions.

Polled Hereford

The Polled Hereford breed originated in Iowa in 1901. An Iowa breeder named Warren Gammon was instrumental in developing the breed. Gam-

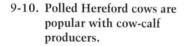
9-10. Polled Hereford cows are popular with cow-calf producers.

mon located four Hereford bulls and ten Hereford cows that were carrying the polled gene and mated them to develop the foundation for the Polled Hereford breed. Polled Herefords have the same basic traits as Herefords, except they do not have horns.

Limousin

Limousin cattle originated in France. Limousin cattle were introduced in the United States in 1968, when semen was imported from Canada.

9-11. A modern Limousin bull. (Courtesy, North American Limousin Foundation, Englewood, Colorado)

The Limousin will be either a light yellow or an orange color. The area around the muzzle will generally be a lighter version of the body color. Recently, black lines of Limousin cattle have been developed. Both horned and polled cattle are prevalent in this breed.

Limousins have a moderate, heavy-muscled frame. The breed is noted for its carcass leanness and large loin areas. Therefore, this breed's carcass is usually very high in *cutability* or the amount of saleable retail cuts that may be obtained from a carcass.

Maine-Anjou

The Maine-Anjou breed originated in France in the early 1800s. It is thought to have developed through crossbreeding English Shorthorns and French Mancelle breeds. Maine-Anjous were originally used as draft animals. Beef traits were developed in the breed through selective breeding. Maine-Anjou cattle were introduced to the United States in 1970, when semen was imported from Canada.

9-12. A young Maine-Anjou bull.

Maine-Anjou cattle are dark cherry red and white in color and have light pigmentation in the skin. They are commonly used in crossbreeding. From this crossbreeding, black Maine-Anjous have been developed. The breed is usually horned and moderate framed. They are generally docile and easy to handle, grow quickly, and have good marbling in their carcasses.

Saler

The Saler breed was developed in central France and recently brought to the United States. Initial herds were established in 1973, when semen was imported.

Salers are a deep cherry red, and sometimes, will have white on their bellies. Recently, black Saler cattle have been developed. Saler cattle are noted for their calving ease, primarily because of their smaller heads and long slender necks and bodies. Most Saler females milk quite well.

Shorthorn

Shorthorns originated in northern England around 1600. Shorthorns were originally developed as a **dual-purpose breed,** meaning they were used for both milk and meat production. The breed was introduced to the United States in 1783, in Virginia.

Shorthorn cattle are red, white, or roan. Shorthorns may be either polled or horned. Frame size of this breed is moderate. The females are good mothers and milk exceptionally well. Shorthorns are generally gentle and have good dispositions.

9-13. A Shorthorn bull. (Courtesy, *Shorthorn Country*, Omaha, Nebraska)

Simmentai

The Simmental breed was developed in western Switzerland and is still quite popular in Europe today. Originally, they had many purposes, including milk and meat production, and use as draft animals. Simmentals were brought to the United States in 1969, from Canada.

Simmental cattle usually have a cream colored face with a light yellow, red, or spotted body. Recently, black strains of the breed have been developed. Simmental cattle may have some color pigment around their eyes.

Horned and polled cattle will be found in this breed as well. The Simmental is large framed, heavy-muscled, and produces a lean carcass.

South Devon

South Devon cattle were developed in the western region of England. This breed of cattle is believed to have originated from a large red breed in France, which was brought to England during the Norman Invasion. The South Devon breed is related to the Devon breed. This breed was developed for dual purposes of meat and milk production. South Devons were first imported into the United States in the late 1930s and early 1940s. However, the popularity of the breed increased in 1970, when a group from Stillwater, Minnesota, started using South Devons for crossbreeding purposes. The South Devon breed will be a medium red color and usually be horned.

9-14. A mature Simmental cow with her calf. (Photograph courtesy of the American Simmental Association, Bozeman, Montana)

Brahman

Brahman cattle were developed in the southwestern part of the United States. The breed is thought to have developed between the mid 1800s

9-15. A modern Brahman cow. (Courtesy, American Brahman Breeders Association, Houston, Texas)

and early 1900s and originated from the cattle from India. *Bos indicus* cattle have a hump on the shoulders and usually have long floppy ears and loose hide. These cattle may also be called Zebu.

Breeders select the Brahman breed because of their hardiness, resistance to insects and diseases, and their heat tolerance, which makes them very popular to the warmer climates of the south and southwestern United States. Furthermore, the females are good mothers and Brahman cattle do quite well on poor quality forages common to range conditions.

Brahman cattle are somewhat unpredictable in their temperament. The Brahman breed is earlier in skeletal maturity and often is used in crossing with other breeds of cattle. Brahman cattle will vary in color from gray to red to black.

Brangus

The Brangus breed is a cross between the Brahman and Angus. This breed was developed at a U.S. Department of Agriculture Experiment Station in Louisiana in 1912.

9-16. A young Brangus bull.

Brangus cattle will have many of the same characteristics as the Angus and Brahman. They are solid black in color and are polled. (There is also a Red Brangus, which is a different breed from the Brangus.) They also have long floppy ears (not as long as the Brahman) and loose hide. Brangus cattle are very adaptable to various climates and have good mothering abilities.

Santa Gertrudis

Santa Gertrudis cattle were developed around 1918, on the King Ranch in Texas. This breed was developed by crossing Brahman and Shorthorn cattle.

Santa Gertrudis are a deep cherry red and may be horned or polled. Because the Santa Gertrudis are ancestors of the Brahman breed, they have many of the same characteristics, including loose skin and long ears. They are also very adaptable to the warmer climates and are more resistant to disease and insects.

Beefmaster

Beefmaster cattle were developed in 1931 in Texas. The breed is the result of crossing Herefords, Shorthorns, and Brahmans. The breed varies in color and reds and yellowish-whites are common. Beefmaster cattle are hardy and milk well.

9-17. A young Beefmaster bull.

Braford

Braford cattle are a cross between Brahman and Hereford cattle. It is assumed that the breed is 3/8 Brahman and 5/8 Hereford. Brafords will have many of the same characteristics of the parent breeds. They are good foragers and are adaptable to many regions.

Tarentaise

The Tarentaise originated in the French Alps in 1859. It was brought to North America in 1973 with importations into Canada. They are a hardy breed. Tarentaise are relatively free of eye cancer and udder burn because of dark pigment around the eyes and on the udder.

Tarentaise are wheat-colored, ranging from cherry to dark blonde. Black pigment is found on the muzzle, eyes, and udder. Females calve easier

9-18. Tarentaise bull. (Courtesy, American Tarentaise Association, N. Kansas City, Missouri)

than most breeds. Bulls are well suited for use with first-calf heifers. Calves are somewhat smaller and vigorous at birth. Bulls may weigh up to 1,800 pounds and cows to 1,150 pounds. Adults are more active than other breeds. The breed is growing in popularity.

ADVANTAGES AND DISADVANTAGES OF BEEF CATTLE PRODUCTION

Producing beef cattle has advantages and disadvantages. The producer has to analyze these and decide if beef production is the right thing to do.

ADVANTAGES

Producing beef cattle may have the following advantages:

- Beef cattle can consume roughages, including most grasses and several weeds, which other livestock may not eat or be able to use. Therefore, little forage is wasted with beef cattle.

- Labor requirements are usually not as great with beef cattle as with other livestock production. Because beef cattle are good foragers, they do not necessarily have to be fed daily.

- Death losses are relatively low, if good management practices are followed.

- Beef cattle adapt well to various sizes of operations. Productive beef herds can be from one or two head to several thousand head.

- There is a good demand for meat. Beef is a mainstay in the American diet. Additionally, meat is an important export to other countries.

- The beef industry creates many jobs. It relies on services, such as herdsmen, trucking firms, processors, researchers, and feed suppliers.

- Beef cattle may provide avocational interests, as well as youth projects through FFA and 4-H, which allow new experiences for both young and old.

DISADVANTAGES

Producing beef cattle has the following disadvantages:

- If cattle are fed to slaughter weight, the producer incurs risk. Grain prices tend to fluctuate and available forage is dependent on the precipitation for a specific area. An example of this would be whenever a region suffers a drought. Grain costs more because yields are lower and forages do not grow readily because of the lack of precipitation.

- Beef cattle do not convert feed and forages into meat as efficiently as other animals.

- Because of the 283-day gestation period of beef cattle, it takes longer to increase herd size. The producer will usually only get one offspring a year from each cow.

- Capital investment is high to start a modern beef operation. Not only does the producer have to invest in the cattle, but also equipment, facilities and herd management.

TYPES OF BEEF PRODUCTION SYSTEMS

Beef cattle production varies among producers. The kind of system depends on the producer's goals. It is also influenced by the resources available and the market in the local area.

COW-CALF SYSTEMS

Cow-calf systems raise calves for sale to other growers. Most cow-calf systems involve high-quality grade cattle. The cows are bred to calve in the late winter or early spring. The calves are sold in the fall. This takes

9-19. This cow–calf producer has a good calf crop underway with grade Charolais cows.

advantage of the warm summer growing season. Most of the calves go to feedlots for finishing into meat animals.

Natural breeding is typically used with grade cow-calf systems. The bull may be put in the pasture with the cows at breeding time. A producer schedules breeding to assure calving at the desired time. One three-year-old bull will breed 25 to 40 cows a year.

Some beef producers use artificial insemination (A.I.). A.I. is more important with expensive, purebred females. This allows the producer to have access to a wider variety of bulls.

Heifers need to be the proper size and age before breeding. Most heifers will have their first calf at two years of age. Since the gestation period of a cow is approximately 283 days, a cow should have one calf a year. Occasionally, cows will have twins or multiple births, but this is not prevalent in beef cattle.

The cow-calf system uses forages for feed. The forages are from grazing pasture grasses, such as fescue, orchard grass, or bermuda grass. If pasture grasses are not available, hay can be fed as a forage.

An additional mineral supplement should also be fed free-choice to meet the need for minerals, which are vital to beef cattle. The calves may also be supplementally fed with grain, which is commonly called *creep feeding.* Creep feeding will allow the calf to gain more weight while it is still nursing the cow. Creep feeding is usually done when the calves are not receiving enough milk from the cow, the supply of forages are limited, or when grain prices are low and it is economically feasible.

Producers of purebred cattle follow a somewhat different cycle. They raise herd replacements and animals for other cattle breeders. Purebred cattle are not raised to go to a feedlot.

BACKGROUNDING SYSTEMS

Backgrounding systems take the calf from the time it is weaned from the cow to the feedlot phase. Backgrounding is essential to add more weight to the calf before it enters the final feeding phase.

During the backgrounding, forages, such as annual grasses, are commonly used to add more weight to the calf. Supplemental grain may also be fed to the calves to increase the rate of gain and prepare the calves to enter the finishing phase earlier. Calves that are backgrounded will also need a free-choice mineral supplement.

FINISHING SYSTEMS

Finishing systems complete the grow-out phase of beef production by taking the calf to slaughter weight. The calf will usually weigh 1,100 to 1,300 pounds at slaughter, depending on the breed.

Some beef producers feed calves to slaughter weight, but most cattle will be finished by commercial feedlots. The feedlots are located in the western and southwestern United States. During the finishing period, calves are fed feed **concentrates.** These are

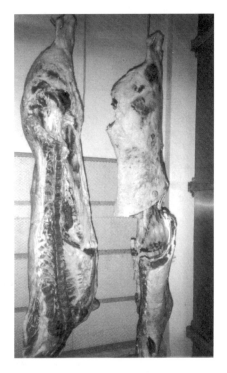

9-20. A finished beef carcass ready for cutting.

9-21. A modern feedlot in California.

feeds high in energy from grains, such as corn, milo, and oats. Free-choice minerals may be provided if the minerals are not part of the finishing ration.

HERD HEALTH

A good herd health program is essential in all phases of beef production. Prevention, including good sanitation and vaccination programs, is key to maintaining good herd health. Diseases and parasites greatly reduce the profitability for the beef producer. A good health program will also aid in marketing beef animals. (Chapter 6 covers animal health in more detail.)

PARASITES

Beef cattle are affected by both external and internal parasites.

Common external parasites are flies, mosquitoes, lice, mites, and ticks. Good sanitation practices with facilities and equipment will reduce problems with these external parasites. In addition, external parasites may be controlled by using insecticides.

Only approved products should be used in controlling external parasites. Some producers consult veterinarians for parasite control. In general, insecticides may be applied by spraying or dipping the cattle. Backrubbers may also be used when spraying or dipping is not an option. A backrubber should be placed in a high-traffic area, such as near feeders and waterers. The insecticide is poured on the backrubber. As the cattle walk under the backrubber, the insecticide is applied.

Internal parasites, such as roundworms and flatworms, must be controlled. Common symptoms of internal parasites are weight loss, poor weight gain, rough hair, and diarrhea.

Anthelmintics are chemical compounds used for deworming animals. Anthelmintics may be administered to cattle orally, by injection, by pour-on, or by adding to the feed or mineral. A regular worming schedule should be followed. Best results are achieved if cattle are dewormed once or twice a year. A veterinarian may be consulted to decide which anthelmintics to use.

DISEASES

Controlling disease is important in a profitable beef operation. Many diseases affect beef cattle. Some producers use a veterinarian to help manage herd health.

Bovine virus diarrhea (BVD) symptoms may include fever, coughing, discharge from the nose, poor gains, rapid breathing, diarrhea, dehydration, rough hair, and lameness. In addition, BVD may cause pregnant animals to abort.

Brucellosis (Bangs) is a reproductive disease. The symptoms may include an aborted fetus, retained afterbirth, and infertility. Government programs have helped control the disease, but it is still a risk to cattle. There is no known cure for brucellosis, and infected animals should be destroyed. All female calves should be vaccinated for brucellosis while they are young. Replacement heifers should be from brucellosis-free herds. Brucellosis may also be spread to humans, causing undulant fever.

Blackleg is a bacterial disease that can live without oxygen. The bacteria commonly live in the soil and may be present for many years. Symptoms of blackleg are sudden death, high fever, swollen stiff muscles, and lameness. Calves should be vaccinated when young and again at weaning.

Infectious bovine rhinotrachetis (IBR, Red Nose, IPV) has several forms. Symptoms may include fever, discharge from the nose, rough breathing, coughing, weight loss, inflammation of the genitals, watery eyes, convulsions, and incoordination. Since the symptoms are quite varied, a veterinarian should be consulted if IBR is suspected.

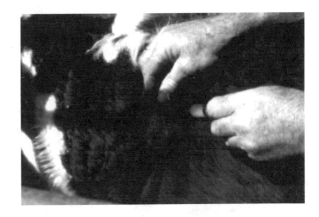

9-22. This calf is being vaccinated for blackleg. (Courtesy, University of Missouri Extension Service)

Leptospirosis is caused by several strains of bacteria. Symptoms may include a rise in temperature, rapid breathing, stiffness, loss of appetite, bloody urine, and abortion. All animals should be vaccinated twice a year to control the disease.

Campylobacteriosis or vibriosis may occur in intestinal or venereal form. The intestinal form usually does not have harmful effects on beef cattle. The venereal form is more serious. The symptoms may include infertility, abortion, erratic heat periods, and poor conception rates due to early embryonic death. Animals should be vaccinated before breeding and the vaccination should be repeated annually.

HERD IDENTIFICATION

Beef animals need to be identified in some way. Proper identification is important in managing the beef herd, as well as specifying ownership. It allows the producer to identify all animals and keep records on production. Several methods may be used.

BRANDING

Branding is a permanent method of identification. Branding may be with a hot iron or by freeze branding. The producer needs to select a brand design that will represent ownership of the cattle. Brands are usually registered with a state agency. A hot iron brand will burn the surface of the skin, leaving a permanent mark. When freeze branding, liquid nitrogen is used with the iron, which freezes the skin pigment. This depigments the hair follicles and allows the hair to grow back white in color. In white-haired cattle, the liquid nitrogen is left on a longer time. This produces a bald spot the shape of the branding iron.

9-23. Cow identified with a brand on the hindquarters.

EARMARKS

Earmarks may be as simple as using a knife or an earnotcher to cut a notch in the ear. Ear tags are commonly used as well. A plastic or metal tag is inserted in the ear of the animal. There is a distinguishing mark on the tag, such as a number, letter, or a combination of both. There are many kinds of eartags available. Select one that is easy to use. Eartags may be torn from the ear or taken out by someone other than the owner.

9-24. Cow and calf identified with an ear tag. The numbers are easy to read and permanently written on the tag.

NECKBANDS

Neckbands or chains may also be placed on animals for identification. Again, animals may tear the bands or chains or someone other than the owner may remove them. In rare cases, an animal may catch the chain or band on an object, which could possibly choke the animal to death.

TATTOOS

Tattoos are a common method of identification used among purebred breeders. An instrument applies letters and numbers by piercing the skin. Ink is then rubbed into the area. A common place for tattoos on beef cattle is the inside of the ear. Tattoos are frequently used. A drawback is that the animal must be confined to read the tattoo. Some are hard to read on dark-skinned cattle.

OTHER METHODS

Several other methods of identification are available, including nose prints, DNA testing, and microchips. Nose prints and DNA testing are used to establish the precise identity of an animal. These procedures are often used with show animals. Microchips may be placed beneath the skin of an animal. A scanner will decode the signal from the microchip. The information identifies the animal.

COMMON HERD MANAGEMENT PRACTICES

Producing quality beef animals is more than turning them out in a pasture. The animals must receive the proper care. Care is known as management practices.

DEHORNING

Dehorning is used with horned breeds of cattle to remove horns. Large horns can be dangerous and injure other animals, as well as people. Several dehorning methods are used. It is preferable to dehorn young calves over larger animals. Caustic chemicals may be placed on the location where horns grow. This prevents horn growth by destroying tissue. It is widely used on younger animals.

Horns may also be sawed or clipped off. This is a surgical procedure that removes the horns. When using this method, all bleeding needs to be stopped afterward. A repellent should be used to keep flies off the wound.

CASTRATION

The beef producer will usually not want to keep all male offspring as bulls. *Castration* removes testicles or destroys their development. Castration makes male calves easier to handle and manage. A calf castrated at a young age is a *steer.* Except for breeding animals, steers often bring higher prices than bulls.

A common method of castrating cattle is with an elastrator. This stretches a specially made rubber band over the scrotum, which will cut off circulation. The scrotum will essentially dry up and fall off. This method should be done on younger, smaller calves.

A knife or a scalpel may be used to cut the testicles from the scrotum. This is a popular method. Bleeding must be stopped and flies controlled. Additionally, castrated calves need to be watched closely for infection.

A Burdizzo clamp is also used. No bleeding occurs with this method. The clamp crushes and severs the cords and blood vessels going to the testicles. This causes them to dry up. When using the Burdizzo, it is important that the cord does not slip out of the clamp and miss being crushed.

FACILITIES AND EQUIPMENT

The facilities and equipment needed for beef cattle depend on the system of production and climate. Factors that assist in planning include: number of cattle, amount of space, location, kind of facilities, herd and property security, environmental conditions, amount of money to invest, and availability of other facilities and equipment.

FACILITIES

Facilities may range from simple to elaborate. Producers should allow for flexibility so they can expand or change beef production systems.

Because beef cattle are primarily range animals, it is not necessary to have a great amount of shelter. However, barns and loafing sheds should be available to provide shade in the summer, if trees or other shade sources are not available. Additionally, barns and loafing sheds may be used to protect cattle from harsh weather. All beef cattle producers need good fences to keep the animals confined.

9-25. A well-designed cattle guard that keeps animals from straying and allows vehicles to cross.

EQUIPMENT

Cattle production requires various equipment. The kind needed depends on the type of animal produced. Water tanks and water fountains need to be available for all cattle.

Because most producers supply salt and mineral free-choice to beef cattle, feeders need to be available. These feeders should be large enough to allow several head to consume salt and mineral mix at the same time.

Feed storage is needed to assure high quality feedstuff and reduce the amount of feed wasted by spoilage and pest damage. Feed may be stored in upright silos, trench silos, metal storage bins, or feed rooms inside barns.

Feeders for hay, forages, and grain need to be large enough to accommodate all cattle at once. This may require the producer to use more than one feeder. This alleviates the dominant, aggressive animals from keeping other cattle away.

Equipment to handle cattle is needed in all beef operations. A good corral makes it easier to work cattle. Corrals should also decrease labor and time, reduce stress on cattle, decrease chance of injury and weight loss, and increase safety for the producer.

Corral designs vary. In planning a new corral, allow for expansion in the future. Sharp turns and corners should be avoided. The corral should be built so cattle can be easily moved and trailers loaded.

9-26. A properly designed corral. (Courtesy, University of Missouri Extension Service)

Working chutes and headgates are used to hold and restrain cattle for tagging and vaccinating. There are many models of chutes and headgates. Select the design that best allows efficient cattle handling.

9-27. A working squeeze chute. (Courtesy, University of Missouri Extension Service)

REVIEWING

MAIN IDEAS

Most beef cattle breeds originated in Europe or Great Britain. Beef cattle are produced by a series of systems: the cow-calf, backgrounding, and finishing. All of these systems are beneficial to the beef producer.

The producer has to follow sound management practices to be successful in producing beef cattle. These practices will increase profitability, loss from disease, and reduce stress to the animals.

Proper facilities and equipment are necessary for a beef cattle operation. Properly planned facilities will decrease labor and make the beef operation more efficient. Designing facilities that will allow for expansion is important to facility design.

Equipment varies from one operation to another but is needed in all beef production. Corrals make it safer and easier to handle cattle, as well as reduce stress on the cattle. Watering equipment is important for most operations. Feeders and feed storage equipment will insure higher quality feed.

QUESTIONS

Answer the following questions using correct spelling and complete sentences.

1. Distinguish between the *Bos taurus* and *Bos indicus* cattle.

2. List advantages and disadvantages of raising beef cattle.

3. What should be considered in selecting a breed of cattle?

4. What beef production systems are used?

5. Name and describe two important herd management practices.

6. What facilities and equipment may be needed in beef cattle production?

7. How are beef cattle identified?

8. Select any three breeds of cattle and provide a brief description of each breed.

EVALUATING

CHAPTER SELF-CHECK

Match the terms with the correct definitions. Write the letter by the term in the blank that is provided.

a. dual purpose

b. cutability

c. polled

d. anthelmintics

e. forager

f. concentrate

g. steer

h. *Bos taurus*

i. *Bos indicus*

j. backgrounding

1. ___ chemical compounds used for deworming animals

2. ___ beef cattle that naturally do not have horns

3. ___ male calf castrated at a young age

4. ___ cattle used for both meat and milk production

5. ___ cattle that are from India and have long floppy ears and loose hide

6. ___ the amount of saleable retail cuts obtained from a beef carcass

7. ___ a system that takes the calf from weaning to the feedlot phase

8. ___ cattle common to the United States of European origin

9. ___ feeds that are high in energy

10. ___ use grasses and plant for food

EXPLORING

1. Prepare a display of the various breeds of beef cattle, including pictures and descriptions of each breed.

2. Construct a bulletin board with all of the components of the beef industry. All components should be identified and pictures should be included.

3. Tour various beef operations in your area and prepare a report on your observations.

4. Make a field trip to a livestock show. Identify the breeds that you see. Describe the conformation of the animals. Note how they are fed, bedded, watered, and groomed.

Chapter 10

SWINE PRODUCTION

What are they: pigs, hogs, or swine? All three refer to the same animal. The word, swine, is more appropriate to use. This is because of the casual ways we use "pig" and "hog."

Contrary to popular belief hogs are not dirty animals. The analogy of "As dirty as a hog" or "Sweatin' like a pig" are incorrect. Hogs are very clean animals. They do not have sweat glands and may wallow in the mud to cool themselves and to keep from getting sunburned.

Swine are popular because of the delicious meat—pork. They reproduce at a high rate, grow fast, require low amounts of labor, and give a fast return on investment.

A big change is underway in how swine are raised. Some small farms have been replaced with large corporate farms. With the new approach, a large volume can be raised efficiently. This keeps the price of pork down and assures a quality food product.

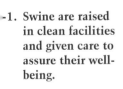

-1. Swine are raised in clean facilities and given care to assure their well-being.

OBJECTIVES

This chapter provides basic information on swine production. The following objectives are covered:

1. Explain the importance of swine production
2. Describe swine as organisms
3. Explain the possibilities of pork production
4. Describe pork production systems
5. Explain important management practices in swine production
6. List nutritional requirements of swine
7. Explain health management practices for swine
8. Describe facility and equipment needs for swine

TERMS

barrow	meat-type hog	prolific
boar	needle teeth	sow
contract production	one-stage production	specific pathogen free
crossbreeding	pedigree	tail docking
farrowing	piglet	two-stage production
feed additive	porcine somatotropine	three-stage production
feeder pig	porcine stress syndrome	type
gilt	probe	ultrasonics

Animals in Our Lives

Horses are often adored by people

A steer can be a companion animal

Grooming a lamb for showing

Koi are among the most colorful freshwater fish

Hogs are interesting animals

Special care is needed in handling this Albino Burmese Python

Selected Breeds of Beef Cattle

Angus bull

Gelbvieh bull

Limousin bull

Brahman cow

Tarentaise bull

Polled Hereford

Selected Breeds of Dairy Cattle

Jersey cow

Holstein-Friesian cow

Brown Swiss cow

Guernsey cow

Ayrshire cow

Milking Shorthorn cow

Selected Breeds of Sheep and Goats

Oxford lamb

Dorset ewe

Suffolk ewe and lamb

Katahdin ewes

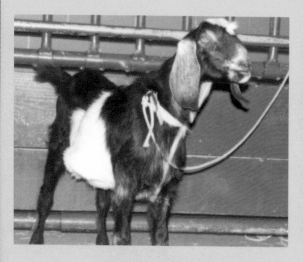

Spanish goat

Nubian goat

Selected Poultry Species

Peacock

White Leghorn rooster and hens

Pekin duck

Ornamental Australian Black Swan

Tom turkey

Canadian goose

Swine Production

Keeping facilities clean reduces disease

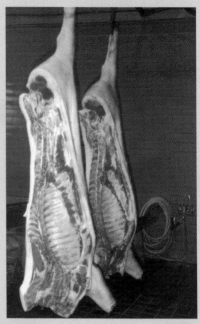

Well-finished, high-yielding carcasses

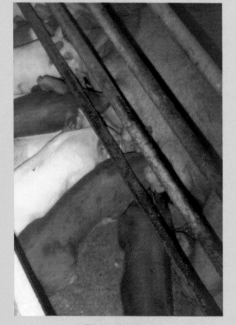

Pigs nursing

Pig being vaccinated

Feeder pigs

Growing market hogs

Animals and Science Connections

Grooming a dog

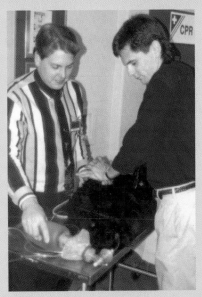

Using CPR on a "dummy" dog

Evaluating boar semen for
artificial insemination

Genie the transgenic pig with a litter of her own
(See Chapter 1)

Culturing animal cells in laboratory research

Fistulated steer for nutrition studies

Exotic Animals

Giraffe

Elephant

Alligator

Flamingoes

Orangutan

Dama gazelle

IMPORTANCE OF SWINE PRODUCTION

Swine have been produced a long time. They were first domesticated in Asia about 9,000 BC. Swine were brought to America by Christopher Columbus in 1493. North America already had wild hogs, often known as boars. The wild animals were scattered throughout North America but had not been important in the swine industry.

The swine brought to America were of European and Asian breeding. Many changes have been made in the animals to have a desired product. Many breeds were developed. Today, the breeds have given way to crossbred animals.

SWINE INDUSTRY

The swine industry closely parallels the production of corn. Three-fourths of the hogs produced in the United States are produced in the Corn Belt. When corn yields are high, corn prices are low and hog production increases.

About 60 percent of the hogs in the United States are grown in confinement. In addition, 12 to 13 percent of hogs are produced under contract. Hogs are the second largest livestock population in the United States, with nearly 70 million head. Iowa, Illinois, and Indiana are the leading swine states. Swine are an important source of agricultural income.

Swine Industry Alternatives

Swine producers have two major alternatives: raising hogs for meat or for breeding stock. Swine produced for meat are selected to provide a high volume of good, lean meat.

Swine raised for breeding stock are often purebred animals. Purebred hogs are produced by breeding purebred animals of the same breed. Careful records are kept on purebred animals. If they meet the standards of an association, they may be registered as purebred animals.

Some breeding hogs are crosses of two or more breeds. These animals cannot be registered as purebred. They are produced to meet specific meat production goals.

Favorable Enterprise Factors

Swine production has many factors that make it a favorable enterprise. Swine are efficient in converting feed to meat. Fewer pounds of feed are needed to produce a pound of pork than for a pound of beef. Less than 5 pounds of feed are needed for a pound of pork. Nine pounds of feed are needed for a pound of beef.

Swine are very prolific. ***Prolific*** means that they will produce a large number of young. Sows will usually farrow 7 to 12 piglets twice a year. They also excel in dressing percentage. They will yield 65 to 80 percent of their live weight. Comparably, cattle only dress out around 50 to 60 percent, and sheep at 45 to 55 percent.

10-2. Sow with litter.

Labor requirements are lower because hogs are good at self-feeding. The capital investment (money) is low because they require little land or buildings, depending on the type of operation. The initial investment is low for hogs. Moreover, the return on investment is relatively short. A return can be made within 10 months.

Unfavorable Enterprise Factors

Several factors make the swine enterprise unfavorable. Hogs are very susceptible to disease and parasites. Hogs have a simple stomach like humans. Because of this, they must consume a large amount of concentrates and a minimum of forages. Hogs also require special attention at farrowing time. Because of the relationship between corn and hog prices, economic conditions can become unfavorable at times.

CORPORATE SWINE PRODUCTION

Swine have typically been produced on small, family-operated farms. The farm might also grow grain crops, primarily corn, as feed for the

10-3. Modern buildings are used in swine production to assure the well-being of the animals and get efficient growth. (This shows the outside of a swine facility in Pennsylvania.)

swine. In recent years, changes have been taking place. These changes have become a major issue with some traditional pork producers.

More swine are being grown in large systems. The producer may be under contract with a company that provides the piglets, monitors production, supplies feed, and markets the hogs. This approach is often known as vertical integration.

The swine are bred for most efficient growth and yield. During growth, careful attention is given to nutrition and disease control. People are not allowed in facilities where the hogs are growing. Producers enter only after bathing and wearing carefully laundered clothing. These precautions allow faster growth to market weight for meat—about 220 pounds.

SWINE AS ORGANISMS

Swine have organ systems similar to most animals. They are mammals, with sows giving birth to litters of pigs. Swine differ from cattle in that they have a monogastric stomach. This means that it has a single compartment. They must be fed feed that is more concentrated. Swine make poor use of forage.

Swine are classified as mammals with hoofs. The scientific name for the species produced in North America is *Sus scrofa domestica*. Other species include pygmy hogs (*Sus salvanis*), bearded pigs (*Sus barbatus*), and the warthog (*Phacochoerus aethiopicus*).

Swine used for food products are young hogs only a few months of age. They have grown rapidly and been protected from disease. Older hogs, especially boars, may develop a strong flavor in the meat.

CLASSIFICATION

Swine are classified by age and sex. Young swine are known as *piglets* or baby pigs. At a young age, male pigs are castrated unless they are kept for breeding. The male castrated at a young age is a *barrow*. A young female that has not had pigs (farrowed) is a *gilt*. An older female is a *sow*. A male hog is a *boar*. Sometimes, male hogs are castrated after they reach sexual maturity. If so, they are known as stags.

The hogs used for high-quality meat are barrows and gilts. Sows and boars may be made into lower-quality meat, such as sausage or cooked food products. Remember, lower-quality meats are wholesome, but may not have the same flavor.

SWINE TYPE

Swine are produced for meat. People want lean meat without much fat. This reflects a change from years ago when fat was used to make lard. Consumer demand has resulted in a *meat-type hog.* Lard-type and bacon-type hogs have declined in use. The meat-type gives the greatest amount of lean meat in high-value cuts, such as the ham.

10-4. Clinics inform people of the desired swine type. Producers learn the type of hog to raise by attending.

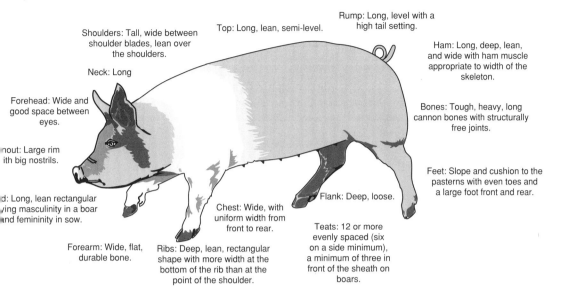

Rump: Long, level with a high tail setting.

Top: Long, lean, semi-level.

Shoulders: Tall, wide between shoulder blades, lean over the shoulders.

Neck: Long

Ham: Long, deep, lean, and wide with ham muscle appropriate to width of the skeleton.

Forehead: Wide and good space between eyes.

Bones: Tough, heavy, long cannon bones with structurally free joints.

nout: Large rim ith big nostrils.

d: Long, lean rectangular ving masculinity in a boar and femininity in sow.

Feet: Slope and cushion to the pasterns with even toes and a large foot front and rear.

Flank: Deep, loose.

Chest: Wide, with uniform width from front to rear.

Forearm: Wide, flat, durable bone.

Ribs: Deep, lean, rectangular shape with more width at the bottom of the rib than at the point of the shoulder.

Teats: 12 or more evenly spaced (six on a side minimum), a minimum of three in front of the sheath on boars.

10-5. Major characteristics of a meat-type hog.

Producers need to know the desired features of a hog. These are useful in describing the desired conformation. Conformation is the type and shape of an animal.

Meat-type Characteristics

Meat-type hogs need to grow fast and efficiently. They need to have plenty of muscle tissue that will be used for lean food products. In general, a meat-type hog is a long animal with deep, well-developed muscles in the hams. Breeding animals should have good bone structure to be able to carry their weight. A meat-type hog will not have thick layers of fat. The major characteristics of a meat-type hog are shown in 10-5.

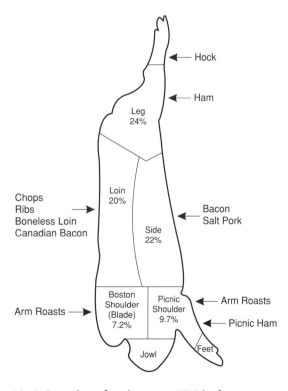

Hock

Ham

Leg 24%

Loin 20%

Chops
Ribs
Boneless Loin
Canadian Bacon

Bacon
Salt Pork

Side 22%

Boston Shoulder (Blade) 7.2%

Picnic Shoulder 9.7%

Arm Roasts

Arm Roasts

Picnic Ham

Jowl

Feet

10-6. Location of major cuts. ("%" is the percentage that part of a carcass is of the entire carcass)

Products

The major meat products from hogs are ham, bacon, loin cuts, and roasts. Other parts produce cuts that are of lower value. Some low-value cuts are ground into sausage. The percentage of the various areas is shown in 10-6.

BREEDS OF SWINE

Several breeds of swine are found in the United States. Swine breeds are very important to the purebred producer. Hogs grown for meat are typically crosses of the purebreds. The breeds of hogs include Duroc, Hampshire, Yorkshire, Berkshire, Poland China, Landrace, Tamworth, and Chester White.

Some producers have detailed breeding systems to get the type wanted. For example, a Duroc boar may be bred to a Yorkshire female. A Hampshire boar may then be bred to female offspring from the Duroc-Yorkshire cross. A Yorkshire boar may then be bred to the females from the second cross.

Most of the hogs produced in the United States are sired by boars of the traditional purebred lines. Some are sired by specialty stock and a few by home-raised boars. In general, there are more differences between animals within a breed than between breeds. Specialized breeding stock suppliers usually have one or more hybrids, crossbred, inbred, maternal female lines, and/or terminal sire lines.

Duroc

The Duroc is light to dark red and has ears that droop over the eyes. It is a popular breed because of prolificacy and hardiness. Durocs are known today as a good meat-type hog. Durocs should not have any white on their bodies.

10-7. Duroc boar. (Courtesy, National Duroc Swine Registry Publications)

Hampshire

The Hampshire has the most striking appearance as a black hog with a white band or belt encircling the shoulders and front legs. The ears standout and do not cover the eyes. Hampshires cannot be registered if the band does not encircle the body and there is too much white on the head and hind legs.

10-8. Hampshire boar. (Courtesy, Seedstock Edge, West Lafayette, Indiana)

Yorkshire

The Yorkshire is a large, white hog. The females are known for their mothering ability. Yorkshires have long bodies that may tend toward a bacon-type of hog. They are sometimes faulted for small hams.

10-9. Yorkshire gilt. (Courtesy, American Yorkshire Club, Inc., West Lafayette, Indiana)

Berkshire

The Berkshire is easily identified by its short, upturned nose. The hog is black with six white points: the feet, some white on the face, and white tail switch. Splashes of white may be in other places on the body. Berkshires are lean and typically have less fat than other hogs.

Poland China

The Poland China is black with white spots on the feet, tip

10-10. Berkshire gilt.

10-11. Poland China gilt. (Courtesy, Poland China Swine, Peoria, Illinois)

of tail, and nose. In the past, Poland Chinas were known as large, fat hogs. Breeders have selected based on more lean and less fat.

PORK PRODUCTION POSSIBILITIES

Producers need to carefully select hogs to establish a herd. Selecting the wrong hogs can result in failure as a hog producer. Three important considerations are: type and pedigree, health, and production testing.

TYPE AND PEDIGREE

Selection based on *type* means to select an animal that is as close to the ideal as possible. A *pedigree* is a record of an individual's heredity. In selecting for type, test for meatiness, porcine stress syndrome, and genetic defects.

Meatiness can be measured with a probe, lean meter, or ultrasonics. All show the amount of backfat present on an animal. The *probe* is a tool for measuring the thickness of backfat. The lean meter is based on the difference in conductivity of electricity in the fat and muscle. Muscle and blood are better conductors than fat. *Ultrasonics* use bursts of high frequency sound. It measures the reflection of the pulses and the time of reflection.

Porcine stress syndrome (PSS) is a nonpathological disorder in heavily muscled animals that results in sudden death loss. A couple of tests can be used to detect PSS, and they are 100 percent accurate.

Litter mates and parents should be noted for any genetic defects that should be avoided. Commonly inherited genetic defects include atresia ani, cryptorchidism, hermaphrodites, PSS, scrotal hernia, and umbilical hernia.

HEALTH

When selecting any animal, always check to see that they are free of disease. Specific swine diseases will be discussed later in this chapter.

Many producers purchase hogs that come from a ***specific pathogen free*** (SPF) herd. SPF pigs are free of disease at birth. They are raised in an aseptic environment. The diseases that SPF animals should be free of are brucellosis, leptospirosis, lice, mange, pneumonia lesions, pseudorabies, swine dysentery, turbinate citrophy, and snout distortion.

PRODUCTION TESTING

Several purebred swine registries, as well as producers, have a production registry and a meat certification program. The most important way to evaluate the productivity of potential animals is through production testing.

Boar Selection. One boar should be selected for every 15 to 20 sows. Because they will be producing so many offspring, their evaluation is extremely important. Table 10-1 summarizes the major characteristics of a good boar.

Table 10-1
Characteristics of a Good Boar

Trait	Standard
Litter Size	from litter with 10 or more farrowed, 8 or more weaned
Underline	12 or more fully developed, well-placed teats
Feet and Legs	medium to large boned; wide stance both in front and in the rear; free in movement; good cushion to both front and rear feet; equal sized toes
Age at 230 lbs.	155 days or less
Feed/cwt gain	275 lb/cwt
Daily Gain	2.00 lb/day or higher
Backfat Probe	1.0 inches or less

The boar should also be docile and have an easy temperament, yet be aggressive enough to breed the herd. The boar should be purchased at six to seven months of age. Use of boars in breeding may begin at eight months of age. Boars should be of good health and have a masculine appearance. The testicles should be well developed. With purebred boars, the standards of the breed should be met.

Female Selection. Gilts and sows should be carefully selected. Only gilts from large litters should be selected. Their backfat should be less than 1.2 inches. The gilt or sow should be sound and free of any flaws or defects that would interfere with normal production. They should be the fastest growing and the leanest of the bunch. Gilts in purebred herds should meet the standards of the breed.

Gilts are often selected from the animals that are being fed for meat production. When gilts reach 180 to 200 pounds (81.6 to 90.7 kg), they should be evaluated for their reproductive potential. If selected as a replacement, they should be removed from the meat animals and fed separately.

SYSTEMS OF THE INDUSTRY

The way swine are raised is often known as a production system. Three production systems are commonly used. These involve a sequence of stages, known as production stages.

PRODUCTION STAGES

The production stages are one-stage, two-stage, and three-stage.

One-Stage Production

One-stage production includes farrow to finish. In this type of operation, a litter is born in a pen and stays there until they are 40 to 60 pounds (18 to 27 kg). They are then sold to someone else to be finished to market weight. This type of operation is not very common.

Two-Stage Production

In a *two-stage production* system, pigs are farrowed, nursed, weaned, and started in one pen until about 60 pounds (27 kg). They are then moved to a finishing unit until market weight.

Three-Stage Production

In a *three-stage production* system, pigs are farrowed in stalls and kept there until weaning. They are then moved to a pen with supplemental heating until they are about 100 pounds (45 kg). Lastly, they are moved to a finishing facility until market weight.

FEEDER PIGS

Feeder pig production is the production of pigs weighing 30 to 60 pounds (13.6 to 27.2 kg). They are sold to farms that will feed them to market weight.

10-12. Crossbred feeder pigs.

CONTRACT PRODUCTION

An agreement between two or more people is a contract. It is usually in writing and signed by all parties. *Contract production* involves a contract between a producer and a buyer before hogs are raised. The contract specifies the kind of hog to be raised and other details. Contract production is becoming increasingly popular. This is because of the high capital investment, the difficulty in financing an operation, and the willingness to forgo large profits in return for security.

SWINE MANAGEMENT PRACTICES

Hogs are unique livestock. Except for a few breeds of sheep, they are the only livestock that produce offspring twice a year. Because hogs reproduce at a younger age and more quickly than other livestock, genetic gains can be made quite rapidly. But, this requires good management. Good management is important because there is little difference between producing a litter of five and one of ten pigs.

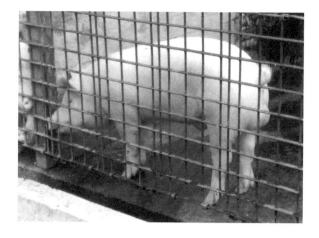

10-13. Boars are penned separately from gilts to prevent unwanted breeding.

NORMAL BREEDING HABITS

Swine reach puberty from four to eight months of age. This wide variance is due to sex, environment, breeds, and breeding lines. Gilts can generally be bred to farrow at 11 to 12 months of age. They should weigh at least 225 pounds when bred.

The heat period or estrus lasts from one to five days and averages two to three days. Signs of heat include restless activity, mounting of others, swelling and discharge from the vulva, increased urination, and loud grunting. The gestation period is 114 days.

10-14. Preparing swine semen for microscopic examination prior to breeding.

THE BREEDING PROCEDURE

Sows can be bred naturally by a boar or artificially. Natural breeding results when the boar and female are placed together for breeding. This is sometimes known as hand mating, which is the most common method. A breeding crate should be used when a heavy boar is to be used on a gilt. This reduces the risk of injury to both.

Artificial insemination (A.I.) is used very limitedly in the United States, yet there is great interest in it. A.I. decreases the risk of disease and increases genetic improvement. However, the main problem with A.I. is that swine sperm cannot be frozen, thus decreasing the opportunities for use.

Thirty to 45 days after breeding, the sow or gilt should be pregnancy checked. The use of ultrasonics has increased detection up to 95 percent accuracy.

CARE OF THE PREGNANT SOW

Many producers are using confinement options to care for pregnant sows. The sows are put in individual pens. The advantages of confinement are a decrease in labor requirements, the use of automatic feeding, freeing land for other purposes, a controlled environment, improved control over disease and parasites, and better management.

Some disadvantages to confining pregnant sows are the increased need for facilities, higher initial investment, possible delay in sexual maturity, lower conception rate, and the requirements of better management.

CARE OF THE BOAR

Care of the boar should not be taken lightly. He will be expected to service dozens of gilts and sows and produce healthy offspring. The boar should have plenty of room to exercise, which will keep him healthy and thrifty. The pasture should be well fenced. If they are kept in confinement, they should have a clean pen with slotted floors or concrete and be kept in individual pens.

Boars that are not being used should be kept separate and away from the rest of the herd. Also, never allow them to grow tusks as this drastically increases their dangerousness.

CARE OF SOW AT FARROWING

Giving birth to pigs is *farrowing*. Care of the sow at farrowing is important to have a strong, healthy crop of pigs. Only 70 percent of the piglets born reach weaning, thus increasing the importance of good, sound management practices at farrowing time. This means that 30 percent die due to disease, mashing by the sow, bad weather, or other conditions. Increasing survival rate makes a farm more profitable.

Signs of Parturition

Signs of parturition include nervousness, uneasiness, enlarged vulva, and mucous discharge. Milk will be present in the teats and the sow will begin to make a nest for the young. She should be placed in a crate or pen by at least the 110th day of gestation. Farrowing crates are commonly used because they reduce the number of young that are crushed.

Sanitation

Sanitation is essential. Before the sow is moved to the crate or pen, she should be scrubbed to remove dirt, manure, and any parasites. Bedding

10-15. Unsanitary conditions for swine.

needs to be clean and fresh. Good sources of bedding are wheat, barley, rye, or oat straw; chopped hay; corncobs; and shavings. The farrowing unit should be scrubbed and cleaned with a disinfectant between uses and left empty for five to seven days.

Environmental Factors

Hogs are very sensitive to their environment. Good ventilation and protection from extreme weather are needed. Their quarters should be kept between 60 and 70°F (15.5 to 21.1°C). Heat lamps or mats need to be placed in the crate or pen for the young to keep warm.

An attendant should check on the sows frequently during and after the birth of the young. This will decrease loss and increase the health and survival of the pigs.

THE SOW AND LITTER

Getting the pig litter off to a good start is important for economic gains. Sanitation is necessary because swine suffer heavy losses due to disease and parasites. A diseased herd can be wiped out leaving the producer in debt and the facilities highly contaminated.

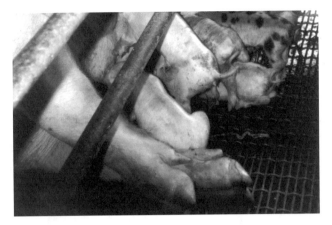

10-16. Newborn pigs with umbilical cords attached
learning to nurse.

Clipping Needle Teeth

When pigs are born, they have eight *needle teeth*. Two are located on each side of the upper and lower jaws. They are sharp and long, much like a short needle. These teeth serve no purpose and will cause injury to the mother as they are nursing and to each other if allowed to grow. These teeth can be clipped off with pliers or forceps. Care should be taken to not injure the pig.

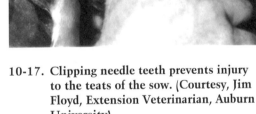

10-17. Clipping needle teeth prevents injury
to the teats of the sow. (Courtesy, Jim
Floyd, Extension Veterinarian, Auburn
University)

Tail Docking

Tail docking is clipping the tail from baby pigs. It is done at the time that needle teeth are snipped. The tail is cut about 1 inch or slightly less from the bone of the tail. Tail docking prevents tail biting that may happen in hogs as they grow in confinement.

Ear Notching

The most common method of identifying hogs is to use a special V-notcher to notch their ears. This is especially important in herds with thousands of hogs. This enables producers to exactly identify the animals, which is necessary when selecting breeding stock and replacements. Plastic ear tags are sometimes used. A disadvantage is that they will tear out easily. Branding and tattooing are also used to some extent, but are hard to read.

Castration

All male pigs being raised for meat should be castrated (removing testicles) before they are weaned. A sanitary and confined environment promotes healing without infection. Males kept for breeding should not be castrated.

10-18. Barrow showing proper castration as a young pig. The testicles have been removed so that no scar is evident.

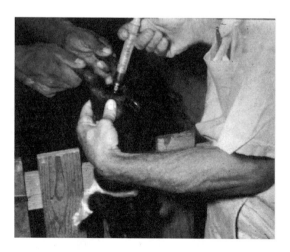

10-19. Pig being vaccinated. (Courtesy, Jim Lytle, Mississippi State University)

Vaccination

Vaccinations should be done before weaning. Pigs are vaccinated to prevent diseases that may occur. The common disease problems of hogs are discussed later in the chapter.

Feeding the Sow

The sow should be fed liberally before parturition to stimulate milk production. During the first three days after giving birth, she should be fed minimal feed. Slowly increase the amount to full feed at two weeks.

CROSSBREEDING

Most meat hog producers use **crossbreeding**, which is the mating of different breeds. Crossbreeding allows for heterosis or hybrid vigor. Heterosis is a biological phenomenon, which causes crossbreds to outperform either of their parents.

Crossbreeding is widely used because it allows for increased production and profits. Roughly 90 percent of commercial hogs are crossbred.

NUTRITIONAL REQUIREMENTS OF SWINE

About 60 percent of all hogs are raised in confinement. Twenty-five to 30 percent of pigs fail to reach weaning age. This is due to many factors of which nutrition is one. Knowledge of feeds and nutrition is important in a swine enterprise because 75 percent of the total cost of production is feed.

NUTRITIVE NEEDS OF SWINE

Each individual varies in the kinds and amounts of nutrients needed. These requirements are influenced by age, function, disease level, environment, and other factors.

The nutrients that swine require include carbohydrates, fats, protein, minerals, vitamins, and water. They are all essential. There are no absolute numbers for any of these nutrients. A factor making this determination difficult to identify is the varying nutritional level of feed, as it is produced in various parts of the country. High performance hogs also require more nutrients than average ones.

Carbohydrates and Fats

Carbohydrates and fats provide energy for the various functions of the body. Cereal grains are the biggest source of energy, mostly carbohydrates. Roughages are a good source of energy, but are not suitable for nonruminants, such as swine. Roughages are too bulky for the restricted size of the digestive system. Corn is the most widely fed concentrate supplying energy. Other sources of energy fed to hogs include barley, sorghum, wheat, and fats and oils.

Protein

Protein is needed to build and repair tissue. Protein is broken down in the stomach and small intestine into amino acids. These amino acids are then recombined into proteins that the animal needs for muscle development and repair of other tissue. Hogs, as nonruminants, are unable to synthesize their own protein, thus increasing the importance of having a well-balanced diet.

Soybean meal is the most widely fed protein. Because of its high palatability to hogs, it should be mixed with grain to prevent overeating.

Young, growing hogs are fed feed higher in protein than older hogs. Protein supplements of 35 to 40 percent protein may be used with corn and other feedstuffs. Commercial feeds with 20 percent protein are used to start pigs at three weeks of age. Growing and finishing feed may have 12 to 16 percent protein. Pregnant sows are fed 12 percent protein feed. Lactating sows and boars are fed 13 to 16 percent protein feed.

Minerals

Hogs are the most common farm animals to suffer from mineral deficiency. Those minerals needed in the largest quantities include salt, calcium, and phosphorus. Cobalt, copper, iodine, iron, manganese, selenium, and zinc are needed in smaller amounts, but are still essential.

Mineral supplements are usually mixed with the feed ration. Newborn pigs are given iron shots shortly after birth.

Vitamins

Vitamins are needed in small amounts, yet are essential for performance of normal body functions. Hogs also suffer from more instances of vitamin deficiency than any other livestock. This is due to confinement and the restricted variety of feeds that they can consume.

Fat-soluble vitamins A, D, and E and water-soluble vitamins biotin, niacin, pantothenic acid, riboflavin, and vitamin B-12 are the most likely to be deficient.

Water

Hogs will generally need $1/4$ to $1/3$ of a gallon of water for every pound of dry feed consumed. Ideally, they should have access to automatic waterers or be hand-watered at least twice a day.

10-20. These swine have free-access to all the feed they want. (Courtesy, Jim Lytle, Mississippi State University)

ADDITIVES

Feed additives are substances added to feed to meet a particular need. Some additives are standard in hog rations. They do not provide nutrients, but provide other advantages. Antibiotics are used to stimulate growth, improve feed efficiency, and control infestations. Piglets do not begin producing antibodies until they are five to six weeks old. Under ideal conditions, antibiotics can increase gain by 10 percent, while decreasing feed consumption by 5 percent.

Sulfas are organic compounds with growth promoting properties similar to antibiotics. Use of sulfas is limited, and they must be withdrawn for at least one week before sale.

Porcine Somatotropin (PST) is a growth hormone for swine. It increases protein synthesis and growth in most tissues. Research shows that PST can increase feed efficiency by 20 to 30 percent, average daily gain by 15 to 20 percent, and improve muscle mass by 10 to 15 percent.

HEALTH MANAGEMENT PRACTICES

Healthy animals vary widely. Some animals may appear healthy, but may actually be suffering from pain or disease. Knowing each individual animal will help to easily detect and correct any ill health. A producer or animal owner will learn to diagnose good and ill health.

The key to personal satisfaction and financial profitability of raising animals is to keep them healthy. Regular practices and management will increase production and decrease losses due to poor animal health. Causes of poor health are parasites, both internal and external, nutrition, and various diseases.

INTERNAL PARASITES

Internal parasites live inside a hog. They may be taken in through the mouth when eating contaminated feed or when eating feed on a dirty floor. Most internal parasites are obtained through feces in the food. Pastures with very short grass also increase the likelihood of ingesting parasites. Hogs on pasture may have increased incidences of internal parasites. Minimize the amount of feed eaten on the ground. Examples of internal parasites include tapeworms, trichina, roundworms, and hookworms.

Roundworms and Tapeworms

Roundworms, tapeworms, and other worms are parasites that can infect all animals. They are found in the intestines and the stomach. Tapeworms can grow several feet long. When they become this large, they use much of the feed that the animal has eaten. Animals with low infestations of small-sized worms may not show symptoms. As the number and size of worms in the digestive tract increases, the animal loses weight, becomes anemic, and may have diarrhea. Sanitation is a good control measure.

EXTERNAL PARASITES

External parasites live outside on a hog's body. They get their food from the blood and tissue of the animal. External parasites can transfer contagious diseases or other parasites to other animals. Examples of external parasites include ticks, fleas, lice, mites, and leeches.

Lice

Lice are external parasites that attack hogs. Lice are small insects that suck their host's blood. Animals may become anemic. Lice also cause animals to itch; they will often be seen scratching or rubbing against trees or posts. Lice usually appear in the winter. Back rubbers and other means can be used to administer pesticides to control lice.

Mange

Mange is a dermatitis caused by one of several types of mites. Mange results in intense itching around the head and neck and wrinkling of the skin. It spreads rapidly. If untreated, it can be debilitating and even fatal. Dipping or high pressure sprays may be used for treatment.

NUTRITIONAL DISEASES

Most nutritional diseases are caused by an unbalanced diet. Animals require specific amounts of nutrients. Too high or too low amounts of any of these may result in a nutritional disease decreasing profitability. Mineral and vitamin deficiencies are quite common in swine.

COMMON DISEASES OF SWINE

Bang's or Brucellosis

Bang's or Brucellosis affects the reproductive tract of the female causing young to be aborted. Prevention of Bang's is by vaccination. Sanitation, testing, and bringing only Bangs-free animals into a herd help to control this disease.

Hog Cholera

Hog cholera is a viral disease that is highly contagious. Hogs will have a high fever, lose their appetite, become weak, and drink lots of water. Considerable effort has been spent eradicating hog cholera in the United States through federal government programs. There is no known treatment. Any hogs with hog cholera are destroyed.

10-21. Keeping facilities clean reduces disease.

Leptospirosis

Leptospirosis is a bacterial disease. Symptoms include high fever, poor appetite, bloody urine, and females will abort their fetuses. Antibiotics are sometimes used to treat leptospirosis. Animals are often vaccinated against this disease.

Pneumonia Lesions

Pneumonia lesions is a specific type of pneumonia that affects swine and can have a major economic impact. Pneumonia itself is not that significant, but the secondary infections that follow can be very costly. Symptoms are a chronic cough and lung lesions, thus the name pneumonia lesions. Lung lesions can result in increasing the incidence of lungworms and migrating ascarid larvae. These decrease the rate of gain and the economic profits. Pneumonia lesions may be found at any time of the year and usually in younger animals. Mortality is low. Control is usually unsuccessful. Use of sulfas and antibiotics helps to decrease secondary infections. Depopulation and the use of SPF animals will lower the incidence of pneumonia lesions.

Pseudorabies

Pseudorabies is a viral disease attacking swine. Adults may serve as unaffected carriers. Suckling pigs usually show clinical signs. Piglets will develop a fever, paralysis, coma, and death, in as little as 24 hours. It may cause adults to abort or to have stillborn pigs. It is spread through contact with nasal and oral secretions. A vaccine is available to prevent pseudorabies.

Swine Dysentery

Swine dysentery or bloody scours is a type of diarrhea caused by *Treponema hyodysenteriae*. It is usually found in young pigs. It is spread by ingesting infected feces. Outbreaks are most common in late summer or early fall. Adding new pigs to the herd, especially ones of unknown origin, can bring in swine dysentery. Once this disease occurs, it will show up periodically, usually once a month. Symptoms include loss of appetite, soft feces, and a slight temperature. The diarrhea will become worse and will get bloody. Because carriers cannot be identified, control is difficult. Antibiotics can be somewhat successful.

FACILITY AND EQUIPMENT NEEDS

Hog production requires good facilities and equipment. The facilities must be properly located, constructed, and maintained.

LOCATION

A good site must be selected for hog production facilities. Location is more important as fewer and fewer people live on farms and understand what is involved. Neighbors may dislike the odors and noise coming from the facilities. Locating buildings downwind of housing areas would be beneficial. Protection from wind and snow and access to good roads are also things to consider in site selection.

BUILDINGS

Swine facilities usually include buildings for farrowing, growing, and finishing. The farrowing barn may contain a nursery. Pasture may be used with gilts and sows depending upon the climate, availability, and time of year.

Swine facilities and equipment focus on temperature control, sanitation, ventilation, and manure disposal.

Ventilation

Well-ventilated facilities help to remove extra moisture in the wintertime. During the summer, it helps to control temperature. Hogs are generally healthier when fresh air is present. Diseases are reduced.

Manure Disposal

Hogs create a lot of waste. Disposing of manure is a major problem for producers with large numbers of hogs. Waste management is needed to maintain good health,

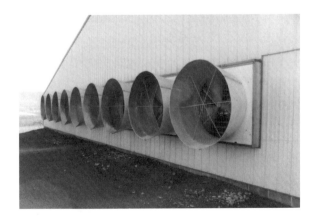

10-22. Ventilation on a confinement building.

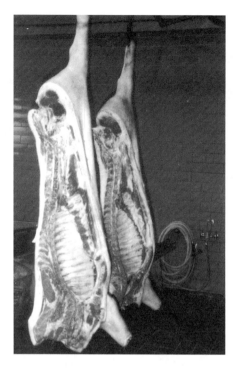

10-23. Well-finished, high-yielding carcasses are the result of swine production.

avoid air and water pollution, and comply with government regulations. Manure pits are usually located under confinement buildings for temporary storage.

FENCING

Some hogs are raised in fenced pastures and other areas. Sows and boars are sometimes put in pastures. Strong, woven wire fencing should be used to make the fence at least three feet high. Temporary fencing may be made of electric wire. Gates need to be made of wood or metal, be of sturdy construction, and have good hinges.

HANDLING EQUIPMENT

Swinging gates make holding and handling swine easier. A series of gates throughout the building allow for easier sorting. Loading chutes are best located at the end of the building.

FEEDING AND WATERING EQUIPMENT

Hogs are hard on feeding and watering equipment. The equipment must be sturdy and not easily damaged. The equipment should also be easy to use and clean.

REVIEWING

MAIN IDEAS

The swine industry is the second largest livestock enterprise in the United States. Over 60 percent of hogs are raised in confinement facilities with over 12 percent raised under contract.

Hogs are good to grow because they are prolific, labor requirements are low, the capital investment is low, and the return can be fast. Hogs are unfavorable because they are susceptible to numerous diseases and parasites. Because they have a simple stomach, they require large amounts of concentrates. A farrowing enterprise requires increased amounts of labor.

The hog industry can be broken down into three systems: one-stage, two-stage, or three-stage production. One-stage production includes farrow to finish. Two-stage production includes farrowing, nursing, weaning, and starting pigs until they are about 60 pounds. A three-stage production system includes moving the pigs between farrowing, nursery, and finishing barns.

Good breeding management of swine is needed to attain high numbers of pigs twice a year. Keeping facilities clean is essential.

Because hogs have a simple stomach, they require concentrates and eat very little forages. Cereal grains, soybean meal, and mineral and vitamin supplements are used in a total mixed ration. Additives, such as antibiotics, sulfas, and porcine somatotropin, may be added to increase efficiency.

A strict health management system should be developed with the help of the local veterinarian. Hogs are very susceptible to many diseases and parasites.

Hogs require good facilities. Selecting the location for a hog facility is an important decision. Locate the facilities downwind of large populations. Building requirements vary based on the type of operation and the management system adopted.

QUESTIONS

Answer the following questions using correct spelling and complete sentences.

1. Why is the swine industry a good or favorable enterprise?
2. What is a meat-type hog?
3. Name five breeds of swine and distinguish between them.
4. List and explain the three different systems of the swine industry?
5. What are the characteristics of a good boar? Good sow?
6. How should the sow be managed at the time of parturition?
7. Describe management practices with newborn pigs.
8. Describe the nutritional requirements of swine.
9. Name three diseases of swine and describe how each affects hogs.
10. Describe the facility and equipment needs of swine.

EVALUATING

CHAPTER SELF-CHECK

Match the terms with the correct definitions. Write the letter by the term in the blank that is provided.

a. prolific   d. probe   g. barrow
b. tail docking   e. specific pathogen free   h. feeder pig
c. pedigree   f. gilt

1. ____ a record of heredity

2. ____ free of disease and raised in an aseptic environment

3. ____ tool for measuring thickness of backfat

4. ____ pig weighing 30 to 60 pounds

5. ____ clipping or cutting the tails on newborn pigs

6. ____ young female swine that have not had pigs

7. ____ produce a large number of offspring

8. ____ male swine castrated at a young age

EXPLORING

1. Visit a swine farm. Note the management of reproduction, nutrition, and health. Interview the manager or a worker about the farm. Determine the breed or cross being raised. Take a camera and make photographs to prepare a poster about the farm.

2. Write a report on the use of porcine somatotropin. Use reference material to help in preparing the report. Contact the animal science department at the land-grant university in your state for information. Your agriculture teacher can provide the name and address of the individual to contact.

3. Make a field trip to a livestock show. Study the breeds of hogs and the conformation of the body parts. Prepare a report on your observations.

Chapter 11

SHEEP AND GOAT PRODUCTION

Sheep and goats look a lot alike. They are closely related and have been domesticated for a long time. Both are ruminants but eat different plants. People who know a little about them can tell them apart easily.

Here are a few distinctions between sheep and goats: goats have a beard, sheep have foot glands, and male goats have a strong smell. Slight differences exist in the horns and skeletons. Goats are more intelligent, independent, and have a better ability to fight and protect themselves.

Goats were one of the first animals to be domesticated. They were once used to plant seed by trampling them into the ground. They were used as a source for food and fiber, and their skin was used for bottles. Sheep and lamb have also long been used for food and fiber.

11-1. Sheep are often pastured on land unsuited for other purposes. (Courtesy, American Sheep Industry Association and National Lamb and Wool Grower Magazine)

OBJECTIVES

This chapter provides basic information on sheep and goat production. The following objectives are covered:

1. Explain the importance of sheep and goat production

2. Describe sheep and goats as organisms

3. Describe different segments of the sheep and goat industry

4. Explain important breeding management practices in sheep and goat production

5. Explain important feeding management practices used in sheep and goat production

6. Explain important health management practices used in sheep and goat production

7. Describe facility and equipment needs with sheep and goats

TERMS

billy	ewe	nanny
browse	farm flock method	orphaned lamb
buck	kid	purebred flock
chammy	kidding	ram
confinement method	lamb	range band method
docking	lamb feeding	wether
doe	lambing	wool
drenching	mutton	

SHEEP AND GOAT PRODUCTION

Sheep and goats are raised for food and clothing. They provide many important products for people. Goats were probably domesticated before sheep. Goats were domesticated about 9,000 years ago and sheep more than 8,000 years ago.

11-2. Goat (on left) and sheep.

SHEEP AND GOAT INDUSTRY

Sheep originated in Asia and Europe. Goats came from the Eastern Mediterranean area and Asia. The first were brought to America by the

settlers some 400 years ago. The numbers of both grew until the mid 1900s in North America.

Australia and New Zealand are the leading producers of sheep. Several European countries produce large numbers of sheep. The per capita consumption of sheep products is greater in New Zealand at 60 pounds a year. The world per capita consumption is about 3 pounds a year.

China and India are the leading producers of goats. These two coun-

11-3. Goats like to climb and play.

tries have over 460 million head. North America has only about 2.5 million head of goats.

United States Production

The number of sheep in North America has been decreasing since the 1940s. This decline is due to lower returns, higher risks than cattle or crops, increased loss due to predators, scarcity and high wages of good sheep herders, and uncertainty in price.

On a global basis, the United States ranks 27th in sheep numbers. In the United States, over one-half of the sheep are found in ten western range states and in Texas. Farm flocks range from a few head to over a thousand.

Products

Sheep and goats provide many products that are important to people. The major products are food and clothing.

Food. The food products from sheep and goats are meat and milk. The meat from sheep is lamb and mutton. *Lamb* is the meat from a young sheep that is less than one year old. Lamb includes both male and female sheep. Lamb is a delicate meat prepared in many delicious ways. *Mutton* is the meat from a sheep that is more than one year old. Mutton has a stronger flavor than lamb and is less desirable as a meat in the United States. Goat is referred to as goat except the younger goat meat known as kid. Many people enjoy barbecue goat.

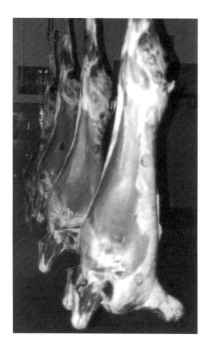

11-4. Lamb carcasses.

Milk is primarily from goats. Goat's milk is easier for people to digest than cow's milk. Goat's milk contains more vitamin A than cow's milk and has smaller fat particles (known as globules) than cow's milk. Goat and sheep's milk are used to make cheese, such as Roquefort cheese from sheep's milk.

Clothing. Clothing is made from the hair or wool and hides of goats and sheep. *Wool* is the soft coat of sheep and is a major fiber used in making clothing. Mohair and cashmere are made from the fiber covering of specific breeds of goats. Sheep and goat skins are used for leather products. **Chammy** is a soft, pliable leather made from sheep and goat skin. It is often used in cleaning and polishing. Because it is very absorbent, chammy is often used in drying automobiles after washing.

11-5. Some sheep produce large amounts of wool. (Shearing is nearly finished on this sheep. All of the wool came from one sheep.) (Courtesy, National Lamb and Wool Grower Magazine)

ADVANTAGES AND DISADVANTAGES OF SHEEP AND GOATS

Many factors make sheep and goat production favorable. Compared to other types of livestock, they are better suited to more arable land for grazing. They are excellent scavengers. Sheep are more efficient at converting feed to meat than cattle. They are dual purpose animals producing wool and meat. Lambs are usually marketed in eight months yielding a fast return on investment.

Goats and sheep can be pastured because they eat different plants. Goats eat **browse** (woody plants) and broad leaf plants. Sheep graze short grass and some broad leaf plants.

Several factors make sheep and goat production unfavorable. The price of wool is low and unstable. Competition from synthetic fibers has hurt the industry. Consumption of lamb is very low. Sheep are susceptible to disease and parasites with little resistance to keep themselves healthy. Sheep and goats are also susceptible to attack from predators. Management of sheep and goats is important because they are dual purpose animals. About 80 percent of sheep production is lamb for meat purposes, while only 20 percent is for wool production.

SHEEP AND GOATS AS ORGANISMS

Sheep and goats are alike in many ways. The differences give advantages in different production situations.

Sheep and goats are mammals with ruminant digestive systems. They have cloven (divided) hoofs. Both are scientifically classified in the Bovidae family. Domesticated sheep are *Ovis aries*. Goats vary in scientific name, with the Spanish goat being *Capra pyrenaica* and other domestic goats being *Capra hircus*.

Sheep are far more important economically than goats in the United States. Greater emphasis in this chapter will be on sheep.

DIFFERENCES BETWEEN SHEEP AND GOATS

Management of sheep and goats is basically the same. However, there are a few differences between them. For example, sheep herders work behind the herd. Goat herders work ahead of the animals to lead them.

Sheep Age and Sex Classification

The names for sheep and goats vary based on sexual classification and age. A lamb is a young lamb of either sex that is less than one year old. A *ewe* is a female sheep of any age. A *ram* is a male sheep kept for breeding purposes. A male sheep castrated before sexual maturity is a *wether*. Its secondary sexual characteristics have not developed at the time it is castrated.

Goat Age and Sex Classification

Goats have similar life cycles to sheep, but the terms applied to age and sex classifications are quite different. A female goat is a *nanny* or *doe*. A male goat is a *buck* or *billy*. Young goats under a year of age are known as *kids*. Most all kids reach puberty by a year of age.

Does are bred to have first kids at two years of age. The process of giving birth is known as *kidding*. Most kids weigh about 5 pounds at birth, but weight depends on the breed of goat.

The gestation period of goats averages 151 days, whereas 148 days is common in sheep. The estrus period lasts 18 to 19 days, which is one to two days longer than sheep. Newborns require greater care than lambs. Kids are either staked near a small A-shaped box or are put in pens or corrals. Goats require little, if any, shelter. If they are well fed, they will not suffer in cold weather.

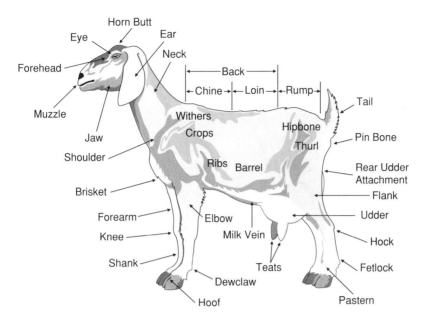

11-6. Major external parts of a dairy goat.

SHEEP CONFORMATION AND TYPE

The desired conformation and type in sheep vary with the product to be produced. Sheep for meat should have large amounts of higher-valued cuts. Sheep for wool should have larger amounts of wool. Producers need to know the major external parts of a sheep and the locations of major wholesale and retail meat cuts.

BREEDS AND CLASSES OF SHEEP

Over 200 breeds of sheep are in existence today. Three-fourths of the sheep raised belong to only six breeds. These top six include Suffolk, Dorset, Hampshire, Rambouillet, Polypay,

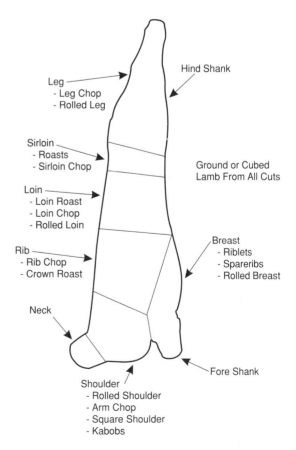

11-7. Location of major wholesale and retail lamb cuts.

and Columbia. Other breeds in North America include Oxford, Southdown, Corriedale, Montadale, Shropshire, Cheviot, and Katahdin.

Sheep may be classed by the type of wool produced or by their breeding use. Type of wool produced is broken down into fine-wool, medium-wool, long-wool, crossbred wool breeds, carpet-wool, and fur sheep breeds. Classification by breeding use breaks down into ewe, ram, dual-purpose, and other breeds.

11-8. Suffolk ewe.

Suffolk

The Suffolk is a medium-wool breed of sheep. It originated in England and is the most popular breed in North America. The head, ears, and legs are black, without any wool on the head and ears. Suffolk are polled, though some males may have scurs.

Dorset

The Dorset is a medium-wool breed. The sheep may be polled or horned, with most being polled. The Dorset is entirely white and has no wool on its face. Ewes breed out of season and are known as good milkers. The Dorset breed originated in England.

11-9. Dorset ewe.

Hampshire

The Hampshire originated in England and is a medium-wool breed. The face, ears, and legs are dark brown and often nearly black. Both males and females are hornless, though males may have scurs. The Hampshire is known as an early-maturing, large sheep.

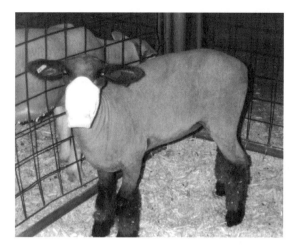

11-10. Hampshire lamb. (The muzzle is used on this show lamb to protect it from eating the bedding material.)

11-11. Oxford lamb fitted for showing.

Oxford

The Oxford is a medium-wool breed that is popular with youth in lamb shows. Originating in England, the Oxford is a cross of Hampshire and Cotswold. The face, ears, and legs may have brownish or grayish coloring. It has a topknot of wool (wool cap) on its head. At maturity, the Oxford is among the largest breeds.

Katahdin

The Katahdin is a breed of sheep with hair rather than wool. Originating in the Virgin Islands, the Katahdin

11-12. Katahdin showing color variations.

was first brought to Maine along with the English breeds of sheep. Colors vary from white to cream or red. The coloring may be spotted or solid. Most Katahdins are polled, though some males may have small horns or scurs. The Katahdin are not tail-docked as are the wool breeds. Docking is not necessary because the hair does not collect feces, as does the wool on other sheep.

GROUPINGS OF GOATS

Goats are in five groups. They are Angora goats, dairy goats, meat goats, Cashmere goats, and pygmy goats.

Angora Goats

There are over one million Angora goats in the United States. However, few people really know what they look like. Texas is the leading producer of Angora goats. They are well adapted to living in areas where the grazing is not suitable for other livestock.

Angora goats are almost totally white. Occasionally a black one will be found. Kids may sometimes be red, but will usually shed their hair and become white. Their coat consists of long locks of hair, known as **mohair**. This fleece covers the entire body, except the face. It is of very fine quality and has a high luster. The average Angora produces up to seven pounds of mohair each year, and is sheared twice a year.

Dairy Goats

Dairy goats produce 1.8 percent of the world's supply of milk. An average doe may produce 5 pounds of milk per day during a ten-month lactation.

Goat's milk has a sweeter taste and contains more minerals than cow's milk. During digestion, goat's milk forms into a soft curd, making it easier for children and the elderly to digest than cow's milk. The presence of bucks around does causes the milk to take on an unpleasant odor.

The major breeds of milk goats are Nubian, Lamancha, Alpine, Toggenburg, Saanen, and Oberhasli.

11-13. Nubian doe. (Courtesy, American Dairy Goat Association)

11-14. Lamancha doe. (Courtesy, American Dairy Goat Association)

Nubian. The Nubian is considered a dual-purpose breed for milk and meat. It has larger, drooping ears than other breeds. The record milk production in a year is 4,420 pounds. Though the amount of milk is less than some other breeds, the milk has a higher fat content.

Lamancha. The Lamancha breed originated in California. A major characteristic is the absence of external ears. The Lamancha has higher milk production than the Nubian.

11-15. Alpine doe. (Courtesy, American Dairy Goat Association)

Alpine. The Alpine is a good milker, with annual production up to 5,700 pounds a year. It has distinctive coloring, with white and darker colors forming unique patterns.

Meat Goats

Meat goats are derived from the Mexican Criollo, thus, they are also called Spanish Goats. Most meat goats are found in Texas. The coloring varies from white to black with some being orange to sandy in coloring.

11-16. Spanish goat.

Cashmere Goats

Cashmere goats have a fine down undercoat. This is the finest animal fiber used. Cashmere produces clothing that is light and soft, yet warm. It also has three times the insulating value of wool.

Pygmy Goats

Pygmy goats are miniature goats used for research and as pets.

11-17. Pygmy goat.

SELECTION OF SHEEP

Management of sheep is more complex because trying to get wool and meat instead of one product from most animals requires greater knowledge and managerial ability. The production of sheep meat makes up about 80 percent of the market and wool only 20 percent.

When selecting an animal, many things must be considered. Each individual must have certain characteristics, including a high rate of gain and high feed efficiency.

11-18. Some producers use trained Border Collie dogs to herd sheep.

Things to consider when selecting sheep include whether to raise purebred, crossbred, or grades; selection of a specific breed for its characteristics; size of the flock or herd; time of year to start; uniformity; health; age; soundness of the udder; and price of the animal.

Usually only experienced breeders need to buy purebred animals. The beginners should start with a crossbred ewe. Selection of breed or class should be based upon the purpose for raising them. Size of the flock should be based upon the space and facilities available. The animal itself should be healthy, sound, and uniform for its age and breed.

PRODUCTION SYSTEMS

There are five types of sheep production systems. They are the farm flock method, the purebred flock method, the range band method, the confinement method, and lamb feeding. Success in any of these types depends upon managing a healthy and productive flock—that is, economically marketing lambs and wool.

THE FARM FLOCK METHOD

The *farm flock method* is the most popular method used in the central, southern, and eastern United States. Production of market lambs is the focus of this method with the production of wool being secondary. Usually, the producer raises other livestock, which compete with the sheep for pasture.

11-19. Farm flock in good pasture. Shade structures and good fencing help assure
well being.

THE PUREBRED FLOCK METHOD

Few farms raise a **purebred flock**. Their main purpose is the sale of
rams and ewes. This requires a high level of management ability. The
producer must raise sheep of ideal type. A full-time caretaker is usually
employed.

THE RANGE BAND METHOD

With the **range band method,** each band of sheep has their own herder
who moves them over a large area of land. The emphasis on lamb or wool
depends upon the region's rainfall and vegetation. In arid regions, such
as in the southwest, wool production is emphasized because there is not
enough grass to finish a lamb.

THE CONFINEMENT METHOD

The **confinement method** is becoming more popular because of parasites
being a major problem. Raising animals in confinement means that they
are raised totally indoors. Other reasons include the success of raising
other types of livestock in confinement, high land prices, and the increase
in young produced.

The confinement method has several unfavorable factors for most
producers. Equipment and building costs are higher. This type of system

is more restrictive than the other methods. More skill and better management practices are required.

LAMB FEEDING

Lamb feeding is a specialized method. After weaning, lambs are sold to feedlots. These feedlots are usually located close to good, cheap food supplies.

11-20. Lambs being fed grain. (Courtesy, National Lamb and Wool Grower Magazine.)

BREEDING MANAGEMENT PRACTICES WITH SHEEP

Commercial and purebred breeders alike strive to have a high lamb crop. The average lamb crop is 106 to 108 percent with the mortality as high as 25 percent from birth to weaning. Breeding and selecting ideal animals is important in meeting these objectives.

NORMAL BREEDING HABITS

Ewes reach puberty at 8 to 10 months of age, while rams reach puberty at 5 to 7 months. Ewes are usually bred to give birth (lamb) at 24 months. Giving birth is known as *lambing*. The time of the year when ewes have lambs is known as the lambing season.

The heat period lasts an average of 30 hours. Ewes show no visible signs of heat other than the acceptance of the ram. Estrus occurs every 16 to 17 days. The gestation period of sheep lasts around 148 days. This varies among breeds.

CARE AND MANAGEMENT OF THE RAM

Rams are normally kept separate from ewes. A dry barn or lot with room to get plenty of exercise is adequate. Rams do not require much, if any, grain. Many rams are pastured. Excess fat is harmful to a breeding ram.

11-21. Katahdin rams on pasture.

CARE OF THE EWE

A pregnant ewe needs feed, water, shelter, and exercise. A ewe that does not lamb is not profitable because lambs are more profitable than wool. Because of this, pregnancy testing is very important. Detection of one or multiple fetuses affects nutritional and health management practices.

11-22. Recently-shorn Dorset ewe.

Care at Lambing

Most death loss of lambs occurs in the first few days after lambing. As parturition approaches, the ewe should be sheared around the udder, flank, and dock; placed in a dry, roomy pen; and the amount of grain should be reduced. Proper care of the ewe and her lamb(s) include use of lambing pens, helping lambs if they are being born in the wrong position, nursing chilled and weak lambs, and examining and treating any health problems.

Orphaned Lambs

If a ewe dies or is unable to nurse, its lambs are orphaned. An *orphaned lamb* needs a foster mother. A ewe who has lost her own young or who is very healthy may accept an additional lamb, if it is first rubbed with the dead lamb. A more effective method is to tie the skin of the dead lamb onto the orphaned lamb. The skin may then be removed a piece at a time after a few days. If the lamb is not accepted, cow's milk or milk replacer may be used to feed the lamb.

MANAGEMENT FROM LAMBING TO WEANING

Between the time that lambs are born until they are weaned, they should be docked and castrated.

Docking

Docking is cutting off all or part of the tail. This is one of the first things done after lambing. It should be done when they are 3 to 10 days old. Docking may be done with a knife, hot docking iron, electric docker, or elastic band. The Katahdin lambs are not docked because they have hair rather than wool. Goats are not docked.

11-23. Ewe with newborn lamb.

Castration

Ram lambs should be castrated when they are young. Many producers dock and castrate simultaneously. Castration may be done by similar methods as docking. Remember to be as sanitary as possible to prevent infection.

NUTRITIVE MANAGEMENT OF SHEEP

Success in the sheep enterprise can be measured by the percentage of lambs raised and the pounds of lamb marketed. Nearly 100 percent of a sheep's diet is made up of roughages. Sheep have a ruminant digestive system, which allows for use of pasture and hay. Sheep are not fed grain except just before and immediately following lambing.

NUTRITIVE NEEDS

Sheep, like other animals require the six basic nutrients of carbohydrates, fats, proteins, minerals, vitamins, and water.

Carbohydrates and Fats (Energy)

Energy, which comes from carbohydrates and fats, is provided chiefly by pasture, hay, and silage. Corn, oats, wheat, barley, or grain sorghums may be added to the feed ration during times of drought, overgrazing, or snow-covered pastures. During the fall, they may have access to stalk or stubble fields.

Protein

Sheep need high levels of protein because they produce wool, which is made up of protein. Alfalfa, clover, soybeans, and green pasture are all good sources of protein. These crops are best when harvested before the protein quantity begins to decline.

When these are not available, a protein supplement may need to be added. Soybean, cottonseed, linseed, canola, peanut, and sunflower meals are good sources.

Range sheep may develop a protein deficiency. Protein blocks may be placed in the pasture to alleviate this problem.

Minerals

Sheep require the macrominerals salt (sodium chloride), calcium, phosphorus, magnesium, potassium, and sulfur. The microminerals that they require include cobalt, copper, fluorine, iodine, iron, manganese, molybdenum, selenium, and zinc.

Phosphorus deficiency may be experienced by range sheep. Self feeding salt and mineral mixtures may be placed in pastures to alleviate any mineral deficiencies.

Vitamins

All of the fat soluble vitamins (A, D, E, and K) are required by sheep. Bacteria in the rumen produce adequate quantities of the water soluble

(B) vitamins. Vitamin A is the most likely to be deficient. Vitamin mixtures may be fed free choice.

Water

The average sheep needs one gallon of water each day. Sheep obtain water not only from obvious sources, but also from feed, dew, and snow. When feeding on lush, green pasture, sheep may go for weeks without drinking water.

HEALTH MANAGEMENT

Keeping sheep and goats healthy means developing a preventive health program. When an animal gets sick, the expense of treating the animal usually outweighs the value of the animal. Therefore, prevention is clearly best.

GENERAL PRACTICES

Some recommendations for prevention of ill health of sheep and goats include: monitor the animals regularly checking for any signs of illness; provide a well-balanced ration; reduce stress whenever possible; control outside traffic from other animals and vehicles; rotate pastures to reduce parasite problems; and keep facilities and equipment sanitary.

A good vaccination and checkup program should be established with the local veterinarian. If an animal becomes ill or has just been purchased, isolate the animal from the rest of the flock or herd. New animals should be isolated for at least 30 days.

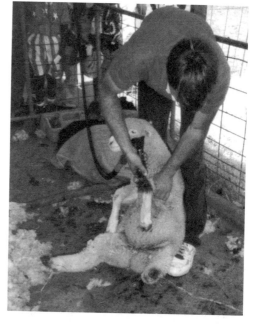

11-24. Sheep should be carefully handled during shearing to prevent injury.

DISEASES OF SHEEP AND GOATS

Common diseases of sheep and goats include: actinobacillosis, actinomycosis, anthrax, blackleg, blue tongue, brucellosis, enterotoxemia, food abscess, foot scald, foot rot, Johne's disease, lamb dysentery, leptospirosis, listeriosis, malignant edema, mastitis, pinkeye, pneumonia, scrapie, shipping fever, and tetanus.

Several of these diseases are discussed in Chapter 6. Other diseases that are mostly seen in sheep and goats will be discussed here.

Actinobacillosis (Lumpy Jaw)

Actinobacillosis causes economic loss to the producer. It affects the jaw and parts of the animal's head. Tumors or lumps of yellow pus appear on the jaw. The muscles and organs of the animal may also become infected, causing parts of the animal to be condemned upon slaughter. The animal will usually not die.

Actinomycosis (Wooden Tongue)

Actinomycosis causes lesions on the animal's head and swelling of the lymph glands. The tongue may have lesions on it as it becomes hard and immobile. The animal finds it more difficult to eat and will lose weight. They will eventually die.

Animals that have actinomycosis should be isolated from the rest of the herd or flock. Contaminated feed usually spreads this disease.

Blue Tongue

Blue tongue is a viral disease that is spread by gnats. Blue tongue weakens the immune system of sheep and goats. Thus, most deaths occur due to secondary diseases.

Signs of blue tongue include a high fever, loss of appetite, and sluggish behavior. The lips, muzzle, head, and ears will become swollen. The mouth will turn red or blue and will develop ulcers. The animal will have trouble eating.

Blue tongue does not have any available treatments. Vaccination at shearing time is the best method of prevention.

Enterotoxemia

Enterotoxemia is a bacterial disease that usually affects lambs and kids. They are commonly found dead with their heads being arched up. This occurs because of the convulsions that they experience shortly before death.

There is no treatment for enterotoxemia. Vaccination, good management, and proper feeding are keys to preventing this disease. Ewes should be vaccinated two and six weeks before giving birth. Lambs should also be vaccinated. Increasing the amount of roughage and chlortetracycline may be necessary.

Foot Abscess

Foot abscess affects the foot's soft tissue. It is usually found when conditions are wet and muddy. Bacteria enter the foot causing abscesses or pockets of pus. The joints and tendons may also be affected.

Put the animals on clean, soft bedding, drain the abscesses, and use antibiotics.

Foot Rot

Foot rot is caused by bacteria, which are different from the one that infects cattle. It is extremely contagious and occurs when the animals are in wet conditions. Foot rot causes animals to lose weight and requires added labor to treat these animals.

Animals will become lame and will have an unpleasant odor about the feet. Keeping their quarters dry and clean and isolating infected animals will help reduce losses from foot rot. A vaccination is available.

Johne's Disease

Johne's disease causes the intestinal walls to become thicker. Animals will appear to have diarrhea, will lose weight, and will eventually die.

There is no effective treatment of Johne's Disease. Purchased animals should be checked for a history of this disease.

Lamb Dysentery

Lamb dysentery is caused by bacteria and affects lambs when they are only one to five days of age. They will lose their appetites, become depressed, have diarrhea, and will suddenly die. Death loss can be very high.

Sanitation and good management practices are the best methods of prevention.

EXTERNAL PARASITES

External parasites cause losses to sheep and goat producers because of the decrease in the quality of wool, mohair, meat, and milk. Controlling of external parasites is done through good sanitation and proper use of pesticides.

Common external parasites of sheep and goats include blowflies, bot flies, lice, mange, and sheep ked or tick.

INTERNAL PARASITES

The biggest problem of raising sheep and goats is the loss caused by internal parasites. Infested animals lose weight and lose milk and meat production. They have poor quality wool and decreased reproductive efficiency.

Signs of internal parasites include weight loss, rough hair coat, loss of appetite, diarrhea, and anemia. Animals should be wormed regularly and rotated between pastures frequently.

Common internal parasites of sheep and goats include coccidia, liver fluke, lungworms, and stomach and intestinal worms.

Drenching

Drenching is using a liquid medication by oral means to control internal parasites. Animals are given this medication through a syringe or gun in the mouth. Make sure to provide the animals with the proper amount and to follow directions.

11-25. Load of wool in bags ready for hauling to market.

FACILITY AND EQUIPMENT REQUIREMENTS

Sheep do not require expensive shelter. They are relatively hardy animals. Some producers raise sheep in confinement facilities using automatic feeders and waterers. This increases costs, but decreases labor requirements. The average farm or ranch does not need confinement facilities.

HOUSING

Barns or sheds for sheep usually open to the south. This provides protection from winter weather. The barn should be well bedded, dry, and well ventilated. Where lambing is to take place, there should be electricity to hook up heat lamps. Troughs, feeders, and waterers should be placed in an easily accessed location.

11-26. Corral designed for working sheep.

Milking does are commonly kept in free stalls. Loose housing may be used for kids and yearlings. Both require extra bedding and care during the cold winter months. Goats kept in loose housing or tie stalls must be dehorned.

FENCING

Fencing is needed to keep predators out, more than keeping sheep and goats in. Sheep need a 60 inch or higher fence. It could be made of woven wire or barbed wire with only 4 to 5 inches between each strand.

EQUIPMENT

Loading chutes, corrals, portable shelters, weigh crates, and pregnancy testing cradles may also be needed, depending upon the operation.

11-27. Youth enjoying competition showing their lambs.

REVIEWING

MAIN IDEAS

Sheep and goats are closely related. They have relatively minor differences in physical characteristics and management practices. The number of sheep in the United States has been declining since the 1940s.

Sheep are used for their meat and for their wool. Meat production is more profitable. Goats are raised for their meat, milk, mohair, and cashmere, or as pets.

Selection of sheep and goats should be based on many things. The main thing to consider is affordability. They should also be sound and healthy and fit comfortably into your facilities and management abilities.

Sheep and goats may be raised by various means. They may be part of a farm flock, a purebred flock, a range band, a confinement system, or they may be fed out in a feedlot.

With the mortality of lambs as high as 25 percent, proper breeding and health management are essential. Unbred ewes that do not have lambs are unprofitable. Lambs should be docked and castrated when young. They consume mostly roughages and very little concentrates. Drenching is a common practice used to prevent internal parasites.

Minimum facilities and equipment are required for raising sheep and goats. However, a good fence is needed to keep out predators.

QUESTIONS

Answer the following questions using correct spelling and complete sentences.

1. How are goats and sheep similar in appearance and in management practices?

2. Describe how sheep are classified?

3. Describe how goats are grouped?

4. What should be considered when selecting a sheep or goat?

5. What are the five types of sheep production systems? Describe each.

6. Which type of production system would best suit the area in which you live? Explain.

7. How should a ewe be managed in terms of reproduction, nutrition, and health?

8. Describe a good health management system.

9. Describe facilities needed on an average sheep or goat farm/ranch.

EVALUATING

CHAPTER SELF-CHECK

Match the terms with the correct definitions. Write the letter by the term in the blank that is provided.

a. ram

b. nanny or doe

c. kidding

d. wether

e. mutton

f. chammy

g. lamb

h. ewe

1. ____ soft, pliable leather made from a sheep skin

2. ____ sheep less than one year old

3. ____ male sheep castrated before sexual maturity

4. ____ male sheep

5. ____ female sheep

6. ____ female goat

7. ____ when a nanny gives birth

8. ____ meat from a sheep that is more than a year old

EXPLORING

1. Visit a local sheep farm. Note the management of reproduction, nutrition, and health. Take a camera and record major observations. Prepare a poster or bulletin board that depicts what you have learned.

2. Visit a local goat farm. Note the management of reproduction, nutrition, and health.

3. Fit a lamb for showing. Use your own lamb or assist someone who owns a lamb. Determine the different procedures in fitting and why these are followed. Write a report on your experiences. Give an oral report to the class.

Chapter 12

DAIRY PRODUCTION

So many of our favorite foods are dairy products! In addition, these foods are good sources of important nutrients. What are your favorite dairy foods? Many people like ice cream, yogurt, and cheese.

Besides dairy foods, milk and milk products are used in many other foods. Cheese is used in pizza, cheeseburgers, and macaroni and cheese. What would these foods be like without cheese?

Milk and the foods made from milk are "nutrient dense." This means that they contain large amounts of essential nutrients compared with calories. This has made milk and dairy foods popular with consumers.

12-1. Ice cream is popular with people of all ages in North America. Dairy products are good sources of calcium.

OBJECTIVES

This chapter provides general information on dairy cattle and dairying. The following objectives are covered:

1. Describe the dairy industry
2. Describe dairy cattle as organisms (including important breeds)
3. Explain important management practices in dairy production
4. Describe how the environment is modified for dairy cattle
5. List the general nutritional requirements needed for body functions of dairy cattle
6. Explain health management practices with dairy cattle
7. Describe facility and equipment needs with dairy cattle

TERMS

alveoli
animal model
cold housing
colostrum
Dairy Herd Improvement
  Program
dry cow
functional type

homogenization
immunoglobulin
linear evaluation
management intensive
  grazing
mastitis
metabolic disorder
milking parlor

pasteurization
predicted transmitting
  ability
progeny
total mixed ration
type production index
udder
warm housing

THE DAIRY INDUSTRY

The dairy industry provides a stable supply of dairy foods to consumers in North America and many foreign markets. Its success can be attributed in part to the ability of the dairy cow to efficiently convert feed to milk. As a ruminant, the cow converts plant nutrients from plants into forms humans will eat. The cow then converts the plant nutrients into the nutrient dense product called milk. Nutrient dense is a food or feed with large amounts of nutrients relative to calories.

EARLY BEGINNINGS

Dairy cows were first brought to the Jamestown Colony in 1611. Early farms had only one or two dairy cows for their own use. Lack of refrigeration made it more difficult for people living in large cities to obtain milk.

Pasteurization, refrigeration, and bottled milk were developed in the second half of the 19th century. This allowed milk to be stored and transported to population centers. Pasteurization destroys bacteria and other tiny organisms. It involves heating the milk to 161°F for 15 seconds. This keeps people from getting disease, such as undulant fever, from drinking milk from diseased cows.

Small family dairy herds began to develop in response to improved storage and handling technology. The dairy industry continued to evolve throughout the 20th century. Compulsory pasteurization, **homogenization,** and addition of Vitamin D to milk became accepted marketing practices in the United States before World War II.

Homogenization is used to keep the fat and milk liquid from separating. Cream rises to the top of milk that is not homogenized. In homogenization, the fat droplets are broken into very small particles so they stay in suspension.

More changes occurred in the mid 1900s. Dairy farmers started using artificial insemination on their cows. Commercial dairy farms became market oriented. This pushed production, per cow, upward.

Table 12-1
Major Events in the Dairy Industry

Year	Event
1611	first cows arrived at Jamestown
1841	unrefrigerated milk shipped to New York by rail
1851	first cheese factory in Oneida, New York
1890	Babcock Butterfat Test introduced
1895	pasteurization available on a commercial basis
1919	homogenized milk marketed
1932	vitamin D fortification of milk introduced
1936	artificial insemination successfully used with cattle
1948	ultra-high temperature pasteurization first used
1975	embryo transfer began to gain acceptance
1994	bovine somatotropin (bST) used on commercial basis

TODAY'S DAIRY INDUSTRY

Nearly 10 million dairy cows are on farms in the United States. These cows annually produce nearly 150 billion pounds of milk.

Each dairy cow, on the average, produces over 15,000 pounds of milk a year. This is three times the production in 1950. One dairy cow, today, provides nearly 25 people with their dairy product needs.

The number of farms with dairy cows has gone down. There are 95 percent fewer dairy farms today than in 1950, but each dairy farm is larger. Most herds today range from 35 to 500 cows, with some regions of the country having herds as large as 10,000 cows. Dairy farms use high-tech milking equipment. The hand labor of milking cows has been replaced with milking machines.

Dairy production is concentrated near population centers. California leads all states in total milk production, even though Wisconsin has more cows. New York is third in both total milk production and number of cows, followed by Pennsylvania and Minnesota. All states have some dairy cattle.

Dairy cattle also contribute to the beef supply. Calves are used for veal. Older cows are used for processed meat products, such as bologna and hamburger.

DAIRY CATTLE AS ORGANISMS

Dairy cattle belong to the family *bovidae*, which includes ruminants with hollow horns. Members of this family also chew their cuds. The following shows the scientific classification of the dairy cow:

Kingdom Animalia: The animal kingdom

Phylum Chordata: Animals with either a backbone (in the vertebrates) or the rudiment of a backbone (in the chorda).

Class Mammalia: Mammals or warm-blooded, hairy animals whose offspring are fed with milk produced by the mammary glands.

Order Artiodactyla: Even-toed, cloven-hoofed mammals.

Family Bovidae: Ruminants have numerous placental attachments and hollow, nondeciduous, upbranched horns.

Genus Bos: Four-footed ruminants, including wild and domestic cattle, distinguished by a stout body and hollow, curved horns standing out laterally from the skull.

Species: Bos taurus: Bos taurus includes the ancestors of European cattle and the majority of the cattle found in the United States.

Box indicus: Bos indicus is represented by the humped cattle (Zebu) of India and Africa and the Brahman breed of America.

DAIRY CATTLE CONFORMATION AND TYPE

Dairy cattle are milk producers. The body structure should contribute to high milk production. Dairy cows may appear angular and lack muscle development in high-value meat cuts. In cows, the udder should be well attached and have the capacity to hold 50 to 70 pounds of milk. Four teats should be shaped and spaced uniformly for machine milking.

Since much feed is needed for the cow to produce milk, the body must be capable of eating and digesting feed. A cow needs a good appetite and the strength to compete with other cows for feed. They also need good body capacity to hold feed for digestion.

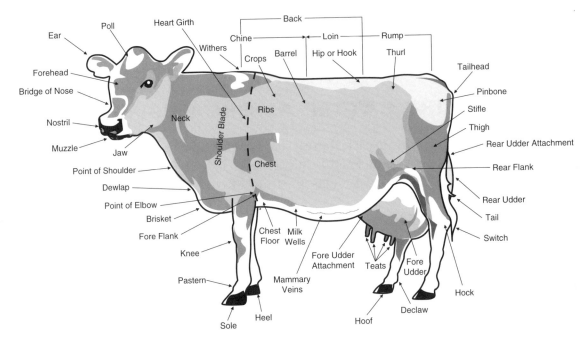

12-2. Major external parts of a dairy cow.

Dairy cows should regularly reproduce. Milk production is associated with the reproductive cycle. Lactation is the secretion of milk by the mammary glands. It begins at parturition and ends when the cow is dried up.

A *dry cow* is one that has stopped producing milk. Most dairy cows produce milk in the herd for 305 days. They are bred to calve each year. With a gestation period of 283 days, some rest from milk production is needed. They are dried-up 50 to 60 days before the next calving—milking is stopped.

UDDER STRUCTURE AND MILK

The *udder* includes four mammary glands or quarters. Each quarter has a teat. A canal through each teat allows removal of milk that has been produced.

The udder should be large and strongly attached to the body. Udders sometimes break away from the cow. This shows problems in producing a large amount of milk. The milk is produced in the *alveoli*. These are tiny structures that remove nutrients from the blood and convert the nutrients into milk.

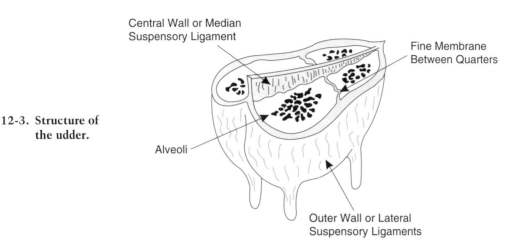

Central Wall or Median
Suspensory Ligament

Fine Membrane
Between Quarters

**12-3. Structure of
the udder.**

Alveoli

Outer Wall or Lateral
Suspensory Ligaments

Milk is 87 percent water. Fat and protein make up a little less than 4 percent each. It has a little over 4 percent lactose (milk sugar). Cows must have the capacity to drink a lot of water to carry on life functions and manufacture milk.

DAIRY CATTLE BREEDS

Seven major breeds of dairy cattle are used. Differences in breeds are milk yield, milk fat, and milk protein. Some variation even exists within each breed. Holsteins produce the most milk by volume. They are followed by Red and White, Brown Swiss, Ayrshire, Milking Shorthorn, Guernsey, and Jersey. The breeds rank in the following order based on milk fat percentage and protein percentage: Jersey, Guernsey, Brown Swiss, Ayrshire, Red and White, Holstein, and Milking Shorthorn.

Jersey

The Jersey is from the Island of Jersey and was brought to North America in 1850. Jerseys are grayish to fawn and near white in color. They are smaller than other breeds. Jersey milk has the highest fat and protein content. The amount of milk produced is less than other breeds. Jersey numbers have been increasing in recent years.

12-4. Jersey cow. (Courtesy, Pete's Photo, Wykoff, Minnesota)

Holstein-Friesian

12-5. Holstein-Friesian cow. (Courtesy, Pete's Photo, Wykoff, Minnesota)

The Holstein-Friesian (commonly shortened to Holstein) breed came to North America from the Netherlands in 1621. It is the most popular breed of dairy cattle. Most Holsteins have distinctive black and white color. They produce the largest amount of milk though low in butterfat and protein. Holsteins are large, with cows weighing up to 1,500 pounds, and bulls, 2,200 pounds.

Brown Swiss

Brown Swiss were brought to North America in 1869 from the Alps of Switzerland. The color is solid light to dark brown. The nose and tongue are black. The Brown Swiss has a calm disposition. The milk has a favorable protein to fat ratio for the present milk market.

Guernsey

The Guernsey was introduced into North America in 1831 from the Island of Guernsey. The color is fawn with white markings. Guernseys are quiet cattle and easy to work with. Their numbers have been declining.

12-6. Brown Swiss cow. (Courtesy, Pete's Photo, Wykoff, Minnesota)

12-7. Guernsey cow. (Courtesy, Pete's Photo, Wykoff, Minnesota)

2-8. Ayrshire cow. (Courtesy, Pete's Photo, Wykoff, Minnesota)

12-9. Milking Shorthorn cow. (Courtesy, Pete's Photo, Wykoff, Minnesota)

Ayrshire

The Ayrshire was brought to North America from Scotland in 1822. Color varies from light to dark red, brown, and white. The Ayrshire is known for well attached udders and grazing ability.

Milking Shorthorn

The Milking Shorthorn breed was designated as a dairy breed in 1968. The coloring is similar to the Shorthorn, with red, white, and combinations of these. The breed is adaptable to a variety of situations.

Red and White

The Red and White evolved from the Holstein. It is similar to other Holstein in milk production, size, and disposition. As the name suggests, the color is red and white.

12-10. Red and White Holstein cow. (Courtesy, Pete's Photo, Wykoff, Minnesota)

MANAGEMENT OF DAIRY HERDS

STARTING A DAIRY ENTERPRISE

Some people like dairy farming. The decision to start a dairy enterprise is a long-term one. Once started, it is difficult to change without serious consequences. As a result, the final decision should be carefully thought out.

Dairying has been one of the most stable agricultural enterprises. It has returned a reasonable profit and allowed long-term financial growth. Personal pride and satisfaction of ownership are additional rewards for most dairy farmers.

The rewards of dairying also come with risks and challenges. The capital investment is sizeable and carries with it risks just as any investment does.

12-11. Large investments and long-term commitments are evident on this Wisconsin dairy farm.

12-12. This California dairy calf facility shows large-scale specialization in dairying.

The final decision is to choose the type of operation best suited to available resources and personal goals. This can include choosing between a commercial (grade) and purebred operation, a family-size (100 or less) or large herd, a total herd (cows and replacement heifers) or a milking herd, and whether the feed will be raised or purchased.

Rather than owning a dairy farm, many people work for other dairy farms. Depending on the size of the facility, the work varies from feeding and milking to managing the herd and recordkeeping.

SELECTION AND CULLING

A key to success in dairying is the ability to select (pick for herd or keep in herd) and cull (eliminate from herd) the appropriate cows or heifers. Criteria used for selection and culling include production factors, genetic makeup, and type.

The dairy industry has been a leader in accurately predicting production capability of sires through *progeny* (offspring) data. This requires accurate measurement of milk production and correct assessment of ancestry and offspring.

Dairy Herd Improvement

Milk production can be assessed by the ***Dairy Herd Improvement (DHI) Program***. DHI is a national industry-wide dairy production testing and record-keeping program. The U.S. Department of Agriculture personnel work with dairy producers to get information and compare the producer's herd with others in the area, state, and nation.

The Dairy Herd Improvement Association (DHIA) is the most common official testing plan. All cows must be properly identified and all DHI rules enforced. Cows are tested every 15 to 45 days. The results of official records are published and used by the USDA to evaluate sires.

The Dairy Herd Improvement Registry (DHIR) includes the standard DHIA requirements plus the added requirements of breed associations. A major difference is additional testing when an individual cow's milk production exceeds specified standards. The records for registered cows in herds enrolled in DHIR are transferred to the respective breed registries for recording.

These rules apply to both the DHIA and DHIR programs:

1. All cows in the herd must be entered into the official testing program.

2. All animals in the herd must be permanently identified.

3. Copies of pedigrees of all registered cows must be made available for DHIR.

4. Testing is done each month with not less than 15 days nor more than 45 days allowed between test periods.

5. Testing is conducted over a 24-hour period.

6. An independent supervisor must be present for supervising the weighing of the milk and the sampling of the milk for milk fat and other determinations.

7. Milk or milk fat records that are above values established by the breed association require retesting of the cow to assure that an error was not made. An owner may request retesting if he or she feels the test does not properly reflect the production of the cows.

8. Surprise tests may be made if the supervisor suspects that they are needed to verify previous tests.

9. Any practice that is intended to or does create an inaccurate record of production is considered a fraudulent act and is not allowed.

Several unofficial DHI testing programs also exist. These programs often involve the owner doing the sampling rather than a neutral supervisor. They may only involve the recording of milk weights. Unofficial records cannot be published or used to promote the sale of cattle.

Progeny Information

Progeny information is useful in selecting dairy animals. Progeny refers to the offspring of animals. Information about the offspring is important. Parents can be studied to predict the traits of offspring.

The dairy industry has benefitted from progeny information. The USDA gathers DHI information in a large, computerized data bank. Sire and dam predicted transmitting abilities are calculated using the animal model.

Predicted transmitting abilities is an estimate of the traits an animal will transmit to its offspring. It includes both genetic superiority and inferiority. In other words, both good and bad traits are transmitted.

The ***animal model*** is the genetic evaluation for dairy cattle production used by the USDA. It uses pedigree data and production related factors.

Linear Evaluation

The type or physical appearance of a dairy cow may be used in culling or selection. Purebred herds are very concerned with outstanding type. It improves the market value and pedigree of an animal. Commercial herds are more interested in functional type.

Functional type refers to the traits of a cow that are good enough to allow her to complete a useful life in a herd. ***Linear evaluation*** is the coding of 15 primary traits of dairy cows. It was started in 1978 and has been well received. Nearly one-half of the herds that use artificial insemination use linear evaluation. With linear evaluation, 15 primary traits of cows between biological extremes are coded. Several numerical scales have been used, but the most common is a 50-point scale with 25 being the midpoint. The most limiting biological extreme is usually given a score of "1."

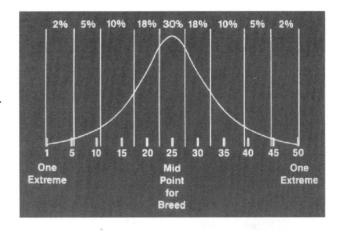

12-13. Linear evaluation uses 15 traits of dairy cows between biological extremes. This shows typical frequencies on a 50-point scale. (Courtesy, Lake-Plains Education Materials, Crookston, Minnesota.)

Most experts agree on the use of linear evaluation. A dairy farmer should select a group of sires that meets specified production standards (goals) for their operation. The selected group of bulls should then be mated to each cow in the herd to maximize type improvement.

Breed associations also calculate predicted transmitting abilities for type (PTAT). A higher PTAT indicates to a dairy farmer that it has a potential to improve the type classification of a particular mating. Dairy

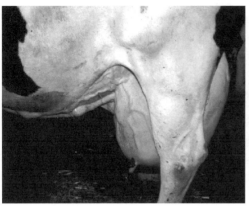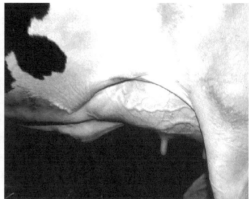

12-14. Fore udder biological extremes. (Courtesy, Lake-Plains Educational Materials, Crookston, Minnesota)

farmers may use PTAT together with linear scores. PTAT is also included in the *type production index (TPI)* which combines several type and production factors into a single figure. The type production index ranks cows based on overall performance. Higher type production indexes are a popular sire selection tool.

Using the ability to judge dairy cattle is often practical with both commercial and purebred herds. This involves comparing animals as a whole based upon their excellence in body type. Four animals are in a class. The animals are compared and ranked. Judging competition is popular throughout the United States. Showing dairy cattle remains popular at county, state, and national shows. Judging complexity increases greatly in a show that has 50 animals per class.

12-15. Dairy judging at the World Dairy Exposition in Madison, Wisconsin.

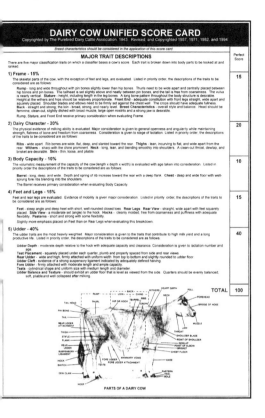 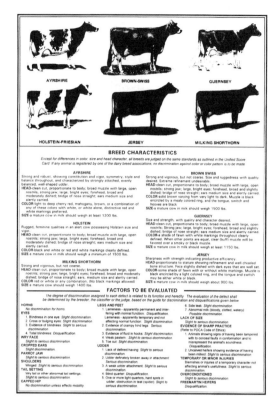

12-16. Dairy Cow Unified Judging Scorecard

FEEDING A DAIRY HERD

Dairy animals have certain inherited genetic possibilities for performance. How well that genetic potential is realized depends upon the environment of the animal. Nutrition is the most important environmental factor.

A great need for energy and protein is created by lactation. A cow that weighs 1400 pounds and gives 80 pounds of milk per day needs 2.5 times more energy for milk production than for maintenance. Feed is the largest production cost in dairying.

Modern dairy technology is used by dairy farmers to feed many times per day. Computerized feeders, mechanized track feeders, or auger-type delivery systems all allow for multiple feedings. *Total mixed rations* (all feeds are mixed together) are also growing in popularity as they provide all feed ingredients in each mouthful of feed a cow eats.

RATIONS

Rations for dairy cattle should be built around high quality roughages. Roughages are usually the lowest cost sources of nutrients. High quality roughages can be bought or grown. Proper soil fertility and harvesting will be needed. The common roughages for dairy cattle are hay, silage, green chop, and pasture. Roughage can be restricted, fed free choice, or as a part of a total mixed ration.

High producing dairy cows cannot get an adequate supply of nutrients from roughages alone. While 50 to 60 percent of a ration may be roughages, the other 40 to 50 percent must be obtained from concentrates. Concentrates refer to high energy feeds, such as grains or protein supplements.

A higher percent of concentrates are usually fed to a dairy cow during the first four months of lactation. The nutrient demand is greatest at this time. Lower producing cows or cows in late lactation are fed a lower percent of concentrates to reduce feed cost.

NUTRIENT REQUIREMENTS

The nutrient requirements of dairy animals can be summarized easily. Maintenance needs are related to body size, but imply no weight gain or loss. Most dairy animals under five years of age also require nutrients for growth. Lactation requires large amounts of nutrients.

Growth rates and nutrients needed for growth vary according to the stage of maturation. Bred heifers and cows require nutrients for reproduction, including both gamete production and fetal growth. The greatest nutrient requirement is for a pregnant lactating cow.

Body condition scores (1 = thin, 3 = average, 5 = fat) are often used to monitor proper nutrition. Ideally, a cow should spend her entire life with a body condition score that fluctuates from 2.0 to 4.0. Peak lactation (60–90 days into laction) requires a cow to use her body reserve to supplment her ration. Her body condition may drop to 2.0 or 2.5 during that phase of lactation. A cow should replenish her body reserve during the last half of lactation, entering the dry period with a body condition score ranging from 3.5 to 4.0.

FEEDING CALVES

Heifer calves must be healthy and well-nourished. It is vital for the calf to become a high producing dairy cow. **Colostrum** fed to a calf ensures that it can use the **immunioglobulins** (antibodies) to develop passive immunity.

Colostrum is the first milk given by a female after giving birth. One-half gallon of colostrum should be fed within the first hour of birth and a full 6 percent of body weight within the first six hours of life. An additional 6 per-

12-17. Dairy calf with all needs met. (Courtesy, University of Minnesota and Northwest Experiment Station, Crookston, Minnesota)

cent of body weight should be fed between the sixth hour and one day of age, for a total of 12 percent of body weight during the first 24 hours.

Colostrum is usually fed to a calf from its mother, but there are times when colostrum from other older cows in the herd should be fed. This is most necessary when cows have been purchased shortly before calving or bred heifers have been raised separately from the older cows and hence do not develop immunioglobulins to the diseases carried by the herd.

Preferably, a calf should be fed whole milk for about five days. From five days until weaning, it can be fed whole milk or a high quality milk replacer. A high quality milk replacer should contain 20 percent fat, 20 percent protein, and less than .5 percent fiber. In addition to milk or a milk replacer, a calf should be fed a calf starter as soon as it will begin to eat. At four to six weeks of age, a calf should be eating 1.5 pounds of calf starter per day.

Nutritionists have found that forage consumption in calves may slow rumen development. Forages should not be fed until after weaning. If hay is fed to a preweaned calf, it should be a leafy, high quality alfalfa or alfalfa-grass hay fed only after four weeks of age.

FEEDING YOUNGSTOCK

Feeding postweaned calves becomes easier due to a functional rumen. Some basic feeding guidelines are still necessary.

Forage can be fed free choice, with concentrates being supplemented at a rate of 0 to 6 pounds, depending on forage quality. Pasture can be the only forage used. Pasture quality should be carefully monitored so that an appropriate supplemental grain mix can be fed.

Just as in lactating cows, body condition can be a useful tool in monitoring nutritional needs. If the nutritional level for youngstock appears correct, but the body condition level and growth rate are less than desired, one should look for other related problems. A common problem is that internal and external parasites may be stealing the animal's nutrients.

FEEDING DRY COWS

Concentrates are not fed when a cow is dried off. For two to three weeks, average to good roughage may be the only feed. Concentrates, as a special dry cow formulation, are often reintroduced to a dry cow about midway through a standard 50 to 60 day dry period at a rate of 4 to 7 pounds per day. Toward the end of the dry period, some producers challenge or lead feed. Lead feeding refers to feeding concentrates that more closely resemble the lactating ration (7 to 14 pounds per day).

A few basic rules of thumb should be followed for a dry cow ration: (1) the calcium:phosphorus ratio should fall in the 1:1 to 1.5:1 range to reduce the incidence of milk fever; (2) a maximum of 35 pounds per day of corn silage should be fed; (3) alfalfa hay or haylage is not a recommended dry cow forage, but if it must be fed, it should never make up more than 50 percent of the forage dry matter; and (4) salt should be reduced to .25 percent of the ration (dry matter basis).

12-18. Ration evaluation helps in feeding dairy cows. (Courtesy, University of Minnesota and Northwest Experiment Station, Crookston, Minnesota)

FEEDING THE LACTATING COW

The most challenging ration to develop is that for a high producing cow. Each cow is unique and should be fed that way when possible. At a minimum, a cow should be fed a ration that is being fed to a group of cows that closely resembles her needs.

Several cow factors must be considered when developing lactating rations. These are age, size, stage of lactation, milk composition, and labor capability. Feed quality, availability, cost, and palatability to the cow must be considered.

High producing cows typically require 1 pound of concentrate (grain) for each 2.4 to 3 pounds of milk produced. Concentrates rarely make up more than 60 percent of the total dry matter (a 50:50 ratio of concentrates to roughages is more desirable). Forages make up the balance of the ration and usually are fed at a rate of 1.5 to 2.0 percent of body weight. Feed protein in a ration should be 19 percent in early lactation and decreased slightly in later lactation. A higher level of bypass protein (undegradeable) is necessary in early lactation.

The fiber level in a ration should be 19 to 21 percent acid detergent fiber (ADF) and 25 to 28 percent neutral detergent fiber (NDF). Acid detergent fiber is the residue of the fiber content of feed. Neutral detergent fiber is cell walls and other feed content that is not readily digested. A maximum of 5 to 6 percent fat level in the ration is suggested. Salt should be fed at .5 percent of the total ration, a calcium/phosphorus mineral source at 1 to 2 percent of the grain mix, and vitamins as needed.

ENVIRONMENTAL MODIFICATION

Modification of a dairy cow's environment has been underway since domestication. Most environmental change is to provide for the well-being of the cow. It includes people who handle the cow, the climate in which a cow lives, societal concerns, and economic considerations. Modification of a cow's environment improves her living conditions over what she would have as a wild animal.

ANIMAL/HUMAN INTERACTION

The major environmental modification for most domesticated animals is increased interaction with humans. This is definitely true with dairy cattle.

A good dairy herdsperson is patient and understands animal behavior. Dairy producers should understand normal behavior to detect and treat animals that are acting abnormally. A caring herdsperson will find that dairy animals bond to kindness.

STRESS MODIFICATION

Many people feel that stress should be eliminated from both human and animal lives. A better viewpoint is that stress should be modified to a desirable level. It should be at a level that prevents boredom, but does not create unnecessary fear. Finding the right level of stress requires knowledge of the past experiences of an animal. A simple analogy can be made between human meals and feeding animals. Humans and dairy animals who are accustomed to eating at precise times have more stress by an hour delay than would humans and animals who were accustomed to variable eating times.

Some potential stressors are under human control. These include regular care, space allocation, transportation, and presence of strangers. These stressors can be easily managed with proper planning.

Other potential stressors cannot be controlled by humans. However, some can be modified. Examples of this type of stressors include weather changes, social order fights among animals, and physiological cycles, such as heat or calving.

Keeping dairy animal stress at an appropriate level requires knowledge of animal behavior, animal history, and coming changes in the life of an animal. A bred heifer raised on a large pasture with little human interaction will require a longer acclimation period to the dairy cow housing facility. A bred heifer raised with close human interaction in a facility similar to a dairy cow housing facility will adapt quickly.

CLIMATE MODIFICATION

Four climate factors affect dairy animals: (1) temperature, (2) humidity or precipitation, (3) wind, and (4) radiation. While dairy cattle have a fairly wide tolerance range to individual factors, combinations of extremes can quickly have an adverse effect on milk production, conception rates, and growth.

A combination of heat and humidity with little wind is especially difficult for dairy animals to cope with. Shade and misting are used to reduce the negative effects. While dairy animals are more tolerant of cold, protection against severe cold, wind, and snowstorm conditions is beneficial.

Understanding an animal as it reacts to the weather is useful. For example, cows eat more before storms and less during hot, humid weather.

Dairy animals prefer to travel in groups or herds to water while grazing if the water is further than 500 to 700 feet away, while they travel individually at closer distances. Hot weather requires greater water intake and disrupts grazing if the water is at a distance that encourages herd travel.

HOUSING

Housing protects dairy cattle and the humans who work with them. The goal is a system that does not allow conditions to get outside the "acceptable" range. Each dairy operation has a level at which additional expenditures will not increase profits.

Cold housing is an unheated building kept cold during the winter. Natural air movement keeps these buildings cold and removes moisture. *Warm housing* refers to buildings kept warm in the winter. Body heat from the animals provides the heat source while insulation helps retain it in the building.

Cold housing for cows usually takes the form of loose housing. Loose housing allows individual cows freedom, while, the herd is handled on a group basis. Loose housing may be an open area with a manure pack or a free-stall system in which cows can enter individual stalls to rest and ruminate. Free stalls often reduce bedding requirements by 75 percent.

Warm housing for cows can also include a free-stall barn with insulation. Northeastern and Midwestern United States dairy farms commonly use stall barns. Stall barns usually consist of two rows of cows confined in tie stalls, comfort stalls, or stanchions. Each cow has an area (usually 4 feet by 6 feet for large breeds) where it spends most of its life. Most often, cows are milked in their stall.

Calves, youngstock, and dry cows are commonly housed in a natural, cold environment. Calf hutches for 0 to 6 week old calves and super hutches or open sided sheds provide shelter for older calves, youngstock, and dry cows. Other sheds are occasionally

12-19. Calf hutches provide a healthy environment for young calves.

12-20. Cows in this lot can go under the shed for shade. Shade reduces the impact of heat.

used, but frequently they lack proper ventilation and temperature control to provide the calf with a healthy environment.

Cattle on open lots in the southern United States or on pasture should be provided with shade and a clean readily available water supply. Northern pastures may also require wind protection or shelter during the early and late grazing season.

MILKING SYSTEMS

Many cows are still milked within their stall in a flat milking barn system. This requires a pipeline to carry the milk from the stalls to the bulk tank.

Milking parlors are more common in the southern and western United States. They are rapidly gaining acceptance in the northern and eastern

12-21. A pipeline brings milk to the bulk tank in this flat barn milking system. (Courtesy, University of Minnesota, Crookston, Minnesota)

dairy areas. A ***milking parlor*** is a concrete platform raised above the parlor floor (pit). Cows enter the parlor in groups or individually and stand on the platform for milking. Milking parlors improve labor efficiency, working conditions, and cleanliness in the milking operation. A wide range of milking parlor designs are used.

FEEDING SYSTEMS

Nearly twenty tons of feed are fed to a cow in one year. Labor efficiency, ability to individualize to each cow or groups of cows, and cost must be considered in selecting a feeding system. Storing and feeding forages can require different equipment from feeding concentrates or can overlap in the case of total mixed rations (TMR).

Storage is needed for silage, haylage, dry hay, and straw. Silages can be stored in traditional upright silos, bunker silos, oxygen-limiting silos, or silage bags. Dry hay and straw can be baled in small square bales, large square bales, or round bales and can be pelleted or put up in loose stacks.

Fresh or ensiled forages are usually fed with mechanized delivery systems or a wide range of silage/green chop wagons. Dry forages are moved mechanically with a tractor or manually in flat stall barns. Mechanical chopping of hay is required in certain feeding systems.

Management intensive grazing is a system that allows the cows to harvest their own forage. Cows are

12-22. Computerized milking parlors with automatic take-off increase labor efficiency. (Courtesy, University of Minnesota and Northwest Experiment Station, Crookston, Minnesota)

12-23. Automated systems can be used to feed forage. (Courtesy, Mississippi State University)

12-24. Concentrates are carefully stored in bins to prevent loss on this dairy farm.

placed in a pasture that can be eaten down to the desirable height in a 24- to 48-hour period. The cows are then moved to another pasture. This concept encourages proper nutrition by forcing the cow to eat both leaves and stems. It also maximizes the number of cows that a pasture can support. Management intensive grazing has the disadvantage of requiring more labor as cattle and waterers must be moved each day.

Concentrates are stored in grain storage bins. A wide range of mechanical feeding equipment is available. Simple push carts, mechanized feeding carts, augers, cable or track feeders, computerized feeders, and lever-operated storage boxes are all available to meet a wide range of housing/feeding arrangements.

Total mixed rations are gaining acceptance as it provides rumen microorganisms with a stable food supply. Chopped hay and silage are weighed and mixed with concentrates in a mixer wagon. Cows can be fed in groups according to production level.

MANURE HANDLING SYSTEMS

Dairy animals produce urine and feces at a rate of about 8 percent of body weight daily. This is about 15 to 20 tons a year for a mature dairy cow. This manure contains nitrogen (N), phosphorus (P), and potassium (K) worth over $100.

A well-designed manure disposal/handling system serves three functions: (1) keep dairy animals clean through frequent removal of manure/waste; (2) provide efficient and economical collection of manure/waste; and (3) dispose of manure/waste in an environmentally sound manner.

Solid manure systems usually require the lowest investment in equipment and facilities. Manure can be hauled on a daily basis or stored and hauled periodically. Daily disposal reduces odor and fly problems. All solid manure systems are labor intensive. Some require daily labor and others require intensive labor for a few days.

Liquid manure handling requires a larger investment in equipment. These systems meet most environmental standards and preserve nutrients. Conventional liquid systems may have above or below ground storage tanks/basins. Water is frequently added to the waste material to allow it to be handled by pumps. A minimum of four to five months storage capacity is needed. Adequate land for manure application must be available to avoid leaching of the nutrients to the underground water supply.

An alternate liquid handling system uses lagoons (a type of pond). Lagoon systems can be aerobic (require oxygen), anaerobic (do not need oxygen), or a combination. Aerobic systems are relatively odor-free, but require larger surface area. Anaerobic systems can handle more types of waste and require less land, but can cause a bad odor problem.

Waste disposal is a big concern. Manure should be viewed as an asset and considered as such when designing a dairy operation. It only becomes a problem when dairy farmers lack the foresight to give it proper planning or fail to recognize its value.

DAIRY CATTLE HEALTH AND REPRODUCTION

A dairy cow must reproduce annually while under the stress of high milk production. A combination of proper nutrition and good health are necessary.

The primary objective of a health program is to prevent and control disease. Knowing the normal and abnormal dairy animal behavior is the first step in proper health and reproductive management. Noticing a cow in heat, a calf with dull eyes, or a cow whose body condition is declining makes it easy to respond.

12-25. Annual calving requires proper nutrition and herd health.

REPRODUCTION

Reproductive goals of most dairy farmers include: (1) a 12 to 13.5 month calving interval, (2) less than 2 units of semen per pregnancy, (3) heifers first calve at 23 to 25 months, (4) a live and healthy calf crop of 95 to 98 percent, and (5) a minimum of reproductive tract disorders. Many dairy farmers today also use embryo transfer, with some considering cloning.

Achieving the reproductive goals requires a good knowledge of animal health. Many diseases that may cause reproductive problems, such as bovine virus diarrhea (BVD), brucellosis (Bangs), and leptospirosis, must be included in a herd vaccination program. Other problems, such as metritis and retained placentas, should be analyzed and treated by a veterinarian.

The danger of many genital diseases, such as trichomoniasis, vaginitis, and vibriosis, can be virtually eliminated by use of artificial insemination. (Some of these diseases are presented in more detail in other chapters of the book.)

Heat detection in cows and keeping good reproductive records are important. Synchronization methods may be used to bring cows in heat or to group the heat cycles of many animals. Most dairy cows are artificially inseminated.

When to Inseminate Based on Signs of Heat

Coming Into Heat	Standing Heat - 12-18 Hours		Going Out of Heat	
Bunts other cows.	Stands to be ridden.		Smells other cows.	
Bellows.	Attempts to mount the front end of other cows.		Attempts to ride other cows, but will not stand.	
Attempts to ride other cows, but will not stand.	Elastic clear mucus discharge.		May still have clear mucus discharge from vulva.	
Non-elastic clear mucus discharge from vulva.	Very restless.			
			Ovum Released ✳	Metestrus Bleeding ✳
Too early to inseminate	Can be inseminated	Best time to inseminate	Can be inseminated	Too late to inseminate

12-26. When to inseminate based on signs of heat. (Courtesy, Gary Stegman, Crookston High School, Minnesota)

SELECTING SERVICE SIRES

Selection of quality sires is important. Dairy farmers have only five or six cow generations to achieve the production goals that have been set. Breeding artificially with bulls that are in the top 20 percent of the breed could be the difference in making or not making a profit.

Semen from five to eight bulls should be adequate for small herds of 60 cows or less. Intermediate size herds (61 to 200 cows) may use 7 to 10 bulls in their breeding program. Large herds (over 200 cows) may use 10 to 15 bulls. Semen is available from many sources. Bulls that meet the standards established by a given herd can easily be selected using predicted transmitting ability data.

METABOLIC HEALTH DISORDERS

Nutrition overlaps most areas of dairy cow health and management. Similarly, several metabolic disorders are directly or indirectly related to nutrition.

Metabolic disorders involve chemical transformation of energy by the body. Most occur at or shortly after calving. This is due to the stress associated with high milk production. The three most common metabolic disorders are milk fever, ketosis, and fat cow syndrome. Other metabolic disorders, such as bloat, retained placentas, and grass tetany, are not as specific to dairy cows.

Milk fever is characterized by low blood calcium and paralysis. It is caused by overfeeding calcium during late lactation or the dry period. Milk fever most often occurs near calving time, when the need for calcium increases rapidly. A low calcium to phosphorus ratio and an adequate amount of vitamin D in the ration reduce milk fever problems.

Ketosis is characterized by a poor appetite and dullness. Blood sugar is low and should be treated with propylene glycol. While ketosis can be a primary problem, it is more often a secondary problem. It results when a cow is struggling with a health problem early in lactation.

Fat cow syndrome develops when cows become overfat in late lactation or the dry period. This leads to other health problems, such as ketosis. Avoid fat cow syndrome by proper feeding.

GENERAL DISEASE PREVENTION AND PARASITE CONTROL

Several diseases are important in dairying. Preventative measures and fast action when problems arise help to keep a healthy herd.

12-27. Flies are pests to cattle and should be controlled. (Courtesy, Mississippi State University)

Mastitis is the most costly disease on dairy farms. It involves inflammation of the mammary gland. Mastitis costs $200 or more per cow annually.

Mastitis is sometimes visible to the human eye. It is often not visible. Laboratory tests are needed to measure somatic cells. A cowside California Mastitis Test (CMT) may be used. Controlling mastitis requires clean yards, barns, and milking equipment. Spread of mastitis is reduced with good sanitary milking practices. Common practices include using individual towels for each cow, pre and post teat dipping, drying of teats, and elimination of overmilking. Cows with mastitis that do not react to treatment during lactation should be dry treated. If a cow is a problem after reasonable treatment, it may be wise to sell her.

REVIEWING

MAIN IDEAS

Dairying is concerned with providing wholesome milk. Producers follow many practices to keep costs down and earn a profit.

Milk production is associated with the reproductive cycle. Cows are milked about 305 days following parturition. They are allowed to be dry 50 to 60 days before the next calving. Cows normally have a calf each year and complete the reproductive cycle.

Dairy cows are selected based on their milking potential. Some differences exist among the breeds. The Holstein is the most popular breed and gives the largest amount of milk. The Jersey gives the milk with the highest butterfat and protein content.

Selecting and culling dairy cows may be based on Dairy Herd Improvement records, linear evaluation, and progeny information. Cows that do not produce or are diseased should be sold for slaughter.

Providing a good environment improves efficiency. The environment should meet the needs of the cows and make high production possible. A ration with high quality feed is needed. An abundance of good water is essential. Preventing and controlling disease requires continual observation of the herd. Reproductive management and a herd health program are needed.

QUESTIONS

Answer the following questions using correct spelling and complete sentences.

1. How have dairy farms changed since dairy cows were first brought to the United States?

2. What are the important breeds of dairy cattle? Distinguish between the breeds.

3. What should be considered before starting a dairy farm?

4. How are cows selected and culled?

5. How does linear evaluation differ from judging?

6. Describe the ration for a lactating dairy cow.

7. How would you react to the statement that "stress is bad for animals"?

8. What are four climatological factors that affect the performance of dairy animals?

9. Identify three types of dairy cattle housing. How does each modify the weather?

10. What are the differences between a flat barn milking system and a milking parlor?

11. What is a total mixed ration (TMR) and what are the advantages of it?

12. What are the three functions of a manure disposal/handling system?

13. What is the overall objective of a health program?

14. List four reproductive goals for dairy farmers.

EVALUATING

CHAPTER SELF-CHECK

Match the terms with the correct definitions. Write the letter by the term in the blank that is provided.

a. immunioglobulins   d. homogenization   g. animal model
b. lactation   e. pasteurization   h. Dairy Herd Improvement Program
c. dry cow   f. progeny

    1. ____ a cow that is not lactating

    2. ____ the genetic evaluation for dairy cattle production

    3. ____ antibodies in colostrum that develop passive immunity in a calf

    4. ____ breaking the fat globules in milk into smaller pieces so that they stay suspended in milk

    5. ____ offspring

    6. ____ heating milk to destroy microorganisms

    7. ____ industry-wide dairy production testing and record-keeping program

    8. ____ the period that an animal secretes or produces milk

EXPLORING

1. Tour a dairy farm. Discuss with the dairy manager how the environment is modified for various age dairy animals.

2. Ask a breed association classifier or an A.I. organization cow evaluator to discuss the linear traits and how they relate to functional type. Refer to Appendix A and complete the practice linear mating problem. Pay close attention to the goals of the dairy family as you complete this problem.

3. Sample several forages and send them to an infrared laboratory for analysis. Discuss the analysis with your instructor.

4. Contact a local veterinarian about preventive health visits. How do these visits differ from emergency calls?

Chapter 13

POULTRY PRODUCTION

Thanksgiving without turkey on the menu would hardly be Thanksgiving! Roast turkey for Thanksgiving dinner is an American tradition. In many homes, the special day would not be complete without turkey. And there are many other poultry products!

Poultry consumption has been zooming! People are eating more poultry than ever. Of course, chicken is the most popular poultry. The average person eats 69 pounds of chicken a year. Much of the increase is because people want economical foods that promote good health.

Poultry provide many products. Eggs are important in many dishes. Feathers are used in making pillows, clothing, and fishing lures. Eggs are used in laboratories to make human medicines, particularly vaccines. Some poultry are kept for ornamental uses and as a hobby. Much of the increased use of poultry is due to changes in how they are raised and marketed.

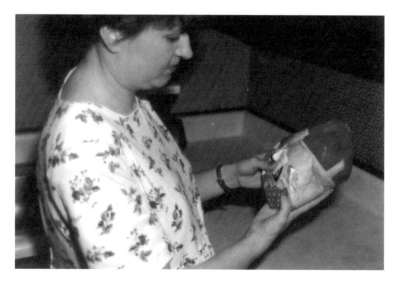

13-1. A carefully processed broiler ready for cooking.

OBJECTIVES

This chapter provides an overview of poultry production. It has the following objectives:

1. List and describe the major kinds of poultry
2. Explain the poultry industry
3. Describe poultry as organisms
4. Explain chicken production systems
5. Describe sanitation and disease control
6. List facility and equipment needs in poultry production

TERMS

albumen	duckling	peacock
broiler	egg injection	poult
candling	gaggle	poultry science
capon	gander	pullet
cock	gizzard	roaster
cockerel	gosling	spent hen
debeaking	incinerator	tom
disposal pit	layer	vertical integration
drake	litter	yolk
dubbing	molting	

KINDS OF POULTRY

Poultry are domesticated birds raised primarily for meat, eggs, and feathers. The products are used primarily by humans. Some products, often by-products, are used as pet food and for other purposes. Eggs are used in making vaccines for humans and other animals. Poultry are also called birds or fowl.

KINDS OF POULTRY

Poultry includes chickens, turkeys, ducks, geese, peafowls, swans, ostriches, and several other species. Chicken is by far the most important. Ducks, geese, and swans spend some time in water, but much of their time on land.

Chickens

Chickens are raised for meat and eggs. The type raised depends on the product wanted. A few other speciality types are raised, such as game chickens and fancy show chickens. The latter are used for hobbies more than anything. Groups of cocks may be seen tethered to small individual houses on a range. These account for only a tiny part of the poultry industry.

13-2. Game chicken roaming free on a Virginia farm.

The meat from a chicken is based on its age and sex. Most chicken is from broilers. A **broiler** is a young chicken six to twelve weeks of age that weighs more than 2½ pounds (1.1 kg). They are tender and easy to cook. Chickens of either sex are used as broilers, which are sometimes known as fryers. About six billion broilers are raised each year in the United States.

Other chickens used for meat include roasters, capons, and spent hens. A **roaster** is a young chicken that is older and slightly larger than a broiler. A **capon** is a male chicken that has been neutered (castrated). Most capons are six to eight months of age and weigh about six pounds. A **spent hen**

13-3. House filled with broilers that are near market size.

13-4. Cocks tethered to A-frame houses in Mississippi.

13-5. White Leghorn cock and hens. (The White Leghorn is primarily for egg production.)

is a hen that is no longer laying. Spent hens go into processed foods, such as soup, or are baked. Older male chickens have tough flesh and are not used much in cooking.

Eggs are produced by mature female chickens, known as hens. A *layer* is a mature female chicken kept to produce eggs. Most hens in laying flocks produce 250 or more eggs a year. A *pullet* is a young female chicken that is being raised for laying. They are less than one year old. Older pullets may have started egg production.

A mature male chicken is a rooster or *cock*. Young male chickens are known as cockerels. A *cockerel* is less than one year of age.

Breeds of chickens are unimportant in commercial poultry production. Specially developed varieties of chickens are raised. These are often developed by a poultry company to meet particular needs. For example, Choctaw Maid Poultry Company uses a Ross variety of breeder chickens for broilers.

The common breeds of chickens are White Leghorn (used for egg production, white egg shell); Barred Plymouth Rocks (meat and eggs, brown egg shell); White Plymouth Rocks (meat and eggs, brown egg shell); and New Hampshires (primarily meat, brown egg shell). The White Leghorn is the smallest, with mature cocks weighing 6 pounds (2.7 kg) and hens

13-6. A Barred Plymouth Rock cock.

13-7. New Hampshire hens. (Courtesy, Mississippi State University)

weighing 4½ pounds (2 kg). The White Plymouth Rock is among the largest, with cocks weighing 9½ pounds and hens weighing 7½ pounds (3.4 kg).

Turkeys

Turkeys are raised primarily for meat. Consumers want birds that have a high proportion of white breast meat. Nearly 300 million turkeys are raised each year in the United States. Turkeys are being raised on fewer farms but the size of the farms has increased. Many commercial producers raise over 100,000 birds each year on a single farm.

A young turkey is a ***poult***. It has not grown to the point where its sex is easily determined. A mature male turkey is a ***tom***, while a female is a hen. Younger and smaller turkeys are roasted whole. Larger turkeys are often made into boneless breasts and other products.

Most turkeys are raised on a range. However, in recent years, a greater number are being produced in houses. Most turkeys are produced under contract with a processing plant. Many turkeys are marketed at 14 to 18 weeks of age, with hens ready about three weeks earlier than toms. Toms weigh quite a bit more than the hens. Toms weigh about 27 pounds (12.2 kg) and hens 15 pounds (6.4 kg) when marketed. Good producers can have feed conversion ratios of 2.5 pounds (1.1 kg) of feed to 1 pound (0.5 kg) of gain. Turkeys have been bred to be larger and have broad breasts.

13-8. White turkeys being raised on a range.

Colors vary from white to bronze, with the broad-breasted white predominating in commercial turkey production.

Ducks

Ducks are raised for meat, eggs, and down and feathers. (Down is the soft feathery covering that grows under the feathers.) Some are kept as hobby or ornamental ducks.

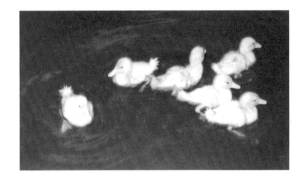

13-9. Ducklings covered with down enjoy a swim.

13-10. Pekin ducks. (Courtesy, Mississippi State University)

Young ducks are covered with down. Feathers develop as they grow. A young duck that still has down rather than feathers is a ***duckling.*** A mature male duck is a ***drake***. Mature female ducks are known as hens.

Duck and duck products are not nearly as widely used as chickens. Some 20 million ducks are raised in the United States each year. White Pekin ducks are most widely found. Ducks grow faster and heavier than chickens. They can also swim, which is not the case with chickens and turkeys.

Geese

About one million geese are raised in the United States each year. Geese are raised for meat, eggs, and feathers and down. Many of these are kept for ornamental purposes. Geese like to eat tender grass in pastures and parks. Some are kept to control weeds and grass, and are known as "weeder geese." They are hardy birds and resist many poultry diseases. A baby goose of either sex is a ***gosling***. A mature male goose is a ***gander***. A mature female is a hen or goose. A flock or group of geese that are not flying is known as a ***gaggle***.

13-11. Goose on a range at Tahoe City, California.

Peafowl

Peafowl are raised for large, beautiful feathers. The males are ***peacocks*** and females peahens. Sometimes, either sex is known as a peacock. Male peafowls may spread their feathers, known as a train. The feathers may be five times the length of the body. Most peafowls are raised for ornamental purposes. The hens lay a few eggs each year.

13-12. Peacock.

Swans

Swans are similar to ducks and geese in their preference for water. Several kinds of swans are found, with colors ranging from white to black. Most swans are kept for ornamental purposes. They are frequently used with pools of water at resorts and similar locations.

13-13. White swan at a California resort.

13-14. Ornamental Black Australian Swan in a Napa Valley pool.

Ostrich and Emu

13-15. Ostrich in a research program at the University of Arkansas at Pine Bluff.

No other birds have created more excitement in recent years that ostriches and emus. They are the largest birds, with the ostrich larger than the emu. An ostrich can weigh as much as 300 pounds and stand 10 feet tall. Ostriches have the longest life span of most birds, living up to 70 years. These birds are raised for feathers, meat, skin, and oil products. The feathers are known as plumes and are used in decorations.

Guinea Fowl

Guineas are raised for food, as novelty birds, and to stock game preserves. They are hardy and often roam free. Their eggs are smaller than chicken eggs and have a thicker shell. Because the eggs are not as easily broken, they are often dyed and decorated in various ways.

13-16. Guinea fowl.

THE POULTRY INDUSTRY

The poultry industry has changed from a family or farm with a few birds for home use to large commercial producers. The typical family no longer has birds. The commercial producer usually raises hundreds of thousands each year.

EARLY POULTRY PRODUCTION

The first poultry production was more than 5,000 years ago in India. This was followed by poultry in Egypt about 3,500 years ago. Poultry were brought to North America by the early European settlers. A few people in Jamestown had small flocks—usually chickens—as early as 1607.

In colonial North America, many families kept small flocks at their homes. The birds usually ran about and often nested in brush and thickets. They ate whatever they could find—insects, seed, and tender leaves. Fox and other animals often preyed on them. Most were kept for eggs. Young males and older hens were used for meat. People often raised only what they needed. Extra eggs or chickens might be traded to a neighbor for a ham or vegetables.

The turkey is native to North America. The Aztec Indians domesticated the wild turkey. Native Americans used the turkey for food and its feathers for decoration before the Pilgrims arrived. Turkeys were exported to Europe before chickens were brought to Jamestown!

Dramatic changes began to occur in poultry production in the mid-1900s. These changes led to the modern poultry industry.

MODERN POULTRY INDUSTRY

Today, poultry production is a large commercial industry. Thousands of birds may be raised in confinement in one house. Science and technology are used to assure well-being of the birds. Exact rations are fed to get rapid growth and high production. The houses are built to keep the birds healthy. Everything possible is done to care for the birds.

The number of small flocks has dropped considerably since 1950. In fact, most people could not raise poultry if they wanted to because of where they live. Urban life is not well-suited to having a small flock of poultry. In addition, people like to be able to go to the supermarket and buy poultry products that have been carefully inspected and are ready to cook.

Providing an abundance of poultry products at low costs required major changes in how they were raised. Good information was needed. A system to market products had to be developed. Consumers had to have a year-round and uniform supply. New methods had to be developed. Much has happened in the poultry industry since Harland Sanders opened the first KFC restaurant in 1956!

The leading states in broiler production are Arkansas, Georgia, Alabama, and North Carolina. California is the largest producer of eggs. North Carolina, Minnesota, and California are leading states in turkey production.

13-17. Modern poultry houses have controlled environments and automated feeding systems. The houses are often 40 feet (12.2 m) wide and 450 feet (138.2 m) long. (Courtesy, Michael Stevens, McCarty Farms, Forest, Mississippi)

Poultry Science

Poultry need certain conditions to live and be productive. Research has studied many areas of poultry. Today's producers know a lot!

Poultry science is the study and use of areas of science in raising poultry. It includes breeding, incubation, rearing, housing, feeding, sanitation, marketing, and other areas. The goal of poultry science is to provide plenty of wholesome poultry products for consumers at a reasonable price.

Poultry science has resulted in improved production. A hen now lays over 250 eggs a year (compared to 100 per year in 1900). Automation allows one worker to care for many birds. A single broiler house may have 15 to 20 thousand birds. One worker can care for 75,000 broilers or 40,000 laying hens in cages.

Most of the information used in poultry science has been developed in the last 50 years. It is based on biology and related areas, such as environmental science.

Vertical Integration

Raising, processing, and distributing poultry has developed into one continuous chain. *Vertical integration* is when one agribusiness is involved in more than one step in providing poultry products. The steps are linked together to assure a continual supply of poultry. The poultry industry has more vertical integration than other areas of agriculture. All areas of poultry are vertically integrated.

The best example is in broiler production. One company vertically integrates by carrying out all of the steps in the process of poultry production. Several steps may be involved. A poultry company may sign agreements with growers, provide the chicks and feed, supervise the grower, process the broilers, and distribute the chicken to buyers. The grower (farmer) provides a poultry house and looks after the chicks. The company may pay the grower a fee for each bird raised. The amount is based on how well the birds grow or other conditions.

Several large companies provide most of the chicken in North America. Holly Farms and Gold Kist are two examples. Some have trademarks and brand labels, such as the Miss Goldy label used by Sanderson Farms.

POULTRY BIOLOGY

Poultry are terrestrial animals with feathers and a backbone. Many have wings that have limited use in flying. Modern poultry does not need to fly, nor do they have the ability to do so. Some species can swim, such ducks and swans. Chickens and turkeys cannot swim. Poultry are classified in the Aves class of vertebrata of animals. Each species has a scientific name. The scientific names of common poultry species are shown in Table 13-1.

Table 13–1
Scientific Names of Common Poultry Species

Kind of Bird	Scientific Name	Kind of Bird	Scientific Name
Chicken	*Gallus domesticus*	Ostrich	*Struthio camelus*
Duck	*Anas domestica*	Pigeon	*Columba domestica*
Goose	*Anser domesticus*	Swan (black)	*Cygnus atratus*
Guinea fowl	*Numida meleagris*	Turkey	*Meleagris gallopavo*

LIFE PROCESSES

Poultry carry out the same life processes as other animals. The structures for doing so vary. Two major areas of difference are food digestion and reproduction.

Food Use

Birds take in food with their beak. Beaks are strong mouths made of material similar to a person's fingernails. It can be used for pecking and breaking foods apart. In large poultry houses, birds are debeaked to prevent damage to each other. **Debeaking** is removing the tip of the beak so that they cannot peck other birds.

Poultry do not have teeth and rely on a **gizzard** to grind food. The gizzard is a strong, muscular organ that may contain grit that the bird has eaten. Grit (grains of sand) helps in the food grinding process.

Reproduction

Birds reproduce by laying fertile eggs. Females must mate with a male for eggs to be fertile. In rare instances, hens lay fertile eggs without mating (known as parthenogenesis). The eggs are incubated for a time to allow the chick to develop.

The length of incubation varies with the species, with most chickens being 21 days. Incubation periods for geese are 29 to 31 days, turkeys are 27 to 28 days, and ducks are 28 days. The length of incubation also varies with temperature and humidity conditions. Eggs go through distinct changes in incubation.

Most eggs in the supermarket are infertile. The hen did not mate with a rooster before laying. Most consumers prefer to have infertile eggs. There is usually no difference to the consumer between fertile and infertile eggs, if they are fresh. A fertile egg that was not refrigerated quickly and stored in a warm environment may show signs of embryo development.

Artificial insemination is widely used in the poultry industry. Some birds, particularly turkeys, are unable to mate because of their mass of flesh. Natural fertilization is still predominate with chickens.

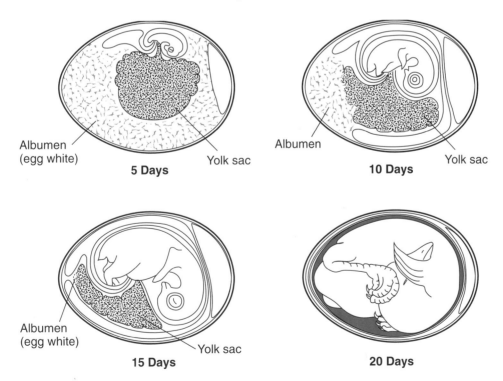

13-18. Stages in the development of a chick embryo.

EXTERNAL PARTS

The external parts of most birds are similar. Some have long spurs near their feet, such as the game chicken. The health and condition of a bird show in its exterior appearance. For example, some breeds have yellow legs. Disappearance of the yellow pigment is a sign that the female has been laying eggs. A large, red comb is a sign that the bird is in good health. Ragged feathers may indicate poor condition.

It is fairly easy to distinguish between most poultry. Some live in and around water, such as ducks and geese. Chickens and turkeys cannot swim. Turkeys have beards and other fleshing structures from their heads. Chickens have serrated combs and wattles. Geese and ducks have rather plain heads. Looking at the heads of poultry is a good way to tell them apart. The heads also vary with the sex of the poultry. Males tend to have larger head features, such as the comb on a rooster.

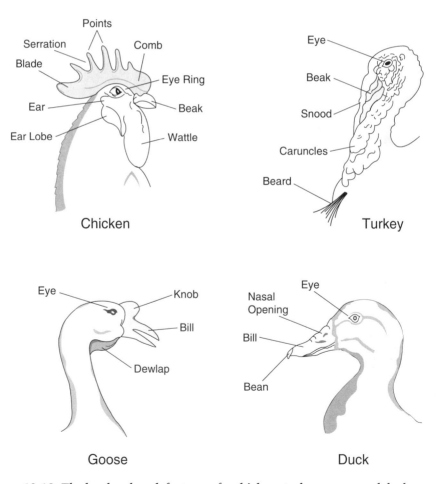

13-19. The head and neck features of a chicken, turkey, goose, and duck.

POULTRY PRODUCTION SYSTEMS

Chickens are the dominant poultry product. Most chickens are produced on large poultry farms that specialize in birds for a particular use. Further, with vertical integration, the producer usually has little investment in the chickens. The producer does have a big investment in the land for the farm, housing and equipment, electricity and water, and labor. The poultry company provides the chicks, feed, medications, and field services to assure that the birds receive proper care.

The chicken industry has four areas: broiler production, egg production, pullets for egg production, and pullets and cockerels for broiler egg production.

BROILER PRODUCTION

The goal of broiler production is to get the most meat as quickly and efficiently as possible. Growers take extra care to keep chicks healthy and growing. Mortality (death) is kept to a minimum—no more than 3 percent while the broilers are growing.

Most broilers are raised in large poultry houses. About six weeks are needed to raise a broiler to market size, or about 3 to 4 pounds (1.7–1.8 kg). The time required is declining as new breeds and better feeding practices are used. Housing is designed to meet the needs of young chickens.

Growers provide an environment that is ideal for broilers to grow. A day or so in the time required to raise a broiler makes a big difference in feed cost. The goal is to get broilers to grow as fast and efficiently as possible. Many growers get a pound of growth on slightly less than two pounds of feed. Young chicks are fed a finely ground meal feed that has 18 to 20 percent protein. As they get older, the percentage of protein is reduced to 12 to 15 percent. Protein in feed is typically from soybean meal and fish and other meat-meal ingredients. The feed also contains ground yellow corn, other grains, minerals, vitamins, and, in some cases, medication.

13-20. Five-day-old chicks learning to eat from a pan.

A small niche market exists for broilers raised on a range. They are outside and allowed to run about inside fenced areas. The efficiency of feed conversion and control over product quality is not as good as in a modern house.

Housing

Housing for broilers is designed to meet the needs of newly-hatched chicks through market size. Poultry houses often hold 20 to 30 thousand birds. A day-old chick needs between one-quarter to one-third of a square foot (0.02 to 0.03 m²) of floor space. Space requirements expand as the chickens grow. A four-week old chicken needs about three-quarters of a square foot (0.06 to 0.09 m²) of floor space. Older, larger birds would need more space. Growers often use a portable fence to isolate baby chicks to a small part of a house. The fence is removed as they grow.

Modern poultry houses have complete environmental control. These controls provide for the well-being of the chickens. Characteristics of environmental control include:

13-21. Wood shavings make excellent litter.

13-22. Economical lighting is provided in environmentally controlled houses.

- Litter—**Litter** is the floor covering in a poultry house. Litter is often made of wood shavings. The litter absorbs moisture from the manure. Litter is normally changed between batches of chickens.

- Lighting—Modern poultry houses rely almost exclusively on electric lights. No natural light may be allowed in some houses. The lighting is timed to give chickens the light needed to eat and grow. Shorter days are made longer by artificial lighting. Most poultry houses use a specially-designed bulb that provides adequate light on a low amount of electricity.

- Temperature—Baby chickens are especially sensitive to temperature changes. Many growers keep house

13-23. Thermostatically controlled gas-powered brooders provide heat for chicks. Note that the brooders are suspended from the ceiling and have large round reflective covers.

temperatures in the 85 to 95°F (29 to 35°C) range for small chicks. Large hood-type brooders with electric or gas heat are used. By six weeks of age a temperature of 70°F (21°C) is adequate. It is lowered as the chickens grow. In warm weather, cooler outside air may be pulled into the house for ventilation.

- Humidity—Humidity control is used to provide the ideal level of water vapor in the air of chicken houses. Heat tends to dry the vapor from the air. Water vapor may be added to the air with mist systems. The humidity level should be 50 to 75 percent in broiler houses.

- Ventilation—Ventilation involves using large fans to bring fresh air into a chicken house. The fans are often on controls to help regulate temperature. Ventilation also removes excess moisture from a house.

13-24. Large fans automatically turn on to ventilate the poultry house.

- Alarm systems—All modern poultry houses have systems to alert people to problems in the poultry house environment. Failures of feeders, waterers, ventilation systems, and other areas are monitored.

13-25. Sensors suspended from the ceiling of a modern poultry house help control various areas of the environment.

- Standby electrical generators—Poultry farms need standby electrical generators in case of power failure. Chickens cannot live long without feed, water, and environmental control.

13-26. A standby diesel-powered electrical generator is available if the electricity fails on this poultry farm.

Feeding and Watering

Most poultry farms use automated feeding and watering equipment. Young chicks may require feed in trays or pans and waterers that are easy to use. The chicks soon learn to eat from automated feeders. After a few days, most producers rely only on automated systems.

Feeder systems often use long troughs that reach from the middle to each end of the house. Feed bins and weighing equipment are located outside at the middle of a long house. A conveyor chain moves the feed throughout the house. A large flock will eat a lot of feed. Plan ahead to be sure there will be plenty. Adequate feed storage must be provided. Feed is usually stored in bins designed to keep the feed from damage and contamination. Bins should be easy to use with access for delivery by trucks.

Watering systems may involve troughs and nipples attached to tubes. The water is carefully metered. The amount consumed is an indication of the health of the flock. Water consumption drops when chickens are sick or in a bad environment. Medications may be added to water. Waterers need to be cleaned each day.

13-27. Feed is stored, weighed, and metered into the poultry house by this automated system.

Most producers use water from wells or municipal systems. Many have standby sources in case one source fails.

13-28. Watering nipple.

13-29. Watering trough.

EGG PRODUCTION

Today's consumers are accustomed to going to the supermarket and selecting carefully graded, high quality eggs. The eggs are in cartons designed to protect from damage. The cartons are filled with eggs graded by size, such as small, medium, and large. The eggs are put in a carton with the small end downward. The larger end should be up to help maintain the quality of the egg. The large end has the egg's air cell.

13-30. Caged layers. (Courtesy, Mississippi State University)

The goal of egg production is to produce high quality eggs as efficiently as possible. Hens are given an ideal environment to produce. Producers strive to get one egg each day from hens. The eggs used for food are not fertile. They will not hatch if incubated.

Managing hens for maximum egg production is essential. Hens need facilities where their well-being is met and the eggs are of high quality. The facilities must keep the eggs clean and prevent cracking. Regularly collecting the eggs or using automated egg collecting systems is needed. Most layers are kept in specially designed cages. Nearly 90 percent of the layers are in cages. Other layers are in houses with slatted or litter floors. A few layers are on a range. The eggs they produce (sometimes known as runabout eggs) are for a niche market. A nest must be provided for the hen in laying.

13-31. Layers in house with litter floor. Note feeders and waterers, with nests along the wall. (Courtesy, Mississippi State University)

Feed and Water

Laying hens need proper feed. A complete commercial feed with 14.5 percent protein is usually fed. The feed should contain 3.4 percent calcium. Egg shells require substantial calcium. Pullets are fed a slightly lower rate of protein just prior to beginning egg production. Pullets usually begin egg production at about 22 weeks of age at a live weight of slightly over 3 pounds (1.5 kg). Peak egg production is reached at about 30 weeks of age. Hens may weigh 5 to 6 pounds (2 to 2.7 kg) and eat 1.3 to 1.7 pounds (0.59 to 0.77 kg) per week. They will drink 2 to 3 pounds (1 to 1.5 kg) of water for each pound of feed.

An average chicken egg weighs 2.2 ounces (0.062 kg). It is made of 66 percent water, 10 percent fat, 13 percent protein, and 11 percent other material, primarily calcium in the shell. The fat is mostly in the egg yolk. Hens that do not get feed with the needed nutrients cannot be productive.

Egg Quality and Color

Egg quality is determined by external appearance and interior condition. External appearance includes size, shape, color, cracks, and blemishes or dirt.

Egg color includes both the shell and internal parts of the egg. The shell is a protective layer of material, primarily calcium. Egg shell color varies. Nearly 95 percent of the market eggs in North America are white. These are normally laid by white chickens. In recent years, a niche market

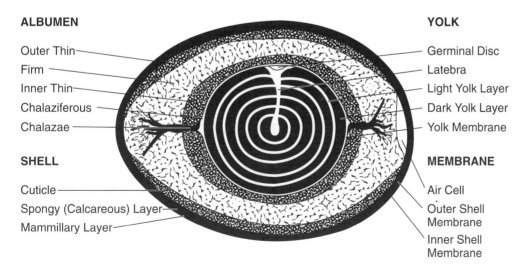

ALBUMEN
Outer Thin
Firm
Inner Thin
Chalaziferous
Chalazae

SHELL
Cuticle
Spongy (Calcareous) Layer
Mammillary Layer

YOLK
Germinal Disc
Latebra
Light Yolk Layer
Dark Yolk Layer
Yolk Membrane

MEMBRANE
Air Cell
Outer Shell Membrane
Inner Shell Membrane

13-32. Major parts of a chicken egg.

has developed for brown and yellow eggs. These are mostly laid by red chickens. The greatest demand for brown eggs is in the New England states.

The inside of the egg has two major parts: albumen (egg white) and yolk (yellow). The **albumen** is the white part that surrounds the yolk. It should be firm and cling near the yolk when broken out. Albumen that spreads out indicates deterioration and aging. The **yolk** is the center part of an egg that varies in richness of yellow color. Hens with diets high in carotene lay eggs with yellower yolks.

Inside structure indicates egg quality. The presence of meat or blood spots is objectionable. An air cell is located inside at the large end of an egg. As an egg ages, the air cell gets bigger. Fresh eggs have small air

13-33. Using a candler to grade fresh eggs. The light behind the egg makes it possible to see any dark spots or other imperfections. (Courtesy, James Lytle, Mississippi State University)

cells. **Candling** is shining light through an egg to see the inside features. Most of an egg is fairly transparent. Dark spots are signs of problems in an egg.

Cracked shells lower the quality of eggs. In most cases, cracked eggs are unacceptable to consumers in the fresh egg market. Cracked eggs go to breaking plants. Here the eggs are made into various egg products.

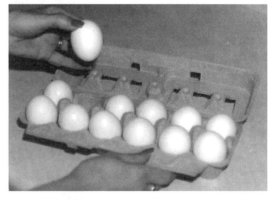

13-34. Consumers want graded eggs of uniform size carefully packaged in protective cartons.

Molting

Molting is the process of shedding and renewing feathers. Hens usually stop laying during molting. If they do not, the laying rate is greatly reduced. Some growers use forced molting. This gives the hen a rest before returning her to production.

A hen may deplete the calcium in her bones during egg production. The rest allows her to restore the calcium and begin high production again.

In the wild, birds molt in the fall so that they have a good protective layer of feathers for the winter. Molting is associated with age. Chickens often molt at about one year of age. Molting takes about four months.

Molting can be induced by decreasing the length of the day. Since most poultry houses have artificial lighting, the light timer is set to make the dark time longer. Hens will molt with eight hours of light. Molting stops and egg production begins in about two months when the light is increased to 14 to 16 hours per day.

PULLETS FOR EGG PRODUCTION

Hens for egg production are raised by pullet producers. The grower takes specially-bred, day-old chicks and raises them similar to broiler chicks. The chicks are sexed, so only females are raised. The sexing is done prior to delivery to the grower. The male chicks are raised as broilers for meat. The birds are bred for maximum egg production. The pullets are known as replacement hens.

Pullets are raised to 20 weeks of age by the pullet producer. They then go to laying farms as starter pullets. Pullets will begin egg production a few weeks after delivery. Pullets start laying regularly at 24 weeks of age and reach maximum production at 30 to 35 weeks.

Some pullets are dubbed. *Dubbing* is removing the comb and wattles. The large comb and wattles on a mature hen can interfere with eating from automated equipment. Dubbing is usually done on a day-old chick by cutting the wattles and comb off. Research has shown that dubbed hens are more efficient users of feed.

PULLETS AND COCKERELS FOR BREEDER FLOCKS

Some growers specialize in raising pullets and cockerels for fertile egg production. The fertile eggs are hatched to become broilers. Varieties of chickens are carefully selected for their potential in producing broilers with large amounts of meat. Many producers raise pullets and cockerels under contract with a poultry company. At 20 weeks of age, the pullets and cockerels are moved to a fertile egg producer for production.

The pullets and cockerels are raised separately. A typical ratio is to raise one cockerel for each five pullets. At maturity, the cockerels and pullets are placed in a combined flock so that mating can occur and the hens will lay fertile eggs.

The initial practices with breeder chicks are similar to broilers in many ways. Growing conditions are used to prepare pullets for high egg production.

SANITATION AND DISEASE CONTROL

Sanitation and disease control are important in successful poultry production. A disease outbreak can quickly wipe out a flock. It is far better to prevent disease than it is to try to treat an outbreak after it occurs.

SANITATION PREVENTS DISEASE

Sanitation is designed to prevent the spread of disease. Many diseases are spread by diseased birds, insects, people and vehicles that travel from one farm to another, carcasses of dead birds, and contaminated feed, water, and litter. Keeping a poultry farm clean is a major part of sanitation.

Removing all litter from a poultry house and spraying the building and equipment with a disinfectant helps prevent disease. This is done after a batch of chickens has been moved to processing or other farms, and prior to bringing a new batch of chickens to a farm.

Many poultry farms restrict the access of vehicles and people. Tires, shoes, clothing, and equipment can transport a disease from one farm to another. Tires of a vehicle must be washed with a disinfectant solution

13-35. Tires on a pickup are sprayed with a disinfectant solution before the vehicle is allowed on a poultry farm.

before going on to a poultry farm. People dip their shoes in disinfectant solution or wear disposable boots.

VACCINATION

Vaccination is used to help poultry develop immunity to disease. The common diseases vaccinated against include Newcastle, Marek's disease, and infectious bronchitis. Most chicks are vaccinated in the nose, eye, or wing. Vaccination is often a part of the debeaking process. Medications are also placed in drinking water.

Day-old chicks suffer a great deal of stress from the handling during vaccination, debeaking, and management. *Egg injection* is a new procedure of vaccinating embryos in the egg. It is used on the eighteenth day of incubation to help the chick be immune to

13-36. Soles of shoes are dipped in the disinfectant solution to destroy any organisms that might be on the shoes before entering the poultry house.

certain diseases at hatching. An egg injection system is used in the hatchery at a high rate of speed. A specially designed needle penetrates the egg shell with a minimum of damage to the shell or embryo. The dosage is delivered inside the egg. One machine and two operators can inject 20 to

13-37. Day-old chicks being de-beaked and vaccinated. (Courtesy, Michael Stevens, McCarty Farms, Forest, Mississippi)

13-38. Egg Injection System. (Courtesy, Embrex, Inc., Research Triangle Park, North Carolina)

30 thousand eggs an hour. Vaccinating during incubation results in the chicks hatching with immunity. Handling and stress are reduced. The chicks are said to be ready for market about two days earlier, if vaccinated before hatching.

DEAD BIRD DISPOSAL

Any birds that die should be promptly disposed of to prevent the spread of disease to others in the flock. Two methods are widely used: incinerator and disposal pit. Every poultry farm should have one or the other in approved condition.

An *incinerator* burns chicken carcasses so that only ashes remain. Incinerators are made of steel and heated with propane, natural gas, or fuel oil.

Disposal pits, also known as compost pits, are dug about 7 feet deep into the earth. Dead birds are dropped into the pit. A cover is kept over the pit. The dead birds are decomposed by microorganisms. Pits should not be located where water supplies would be contaminated.

13-39. Approved commercial incinerator on a poultry farm.

SELECTED DISEASES OF POULTRY

Poultry are affected by some 33 pathogenic diseases and 10 parasites. In addition, nutritional diseases and deficiencies affect poultry. These cause huge losses among poultry producers. In some cases, the birds die. In others, the birds are less efficient. Either way, the owner has a loss of money from the disease. Further, who wants to eat a diseased chicken?

Diseases

Pathogenic diseases are caused by bacteria, viruses, or other living agents. Examples are:

- Marek's disease—Marek's disease is also known as range paralysis. It is caused by a herpes virus. Symptoms include diarrhea, paralysis of legs or wings, loss of weight, and death. No treatment is available. Prevent Marek's disease by vaccination of day-old chicks or embryo vaccination. Some genetic resistance has been developed.

- Newcastle disease—Newcastle disease is caused by a virus. Affected birds gasp, wheeze, twist necks around, may be paralyzed, and lay soft-shell eggs, if any are laid. No treatment is available. Vaccination prevents Newcastle disease.

- Infectious bronchitis—Infectious bronchitis only affects chickens. Young birds wheeze and gasp and have nasal discharge. Older birds stop laying. Sanitation and isolation are effective in prevention. A vaccine is available. Infectious bronchitis is caused by a virus.

- Fowl cholera—Fowl cholera affects many different birds. Caused by bacteria, birds get a fever, purple-colored heads, yellowish droppings, and sudden death may result. Some growers have success in treating fowl cholera with antibiotics and sulfonamides. Vaccination can be used to prevent fowl cholera.

Parasites

Sanitation and control of pests, such as flies and wild birds, help in preventing parasites in poultry. All poultry may be affected. Examples of parasites are:

13-40. Broilers are carefully inspected for wholesomeness during processing. (Courtesy, James Lytle, Mississippi State University)

- Coccidiosis—Coccidiosis is an internal protozoan parasite disease. It is transmitted by the droppings of affected birds, including wild birds. The droppings may be bloody, and affected birds may be sleepy, pale, and listless. Anticoccidials may be put in feed or water.

- Large roundworms—Birds with large roundworms are droopy, emaciated, and may have diarrhea. The worms may reach lengths of 3 inches (7.62 cm) in the intestines. Sanitation and rotation of range and yards help control large roundworms. Dewormers can be used to treat affected poultry.

- Mites—Mites are external parasites that cause birds to be droopy, pale, and listless. Approved insecticides can be used to treat mites. Mites are prevented by sanitation and controlling access of wild birds to the flock.

- Tapeworms—Tapeworms are internal parasites that use snails, earthworms, beetles, and flies as intermediate hosts. Pale head and legs and poor body flesh indicate tapeworm infestation. Dewormers are available.

REVIEWING

MAIN IDEAS

Poultry are domesticated birds raised for meat, eggs, and other products. Chicken is by far the most important poultry. Turkey is second in importance.

Broilers are the major source of chicken meat. A broiler is a young chicken 6 to 12 weeks old. Most commercial broiler producers can grow them to the weight of $2^1/_2$ pounds or more in six weeks or a little less.

Egg production, pullet production, and breeder pullet and cockerel production are also specialized areas of the chicken industry.

Vertical integration is used to link the poultry industry. Most producers specialize in one kind of bird, such as broilers or layers. They get the baby chicks or pullets from specialized producers.

Housing and equipment are used to provide a good environment for chickens. The well-being of birds is a major part of the environment. Birds are managed to efficiently grow and produce.

Sanitation is an important part of reducing disease problems. A method for properly disposing of dead birds is essential on every poultry farm. Many farms use incinerators.

Chicks are vaccinated for disease. A new system of egg injection has been developed to vaccinate embryos on the eighteenth day of incubation. This reduces the handling stress newly-hatched chicks often experience.

QUESTIONS

Answer the following questions using correct spelling and complete sentences.

1. What are the major kinds of poultry? Briefly describe the two most important.

2. Explain how the poultry industry has changed chicken production.

3. How does the digestive system of poultry vary from other animals?

4. What is the reproductive process in poultry?

5. What are the distinguishing features of the heads of chickens, ducks, turkeys, and geese?

6. What are the four production systems with chickens? How do these relate to vertical integration?

7. What are the housing and equipment needs in raising broilers?

8. How does the feed vary from young chicks to older broilers?

9. How are layers managed for production?

10. What are the major parts of an egg?

11. What is molting? Forced molting?

12. Why is sanitation important in controlling disease?

EVALUATING

CHAPTER SELF-CHECK

Match the terms with the correct definitions. Write the letter by the term in the blank that is provided.

a. egg injection   c. disposal pit   e. debeaking   g. tom   i. layer
b. incinerator   d. litter   f. gosling   h. cockerel   j. capon

1. ____ mature female chicken kept for egg production

2. ____ castrated male chicken

3. ____ mature male turkey

4. ____ young male chicken

5. ____ floor material in a chicken house

6. ____ removing the tip of the beak

7. ____ vaccinating chicks as embryos through the egg shell before hatching

8. ____ equipment to burn the carcasses of dead poultry

9. ____ a young goose of either sex

10. ____ pit dug into the earth for disposing of dead poultry

EXPLORING

1. Invite a poultry technician or other qualified person to serve as a resource person in your class. Have them explain the nature of the poultry industry, particularly their role in providing poultry products.

2. Incubate a small batch of eggs. Use a laboratory educational incubator. Get the fertile eggs from a local producer or person with a small farm flock. Carefully monitor temperature, humidity, and other conditions. Follow instructions with the incubator to be sure the eggs are properly incubated. Keep records on all processes and activities.

3. Use a whole broiler from a local supermarket. Carefully cut the broiler into different pieces based on the natural joints of the bird. Study muscling and bone structure. CAUTION: Be careful with sharp knives. Practice proper sanitation. Wash your hands and all equipment thoroughly. Bacteria, especially salmonella, may grow in water or blood from the chicken.

Chapter 14

AQUATIC ANIMALS: FISH, MOLLUSKS, AND CRUSTACEANS

More and more people like to eat fish. Consumers have a wider variety of aquatic foods than ever before. The amount of fish and other species eaten has increased in recent years. People are now more concerned about foods that promote good health. This has resulted in greater demand for many aquatic animals.

Demand is greater than the supply of wild aquatic animals. Streams, lakes, and oceans cannot produce enough. Some waters have been over-fished, resulting in no more wild catch. To meet the demand, a new industry has emerged in the United States: aquaculture. People have tried aquaculture for hundreds of years in some countries. It has become a major food source in the United States only since 1970.

Raising aquatic animals requires different skills from livestock and poultry, which are terrestrial animals. However, the basic needs of aquatic animals are similar to those of other animals.

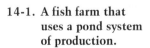 14-1. A fish farm that uses a pond system of production.

OBJECTIVES

This chapter provides important background information on aquaculture. It has the following objectives:

1. Explain aquaculture and list requirements of aquacrops
2. Distinguish between water on the basis of salinity and temperature
3. Describe fish, mollusks, and crustaceans as organisms
4. Explain fish production systems
5. Describe management practices in the fish production cycle
6. List the nutritional and feeding requirements of fish
7. Explain important health and predator management practices
8. Describe harvesting procedures with fish

TERMS

aquacrop	freshwater	production intensity
aquaculture	fry	raceway
Arthropoda phylum	harvesting	sac fry
brackish water	hatchery	saltwater
broodfish	incubation	smolt
cage	mariculture	spawning
Chordata phylum	Mollusca phylum	spawning nest
crustacean	mollusks	spring
dissolved oxygen	operculum	surface runoff
expressing	pen	water facility
fin	plankton	water quality
fish	pond	well water
fingerling	predator	
food fish production	production cycle	

AQUACULTURE

Aquaculture is the production of aquatic plants, animals, and other species, such as algae. Aquatic plants and animals are species that grow naturally in water. Terrestrial species grow on the land. Most aquaculture

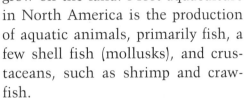

in North America is the production of aquatic animals, primarily fish, a few shell fish (mollusks), and crustaceans, such as shrimp and crawfish.

4-2. Crawfish are raised in many countries, with Louisiana being the major state in the United States.

Important skills are needed in aquaculture. The needs of each species are different. Species must be reproduced and raised to grow-out size. Appropriate water must be provided and managed to keep it in good condition. Feeding fish properly and controlling disease are important. The crop must be harvested and marketed so that the producer has a profit.

AQUACROPS

Aquacrops are commercially-produced aquatic plants, animals, and other species. "Commercial" means that the aquacrop is produced to be marketed for income to the producer. This is different from a small farm pond used for occasional sport fishing where no money is gained.

Aquacrops are produced for several purposes. The most important purpose is for human food. These are known as food aquacrops. Other uses of aquacrops include bait, for the sport fisher; sport fishing or recreation; ornamental fish or pets; and feed, which is using a fish, such as goldfish, as feed for other fish crops.

4-3. Harvested fish ready for processing at an Arkansas plant.

The kinds of aquacrops produced vary with location. Climate, water, consumer demand, and other factors are important. Oysters are grown in protected saltwater areas, such as Wilapa Bay off the coast of Washington. Catfish are raised in "pure" freshwater in warmer areas, such as the delta area along the Mississippi River. Trout in cool freshwater in areas with mountain streams or in the north, such as Pennsylvania and Idaho.

Important Species

Different aquatic animals are raised, with some more important than others. Species chosen are adapted to the available water and market. Some species are in more demand as food for humans. Species with high demand include trout, shrimp, oyster, salmon, and catfish.

Examples of the most important cultured species are listed in Table 14-1.

Table 14-1
Important Cultured Aquatic Species in North America

Species (scientific name)	Climate/Culture	Use
alligator *Alligator mississippiensis*	warm; freshwater	food;leather
bullfrog *Rana catesbiana*	warm; freshwater	food; lab use
catfish (channel) *Ictalurus punctatus*	warm; freshwater	food; sport
Chinese waterchestnut *Eleocharis dulcis*	mild; freshwater	food
coho salmon *Oncorhynchus kisutch*	cool or cold; saltwater/freshwater	food
crawfish (red swamp) *Procambarus clarkii*	warm; freshwater	food
prawn (freshwater) *Macrabrachium rosengbergii*	warm; freshwater	food
minnow (fathead) *Pimephales promelas*	cool or mild; freshwater	bait

(Continued)

Table 14-1 (Continued)

Species (scientific name)	Climate/Culture	Use
oyster (American) *Crassostrea virginica*	cool to warm; saltwater or brackish	food; pearl
red drum (red fish) *Sciaenops ocellatus*	warm; saltwater	food; sport
salmon (chinook) *Oncorhynchus tshawytscha*	cool or cold; saltwater/freshwater	food
shrimp (western white) *Penaeus vannamei*	mild; saltwater	food
hydrid strip bass *Morone chrysops x M. saxatilis*	mild; freshwater	food; sport
tilapia (blue) *Tilapia aurea*	warm; freshwater	food
trout (rainbow) *Oncorhynchus mykiss*	cool or cold; freshwater	food; sport

General Requirements

Aquacrops differ from terrestrial crops. A major difference is the water environment. Farmers who are very good in producing terrestrial crops will find that water crops require new skills.

Six requirements must be met:

1. A suitable species must be selected. (The species should be adapted to the water and the available growing conditions.)

2. Water that is appropriate for the species must be available. (Without the "right" water, production will be difficult or impossible.)

3. The nutritional needs of the aquacrop must be met.

4. Practices must be followed to keep diseases under control.

5. The water must be properly managed for the aquacrop to grow.

6. A market must be available. (Large amounts of aquacrops require mechanical harvesting equipment and automated processing plants. A system must be in place to get products to consumers.)

14-4. Four species of fin fish cultured in North America: channel catfish (top left), hybrid striped bass (top right), rainbow trout (bottom left), and coho salmon (bottom right). (salmon courtesy, American Fisheries Society)

Production Intensity

A pond, stream, lake, or ocean naturally supports a low density of production. Foods for aquatic species grow in most streams and bodies of water. These food sources include tiny plants and animals, known as **plankton,** and larger plants and animals. Aquafarmers want to have a much higher production level than exists naturally. Ponds will naturally support the growth of 100 to 200 pounds of fish a year. This production level is far below the amount needed in fish farming.

Production intensity is the number or weight of aquatic species in a volume of water. Scientists call intensity biomass. The intensity may be low, as in natural water areas, or high, as in commercial fish ponds. Ponds may be stocked at 6,000 to 8,000 or more fish per acre in intensive production. This level of stocking may produce 8,000 pounds of fish per acre in a year. Careful attention is needed in managing the water, feeding, and controlling disease. Tanks and raceways are stocked at higher intensity than ponds.

WATER ENVIRONMENTS

Aquatic animals vary in the kind of water environment that they need. Water is the limiting factor with most aquatic animals. Two features of water are: salt content (salinity) and temperature.

Water Salt Content

Aquaculture is in three different water environments based on salt content: freshwater, brackish water, and saltwater. Species must be carefully selected for the different water environments.

Freshwater. *Freshwater* is water with little or no salt content. Salt is measured as parts per thousand (ppt) or the number of parts of salt in a thousand parts of water and salt solution. Freshwater typically has less than 3.0 ppt salt. The water from most wells, streams, and surface runoff is freshwater.

Species that are adapted to freshwater will not thrive, and may die, when put into water that is not freshwater. Most of the growth in aquaculture in North America has been with freshwater species, such as trout, catfish, and hybrid striped bass.

Brackish Water. *Brackish water* is a mixture of freshwater and saltwater. It is found in areas where freshwater streams run into saltwater oceans and seas. The salt content is higher than freshwater, but lower than saltwater. Only a few species are cultured in brackish water, including mullet and crab.

Saltwater. Much of the surface of the earth is covered with saltwater in oceans, seas, and lakes. Water that has more than 16.5 ppt salt is *saltwater.* The salt content of oceans and seas is typically 33 to 37 ppt.

Many different species grow in saltwater. Salmon, oysters, shrimp, and red fish are a few examples. Some of these are produced in large cages or other devices that confine the animal. Producing aquacrops in saltwater is *mariculture.*

14-5. The streaks in this water were caused by melting snow (freshwater) that has not mixed with the saltwater off the coast of Alaska.

Water Temperature

Water temperature influences the growth of aquatic animals. Some species will live only in water that has a temperature within a suitable range. Usually, species can be classified as warm water, cool water, and cold water. Most cultured species grow in either warm or cool water.

Water temperature is important in the metabolism of fish. When the temperature is too cool or too warm, the fish will stop eating and, consequently, stop growing. The species selected must be appropriate for the water temperature.

14-6. This stream in the Smoky Mountains of Tennessee has cool water and is well known as a good place to catch trout. The trout are stocked in the stream from hatcheries to enhance the natural fish population.

Cool Water. The predominant cool water species cultured in the United States are trout and salmon. These species grow best at a water temperature of 50 to 68°F (10 to 20°C). The rate of growth slows below and above this temperature range. The species will not survive long in water that is warmer than 80°F (26°C).

Warm Water. Catfish, tilapia, crawfish, and hybrid striped bass are the predominant species of cultured warm water fish in the United States. Catfish grow best in water that is 75 to 85°F (24 to 29°C), but will survive in water below or above the temperature range. Tilapia prefer a similar temperature and will not survive in water below 50°F (10°C). Crawfish prefer water that is 65 to 85°F (18 to 29°C). Hybrid striped bass prefer 77 to 88°F (25° to 31°C), but will survive in water as low as 40°F (4°C).

Other Water Factors

Besides salinity and temperature, water must be suitable in other ways. Water quality is affected by pollution, pH, mineral content, gases, and the presence of other living plants and animals.

AQUACROPS AS ORGANISMS

Aquacrops carry out the same life processes as other organisms. Although, the way they do this may vary a great deal. The major emphasis in this chapter is on freshwater fish. Short sections are included on crustaceans and mollusks.

FISH

Fish are vertebrates, or animals with backbones. Fish are classified in the phylum of *Chordata,* meaning that they have a spinal cord. As organisms, they typically have bony skeletons, a skull, teeth (though not always well developed), nasal open-

14-7. **Tilapia are increasingly being cultured in the United States. This tilapia is in an aquarium in North Carolina.**

ings, swim bladder, and gills. A major distinction from other animals is that fish do not have lungs. Fish use gills to get oxygen from the water.

External Anatomy

Most fish have a streamlined body shape. The body is covered with skin or scales. The color of the fish matches their natural surroundings. *Fins* help fish move and keep their balance in the water. Some fins are sharp and can inflict painful wounds in human skin.

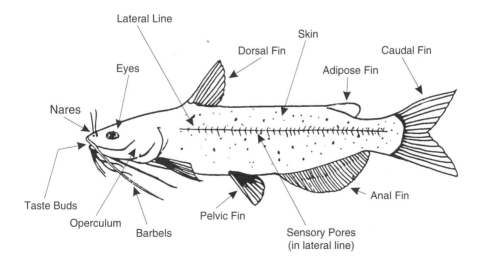

14-8. **Major external parts of a fish.**

The body of a fish is divided into three parts: head, trunk, and tail. The head extends to the back of the *operculum* (gill cover). The trunk begins at the back of the operculum and extends to the anus. The tail extends from the anus to the tip of the caudal fin.

Internal Anatomy and Systems

Fish have internal anatomy and systems similar to most other vertebrates. The major systems are:

Nervous system—Consisting of the brain and spinal cord and branching spinal nerves, the nervous system sends impulses or commands throughout the body.

Circulatory system—The circulatory system consists of the heart and blood vessels. Blood contains red and white cells, similar to other animals.

Sensory system—The sensory system collects information from the fish's environment. Skin and eyes are important to fish. Some fish also have barbels, or feelers, about the mouth to help in sensing danger, food, and movement.

Skeletal system—The skeletal system consists of bone and cartilage. The skeleton gives the body shape and protects the internal organs.

Muscular system—The muscular system is the largest system in the body of fish. Muscles make it possible for fish to swim and move in other ways. The large muscles along the trunk and tail are highly desired as food.

Respiratory system—Fish have gills that remove dissolved oxygen from the water and release carbon dioxide into the water. The blood flows through the gills on its way to other parts of the body.

Digestive system—The digestive system prepares food for use by the body. Fish take in food through their mouths. Digestion is in the stomach and intestines. Wastes pass from the body through the anus.

Reproductive system—Most fish spawn. *Spawning* (fish reproduction) is the process under which new fish are created. The female fish lays eggs that are fertilized by sperm from the male. After a few days of incubation, the eggs hatch as tiny young fry. A few species give birth to live young.

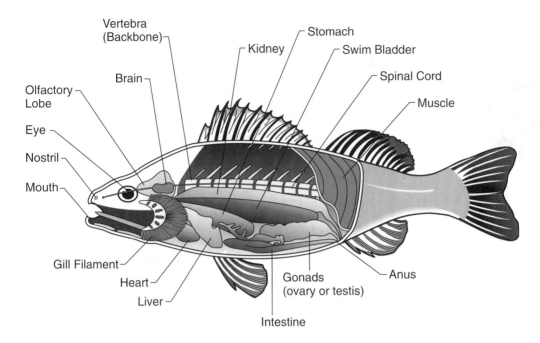

14-9. Major internal parts of a fish.

Swim bladder—The swim bladder is a small balloon-like structure that helps a fish with buoyancy. Buoyancy allows a fish to stay at a particular depth in the water. The deeper a fish swims, the smaller its air bladder becomes.

CRUSTACEANS

Crustaceans are in the *Arthropoda phylum.* Arthropods have exoskeletons, segmented bodies, and jointed legs. The major crustaceans used for human food include shrimp, prawns, crawfish, crabs, and lobsters. Shrimp and crawfish are more common as aquacrops. New methods have been developed for prawn culture. Freshwater prawns are now being cultured. Lobsters are primarily harvested from the wild. These crustaceans are decapod crustaceans, or decapods, because they have five pairs of limbs.

Crustaceans have exoskeletons that cover their bodies. The material is much like cartilage or the human fingernail. Crustaceans are in the same scientific order as insects. They have many other body features similar to insects, including three major body parts: head, carapace (similar to the thorax in insects), and abdomen. Decapods have five pairs of limbs, some of which are used for walking. A few decapods form one pair of

14-10. Freshwater prawns are now raised in some
       locations.

14-11. Basket of harvested cultured prawns.

strong pincers, such as the prawn and
crawfish. Crustaceans have body systems similar to fish, but the systems
are not as complex.

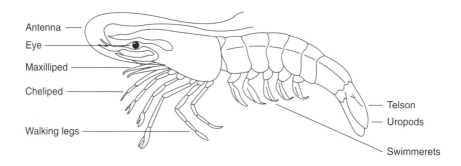

14-12. Major external parts of a shrimp.

MOLLUSKS

Mollusks have thick, hard shells. They are in the *Mollusca phylum*.
Some shells are hinged and easily opened for the organism to capture food.
Mollusks with hinged shells are bivalve mollusks. Oysters, mussels, and
clams are examples. The major muscle in a bivalve is the abductor muscle,
which opens and closes the shell.

Mollusks have simple organs and systems. They have gills for respiration. A heart forces blood over the gills and throughout the body. Most
mollusks are filter feeders, meaning that they remove food particles from
the water they live in and use the particles for food. Quality water is
essential; mollusks will eat pollution if it is in the water!

14-13. The oyster is a bivalve mollusk.

14-14. The mussel has a smooth bivalved shell.

The most valued aquacrop from mollusks is the pearl. Formed naturally by oysters, pearls are cultured by implanting a tiny grain of sand or other irritant in the oyster.

FISH PRODUCTION SYSTEMS: FACILITIES AND EQUIPMENT

Aquacrop production **systems** vary with the kind of water facility used. *Water facilities* are the structures in which aquacrops are grown. Several kinds of water facilities are used: ponds, raceways, tanks, vats, aquaria, and pens and cages. Pens and cages are primarily used in natural bodies of water to contain the fish crop.

PONDS

Ponds are water impoundments made by building earthen dams or levees. Ponds may range from less than an acre to over 50 acres in size. Many producers prefer ponds that are about 20 acres in size. The water in ponds does not flow and must be managed to keep fish growing.

14-15. Small ponds used in freshwater prawn production in Mississippi.

Ponds should be designed for easy management and harvesting. Trees, stumps, and other obstacles are removed from ponds. The bottoms of ponds must be smooth, which makes harvesting easy. Levees should be wide enough for vehicles to travel on them and free of trees and utility poles. Ponds are typically rectangular and have water that is 3 to 5 feet (1 to 1.7 m) deep.

RACEWAYS

A *raceway* is a water impoundment that uses flowing water. They are typically long and narrow. The water may be 3 to 4 feet (1 meter) deep. More fish can be stocked in flowing water than in still water. Flowing water has more oxygen and carries waste away. Raceways vary considerably, with some built of earth, and others, of metal or concrete. Using pumps to keep the water flowing is expensive because of the cost of electricity to run the pump motors. Raceways also require more water and a method for disposing of used water.

14-16. Raceways used to raise trout in Virginia.

14-17. Tanks used to raise tilapia in Washington.

TANKS, VATS, AND AQUARIA

Tanks are made of many different materials, such as fiberglass, concrete, and metal. Large round tanks are often used for intensive fish culture.

4-18. Large tanks used to raise fresh-water prawns in a specially designed greenhouse-type facility in Mississippi.

Vats, often made of concrete, are similar to tanks and often known as tanks. Aquaria are smaller tanks made primarily of glass.

The water in tanks may flow, much like a raceway, or it may be managed, as a pond. Intense production requires flowing water that has plenty of oxygen.

PENS AND CAGES

Pens and cages confine aquacrops in small areas in large bodies of water. **Pens** are made so that the bottom of the pen is joined with the earth in the pond, lake, or stream. Posts that support the pen may be imbedded in the

14-19. Round net pen used for raising salmon in a freshwater stream at Skagway, Alaska.

14-20. Square net pens used for raising salmon in Wilapa Bay, Washington.

earth. Pens are often used with broodfish where the nesting container rests on the mud at the bottom of the water. Water where pens are located is often only 2 to 3 feet (0.6 to 0.9 m) deep.

Cages float on the surface of the water and do not touch the earth at the bottom. They have frames covered with net-type material, such as wire or plastic. Cages can be in water that is deep and open, such as a large lake, in which many other species grow. Cages help confine the aquacrop so that it can be managed and harvested.

MANAGEMENT IN FOOD FISH PRODUCTION

Aquafarmers must manage the environment for fish. This is more complex because the water must be kept in good condition.

Water is the environment in which aquacrops grow. In the right water, they will grow well. The producer must keep the water so that the aquacrop grows efficiently.

PRODUCTION CYCLE

The production of aquacrops follows a cycle. A *production cycle* is the complete production of a crop. It begins with seedstock and ends with the crop reaching market size. In nature, these cycles often follow the seasons of the year. Reproduction may be in the spring, with the young growing all summer. A couple of years, or more, may be needed for the young to reach adequate size.

The production cycle described here is primarily for freshwater fish. It has four phases: broodfish and spawning, hatching and raising fry, producing fingerlings, and growing fish to food size.

Broodfish and Spawning

Young fish are used to stock food fish growing facilities. *Broodfish* are the sexually mature fish kept for reproduction purposes. Female fish produce eggs and the males produce sperm that fertilize the eggs. The releasing of eggs by the female and subsequent fertilization by the male is spawning.

Broodfish grown in warm water, such as catfish, are often kept in separate ponds from young, growing fish. Sometimes, broodfish will eat young fish if they are together. Broodfish are usually larger than food fish and three or more years old. At spawning time, pairs of broodfish may be isolated in pens or left in small ponds that contain spawning nests. A **spawning nest** is a container large enough for the pair of fish and provides some protection for the spawn (egg mass).

The size of eggs varies with the species of fish. Some eggs, such as the catfish, may be the size of a mustard seed or small pea. The red drum eggs are about the size of the head of a straight pen. Other species, particularly the smaller bait, fish may have tiny eggs.

Female fish may produce thousands of eggs at one spawning. For example, a female channel catfish may produce 2,000 or more eggs per pound (0.45 kg) of body weight. This means that a fish weighing 5 pounds (2.27 kg) might produce 10,000 or more eggs at one spawning.

Spawning varies among the species. Catfish spawn when the weather begins to warm in the spring. Trout spawn a little later in the spring or in the early summer. Some species spawn only once and die, such as the coho salmon, while others can be used in repeated years, such as the tilapia.

How fish respond to spawning also varies. Some female fish, such as the catfish, leave the eggs in a nest where the male tends them for natural hatching in 5 to 7 days. Other fish, such as some tilapia, incubate the eggs in their mouths, and this is the reason they are called mouth brooders. Producers of mouth-brooding fish check the mouths of females for eggs and remove them for artificial hatching.

14-21. A spawn is the egg mass produced by one female fish. This shows a fertilized catfish spawn from one female being collected from spawning nests located in the water. The spawn will be taken to a hatchery. (Courtesy, Progressive Farmer Magazine, Birmingham, Alabama)

14-22. Male coho salmon at a Washington hatchery.

Spawning can be encouraged by regulating the environment of fish, such as the water temperature, and by injecting the fish with a hormone. Of course, these are often practical only in laboratory situations.

In some cases, the eggs are artificially removed from the female fish. With fish that die after spawning, the abdomen may be cut open and the eggs removed. With some species, particularly trout, the abdomen of the female is carefully squeezed to force the eggs out into a container. This procedure is known as *expressing.* Females that have eggs forced from them by gentle pressure on their abdomens have been expressed. Sperm, also known as milt, may be expressed similarly. The milt is placed on and mixed with the eggs in a container so that fertilization can occur.

Hatcheries and Fry Rearing

A *hatchery* is a place where eggs are artificially incubated and hatched. *Incubation* is the time between spawning and when the young fish are hatched from the eggs. A favorable environment must be maintained for the eggs during incubation. The water must be within an appropriate temperature range and the water around the eggs must typically be moving.

14-23. Salmon hatchery. (Salmon eggs require about 60 days to hatch.)

Hatcheries must have the proper equipment. Eggs are incubated in jars, trays, or troughs, depending on the species. The conditions provided must be ideal for maximum hatch and survival of the young fish. Red drum eggs hatch in about 30 hours at a water temperature of 72°F (22°C). Channel catfish eggs hatch in 5 to 9 days at water temperatures of 70–85°F (21–29°C).

14-24. Incubator used with trout eggs at a Virginia hatchery.

4-25. Basket of coho salmon eggs in a hatching trough. The light-colored eggs are not developing embryos and need to be removed from the basket by a process known as egg-picking.

Young fish are *fry* or larvae when they hatch. Fry are very small and delicate. Conditions must be provided so fry survive and grow free of disease. At first, a fry may have the yolk sac from the egg still attached for a few hours or a day. These are *sac fry*. (Newly hatched salmon are known as alevin.)

Feeding fry is essential. Since they are very small, they cannot eat the commercial feeds made for larger fish. Some feed manufacturers produce a finely-ground meal that is suitable for certain fry. Fry may be fed very finely ground organ meat (heart or kidney), yeast, or natural foods. Some aquafarmers grow plankton for the fry. In a few cases, the yolks of chicken eggs may be made into a slurry and used as fry feed. Once the fry have grown sufficiently, they are moved to other growing facilities. With food fish, fry are about 1 inch (2.54 cm) long when moved to a fingerling area.

Fingerling Production

A *fingerling* is an immature fish. They are between the size of a fry and the size of a fish for stocking in a food fish growing facility. Salmon

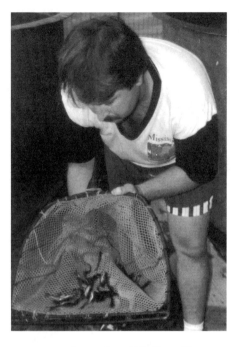

14-26. Channel catfish fingerlings.

this size may be known as **smolts.** A smolt has other characteristics, such as moving from the freshwater where it hatched to the saltwater where it will grow to maturity. A smolt's color changes from dark to light as it grows.

Fingerlings may be grown in tanks, small ponds, or other facilities. The rate of stocking is quite high because they will not grow to maturity in the facility. Fingerlings must be fed a nutritionally-complete feed. Since fingerlings are young and rapidly growing, their feed must have a higher percentage of protein (about 36 percent) than feed for older fish. Commercial feed manufacturers often have a specially-formulated fingerling feed. Feed particle size is small so that the young fish can eat the feed. Disease control is essential in order for the fingerlings to survive.

With the channel catfish, a fingerling ranges from 1 to 8 or 10 inches (2.54–20 or 25 cm) in length. One growing season is needed for fingerlings to reach adequate size. Fingerlings should be at least 5 inches (12.7 cm) long when stocked in growing ponds.

Food Fish Production

Food fish production is raising fingerlings into fish that are large enough for harvesting. The harvest-size fish are food fish.

Fingerlings are stocked in growing ponds, raceways, or tanks. Appropriate feed is needed at the rate of about 3 percent of the weight of the fish each day. The feed may have a protein content of 28 to 32 percent. Of course, the intensity of production has a big impact on the rate of feeding and other management details.

Fish may reach harvest size in six months or more depending on the climate and other conditions. Fish are considered food size when they are three-fourths pound or larger. Most processing plants are designed for fish within a certain size range. The fish that are too large may not process well with the equipment in the processing plant. With catfish, the processing plants most efficiently handle fish that are 1 to 2 pounds in size.

All areas that affect the growth of a fish must be considered. This includes feeding, water quality, and disease and predator control.

WATER MANAGEMENT

Water must be well managed for the aquacrop to survive and grow. This involves several areas, beginning with the source of the water and includes maintaining oxygen and disposing of used water.

Sources of Water

14-27. Food-size channel catfish.

Water from several different sources may be used in aquaculture. The source of water depends on the local climate and requirements of the species of fish to be grown.

Surface runoff is excess water from precipitation. It may collect behind dams or in streams, lakes, or oceans. Runoff may be okay if it is free of pollution and dependable in supply. Water must be available throughout the growing season for an aquacrop. Runoff tends to be seasonal and available only during certain times of the year. Legal regulations may restrict pumping water from a stream.

Well water is water pumped from aquifers deep in the earth. The water is often of high quality, but is expensive to get. Deep wells, pumps, and electricity to operate the pumps are costly. Occasionally, the water may be too cool for pumping directly into fish facilities. It may need to be pumped and stored in open ponds for a few days so that the sun can warm it. Permits are needed to drill water wells for aquaculture.

Springs are natural openings in the earth that provide water similar to well water. The water may be of good quality for aquacrops. Some springs dry up at certain times of the year and are not dependable sources of water for aquaculture.

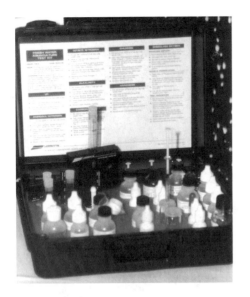

14-28. Water quality test kit.

Other sources of water include industrial waste water and municipal water systems. Waste water must be free of dangerous substances and of the right temperature. Municipal water systems may treat the water with various chemicals that are harmful to fish. This source of water is too expensive for large-scale aquaculture but, with proper care, can be used for small aquaria.

Water Quality

Water quality is the suitability of water for a particular use. In aquaculture, water must have sufficient dissolved oxygen and be free of pollutants. In addition, acidity, alkalinity, hardness, nitrogen compounds, and carbon dioxide are important. The major emphasis in this chapter is on dissolved oxygen.

Dissolved oxygen (DO) is oxygen in the water that is available for living organisms to use. It is in a free, gaseous form and not the oxygen in a molecule of water. DO is measured as parts per million (ppm) in water. For example, 5.0 ppm DO in water means that there are 5.0 parts DO in 999,995.0 parts water, or a total of 1 million combined parts.

Cooler water holds more DO than warmer water. Fish also are more active in warm water and produce more wastes. Aquafarmers must carefully monitor water in warm weather. DO is highest about noon on a bright,

14-29. Paddle wheel aerator being used to increase DO in water.

sunny day and lowest just before sunrise. This is because tiny plants in the water, known as phytoplankton carry on photosynthesis and produce oxygen. At night, when there is no sun, the phytoplankton does not produce much oxygen.

Most aquacrops need 5.0 ppm DO or above to grow, though some will survive below that level. DO is frequently measured with an oxygen meter. If DO is low, the water is oxygenated by splashing it into the air, injecting pure oxygen into it, or in other ways. Fish can die quickly if the DO level drops too low.

NUTRITION AND FEEDING

14-30. Dissolved oxygen meter.

Aquacrops must have the appropriate nutrition to grow. Nutrient deficiencies may result in disease outbreaks in the aquacrop. Most aquacrops have nutrient needs similar to other animals.

FISH NUTRITION

Fish need a ration (daily feed amount) that meets nutritional needs. The nutrient requirements of fish vary, but are very similar. Fish need the following:

- Protein—Proteins are needed for growth. Research has provided considerable information about the protein needs of fish. Most commercially-produced feeds have the needed protein. The protein must partially come from an animal source, such as fish, shrimp, beef, or chicken scraps, that have been properly prepared in the feed. Feed usually contains about 28 percent protein for growing fish.

- Carbohydrates—Carbohydrates provide energy. Most fish feed contains grains, such as corn and soybeans, which are good sources of carbohydrates. The amount of carbohydrate in the diet of a fish is about 10 percent.

- Fats—In fish rations, fats usually come from animal wastes, such as by-products from a processing plant. The fat content of feed is 4 to 15 percent.

- Minerals—Fish need about the same minerals as other animals. These include calcium, iron, silicon, manganese, sodium, and seven others.

- Vitamins—Fish need vitamins or they will develop severe deficiency symptoms. Eleven vitamins are essential for fish.

FEEDS AND FEEDING

The feed that aquacrops need depends on the intensity of production. If only a few fish are in an acre of pond, adequate natural food may be available. This is not true with aquaculture, however. High stocking rates require appropriate feed.

Natural Food

Plankton, plants, and animals naturally grow in ponds, streams, and other water impoundments. These serve as food for fish and other aquacrops. Depending on the stocking rate, these can provide some of the needed food. The use of natural foods also depends on the species of aquacrop.

Crawfish eat vegetation in the water. Most crawfish growers plant rice in the crawfish ponds. The rice (stem, leaves, and grain) serves as food for the crawfish.

Manufactured Food

Several commercial feed mills produce products for feeding aquacrops. These products are often specially designed for the intended species.

Feeds are manufactured in several forms, as follows:

14-31. Bag with label providing feed analysis on a manufactured minnow feed.

- Meal—A meal is a finely ground feed used with small fish.

- Pellets—A feed product made by binding meal into larger particles. Pellets range in size for different sizes of fish. Pellets are made to float, sink, or remain relatively neutral in the water. Many food-fish farmers prefer floating pellets because the fish have to come to the surface of the water to get food. This helps the farmer observe the aquacrop that otherwise may be unseen.

14-32. Floating pellets (on left) and meal fish feed.

- Other forms—Crumbles, blocks, and other forms may be used with fish feed. These forms are for specific species of aquacrop.

Frequency and Rate of Feeding

Frequency is how often fish are fed. It ranges from feeding every hour to once or twice daily to weekly. How often fish are fed depends on the size of the fish, climate, and rate of growth. Fry are fed several times during the day, often each hour. Fingerlings may be fed less often. Food fish are fed once or twice a day. Broodfish may be fed once a day or two or three times a week. In the winter, broodfish are fed less often.

The rate of feeding depends on the size of the aquacrop and water temperature. Small fish are fed at a higher rate than larger fish proportionate to body weight. A general rule of thumb is to feed fish 3 percent of their body weight each day. This depends on the water temperature and other factors. Most fish eat more in water as it warms. The rate of feeding is regulated to keep good water quality. Over feeding fouls the water. Uneaten food particles float and settle to the bottom.

14-33. Tractor-pulled mechanical feeder that blows feed out into the water.

HEALTH AND DISEASE AND PREDATOR CONTROL

Aquacrops must be kept in good health and diseases controlled. Diseases and predators can rapidly wipe out a fish population.

HEALTH

Health is the absence of disease. Aquacrops that receive adequate feed and where the water quality is maintained are less likely to have health problems.

Prevention of disease is the best control. Fish are difficult to treat once an outbreak of disease occurs.

Kinds of Disease

Diseases are infectious and noninfectious. Infectious diseases are caused by germs or pathogens. Bacteria, fungi, viral, and parasitic diseases may occur. Parasites include worms, flukes, and protozoa. These diseases attack the tissues of aquacrops and cause loss of growth and, sometimes, death.

Noninfectious diseases are caused by organs that do not work properly, the environment in which an aquacrop lives, or the food it gets. Sometimes the organs may not operate properly. This may be related to water quality, such as a sudden change in pH may cause blood problems. Nutritional disease results when the aquacrop has an inadequate diet. Environmental diseases result when the water contains gases, chemical residues, or other materials that are damaging to fish.

Disease Control

Keeping fish healthy is the best way to control disease. When fish for stocking are bought, only disease-free fish should be brought on a farm or put in a tank. One fish with a disease can infect an entire population!

A few chemical treatments are available for fish. Some involve putting medications in the feed of the fish. Other treatments involve dipping fish into a chemical solution or bathing the fish in a solution. Valuable broodfish can be injected with medications. The U.S. Food and Drug Administration

(FDA) regulates the medications that can be used on fish that will be used for human food in the United States.

PREDATORS

A *predator* is an animal that hunts and kills other animals. Large birds can prey on aquacrops in ponds or other growing facilities. For example, the cormorant (a water bird) can cause large losses in a fish production facility by eating a pound or so of fish a day. Other predators include alligators, turtles, frogs, other fish, snakes, and insects.

With ponds, keeping the grass mowed around the edge of the water will discourage some predators. Air cannons are sometimes used to frighten protected species away. Small ponds and tanks can be covered with netting to keep predators away. Constructing ponds with a minimum area of shallow water where birds can stand will help keep some species away.

HARVESTING

Harvesting is important in the economic success of aquaculture. *Harvesting* involves capturing the aquacrop so that it can be hauled and sold to a processing plant or other facility.

How an aquacrop is harvested depends on the crop and the water facility. Large seines may be put around ponds and the fish confined to a small area in the pond. This is accomplished by drawing the seine through the water. After seining, the fish are dipped and loaded into a haul truck. Frequently, the basket used to dip the fish has a scale for weighing the fish as they are being loaded.

14-34. Seine being placed around a pond for harvesting.

14-35. Careful records are kept as seined fish are dipped from the water.

Fish in tanks or raceways may be harvested similarly to those in ponds. A net or grading device may be put across the facility and moved toward one end where the fish are confined. The confined fish are dipped from the water.

Sometimes, the water level is lowered so the aquacrop is easier to confine. This creates the need for additional water to again fill the facility. Water is expensive!

Harvested aquacrops must be kept alive until they reach the processing plant. This requires hauling in a tank of water with oxygenation equipment.

14-36. Cultured oysters being shucked at processing plant in South Bend, Washington. (Shucking is opening the tight shell and removing the oyster in one piece.)

REVIEWING

MAIN IDEAS

Aquaculture is the production of aquatic organisms. Plants, animals, and other species, such as algae (kelp), are produced. The major emphasis in the United States is on aquatic animal production. The most widely produced aquacrops are: Fish (known as fin fish in the phylum Chordata), crustaceans (in the phylum Arthropoda, including shrimp, prawns, and crawfish), and mollusks (in the phylum Mollusca, including clams, oysters, and mussels). Freshwater fish farming has grown most rapidly in recent years, particularly catfish, trout, tilapia, and hybrid striped bass.

Water environments are very important in aquaculture. Freshwater, saltwater, and brackish water are used, as appropriate to the species. Water is obtained from runoff, wells, and waste from industry. Water quality is the suitability of water for a particular aquacrop. A critical factor in any aquaculture is dissolved oxygen (DO). Most aquacrops prefer a minimum DO level of 5.0 ppm. The common water facilities (impoundments) include ponds, raceways, tanks, vats, and aquaria.

Production cycles of aquacrops include broodfish and spawning, hatcheries and fry rearing, fingerling production, and food fish production. Nutrition and feeding are important at all stages. Controlling diseases and predators is essential. Harvesting is a part of the final production cycle with food fish.

QUESTIONS

Answer the following questions using correct spelling and complete sentences.

1. What species of aquacrop are commonly produced in the United States? List any five and give one item of descriptive information about the requirements of the aquacrop.
2. What are the general requirements for aquaculture?
3. What is production intensity? Why is it important in the production of aquacrops?
4. Distinguish between the kinds of water based on salinity and temperature.
5. Draw a fish and label its external anatomy.
6. What kinds of water facilities are used? Describe each.
7. Name and briefly describe the four phases in the production cycle of aquacrops.
8. What are the major sources of water?
9. What is water quality?
10. What is dissolved oxygen? Why is it important?
11. What are the major areas of nutrition and feeding of aquacrops?
12. What are the major areas of health management with fish?
13. How do predators cause losses?
14. What is harvesting? How is it done?

EVALUATING

CHAPTER SELF-CHECK

Match the terms with the correct definitions. Write the letter by the term in the blank that is provided.

a. predator
b. aquaculture
c. aquacrop
d. pond

e. production intensity
f. freshwater
g. brackish water
h. saltwater

i. fish
j. crustaceans

1. _____ Arthropoda phylum

2. _____ contains little or no salt

3. _____ mixture of freshwater and saltwater

4. _____ water with more than 16.5 ppt salt

5. _____ production of aquatic plants, animals, and other species

6. _____ the number or concentration of aquatic species in a volume of water

7. _____ commercially-produced aquatic plants, animals, and other species

8. _____ Chordata phylum

9. _____ water facility made with an earthen dam or levee

10. _____ animals that hunt and kill other animals

EXPLORING

1. Set up an aquarium or fish tank in the classroom. Select a species that will survive in the environment. Goldfish are very hardy. Develop a plan for maintaining the water facility and caring for the fish. (Reference: *Aquaculture: An Introduction* available from Interstate Publishers, Inc.)

2. Tour an aquafarm or hatchery in the local area. Determine the species produced and the management needed for successful production. Also, learn how the aquacrop is marketed.

3. Invite an aquaculturist or fisheries specialist to serve as a resource person in class and discuss the locally-adapted species and how to culture them.

Chapter 15

HORSES

Most horses are owned and cared for because of their recreational and personal value. This can be rewarding and educational. In the past, horses have been useful in many ways.

Horses have served various purposes over the years. They have been used as a source of food. They have traditionally been used for military purposes. The racing horses of today serve as the largest spectator sport in the world. Horses are used in agriculture and commercial enterprises and for recreation and sport.

In the 1800s and early 1900s, horses were used for power and as a major mode of transportation. The United States had 20 million horses in 1900. This dropped to 9 million in 1977 and to less than 6 million today. In 1908, Henry Ford produced cars so cheaply with the assembly line that horses were replaced for transportation. The mechanization of agriculture, with the advent of steam engines and tractors, had a huge impact on the decline of horses.

Horses are a major industry in the United States. On the average, owners spend over $1,000 a year on each horse. Today, most horses are for pleasure. A few are used in working cattle and in providing power for farming. Horse owners get much pleasure from their animals.

15-1. Horses give their owners a lot of fun. This young woman enjoys her Quarter Horse. (Courtesy, DeShannon Davis, Meridian, Mississippi)

OBJECTIVES

This chapter will provide a broad insight into the care and management of light horses. The following objectives are included.

1. Describe horses as organisms (classes and important breeds)
2. Explain breeding practices with horses
3. List the nutritional requirements of horses
4. Explain important health management practices with horses
5. Describe facility and equipment requirements of horses
6. Describe types of equitation

TERMS

draft horse	gelding	polo mount
driving horse	hand	pony
equitation	horsemanship	racehorse
farrier	hunting and jumping horse	riding horse
filly	jog	stallion
floating	light horse	stock horse
foal	lope	stud horse
foaling	mare	walk
gait	plug	

HORSES AS ORGANISMS

Horses are similar to other animals in many ways. Yet, they have differences that make them unique. The scientific name of horses is *Equus caballus*. Horses are sometimes called equine, which is based on the scientific name. Riding and managing horses is **equitation.**

AGE AND SEX CLASSES OF HORSES

A mature female horse over four years of age is a **mare,** while a mature male horse over four years of age is a **stallion.** In Thoroughbreds, both classes begin at five years of age. A mare that has never been bred is a maiden horse. A sire is the male parent. A female parent is a dam.

The gestation period in horses is longer than other livestock at 336 days. Giving birth (parturition) is **foaling.** A young horse of either sex that has not been weaned is a **foal.** A **gelding** is a male horse castrated before reaching sexual maturity. A **filly** is a female horse under three years of age (except Thoroughbreds, which are under four years of age). A **stud horse** is a male kept specifically for breeding purposes. A stag is a male castrated after reaching sexual maturity.

Horses are known as intelligent animals. They can be taught and directed in many ways. Some horses fail to meet the notion of a beautiful, intelligent animal. A horse that has an ugly head or behaves oddly is known as a jughead. A horse with a large, coarse head is called a hammerhead. A **plug** is a horse with poor conformation and common breeding. Older horses that show signs of aging may also be known as plugs.

SIZE CLASSES OF HORSES

Horses are classified by size as either light horses, ponies, or draft horses. Draft horses are suited to drawing loads and performing heavy work. This chapter includes light horses. Draft horses are covered in Chapter 16.

Horse classification is based on height. Height is the distance from the highest point of the withers to the ground. Horses are measured by hands. One **hand** is equal to 4 inches (10.2 cm). If a horse is 15.2 hands, then it is 61 inches (155 cm) tall.

A *light horse* stands 14.2 to 17 hands high and weighs 900 to 1,400 pounds (408–635 kg). They are used mostly for riding, driving, and racing. A *pony* stands under 14.2 hands high and weighs 500 to 900 pounds (227–408 kg). A *draft horse* stands 14.2 to 17.2 hands high and weighs over 1,400 pounds (635 kg). Donkeys and mules are again becoming popular. Mules are being used for trail rides and working with cattle.

TYPES AND USES OF LIGHT HORSES

Specific breeds of light horses have been produced for many specialized purposes.

15-2. Four breeds of horses are: Quarter Horse (top left), Paint (top right), Appaloosa (bottom left), and American Saddlebred. (Paint courtesy, BarLink Paint Horses, Madras, Oregon)

Riding Horses

A *riding horse,* as the name implies, is a horse ridden for pleasure or work. These horses have a definite utility value. Types of riding horses include gaited horses, stock horses, polo mounts, hunters and jumpers, and ponies. The most popular riding horses are the Quarter Horse, Arabian, Appaloosa, Morgan, Thoroughbred, American Saddle Horse, and Tennessee Walking Horse.

Gaited Horses

Gait is the way a horse moves—walks or runs. It is associated with the rhythmic movement of the feet and legs. The run or gallop is a fast gait. The two hind feet leave the ground at different times followed by the two front feet. The walk is a gait in which each foot leaves and touches the ground at different intervals.

Three-gaited horses are known for their gaits of walking, trotting, and cantering. They are used chiefly for pleasure riding. After the development of New England towns and the opening of roads, these horses greatly increased in popularity. Gaited horses lack the style and action of a stock horse, yet they have their own unique style.

Other gaited horses are five-gaited saddle horses. The five-gaited horse must also perform a slow gait and the rack or single-foot. The slow gait is either the running-walk, fox trot, or slow pace. These horses were developed on the southern plantations by people who spent long hours on horseback. They chose animals that had an easy, springy step. Gaited horses are ideal for the amateur rider because of their easy disposition and easy gaits.

Stock Horses

Stock horses are usually of mixed breeding and descend from the Mustang. They are the most popular type of horse in the United States. They are short coupled, well muscled, and deep bodied. The stock horse needs to be hardy, agile, sure footed, fast, short coupled, deep, powerfully muscled, durable, and must have good feet and legs. Stock horses must also possess "cow sense."

15-3. Horses are useful in managing cattle.

Polo Mounts

The **polo mount** was developed for the game of polo. Polo is played by four people on horseback who try to drive a wooden ball between a goalpost. A polo mount is of smaller size than hunters and Thoroughbred types. They must be quick and clever in turning. They must be able to dodge, swerve, and turn while running. This sport is very expensive because up to six years is required to train a polo mount and a person may use four to six mounts during a single game.

Hunters and Jumpers

Hunting and *jumping horses* are used in fox hunting. They are large, clean-cut horses that perform well in cross-country riding and jumping. Again, Thoroughbred blood is predominantly found in these breeds. They must be of good size and height to jump tall fences and wide ditches. They also require stamina and good conformation to keep up with the pack as the hounds track the fox.

15-4. Horses may be trained to jump.

Ponies

Ponies are typically raised for children. They are not only unique in their small size, but

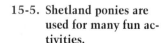

15-5. Shetland ponies are used for many fun activities.

must also possess many other characteristics. They must be gentle, sound in feet and legs, symmetrical, have good eyes, and possess endurance, intelligence, patience, faithfulness, and hardiness. The most popular breeds of ponies are the Welsh, Shetland, and Pony of the Americas.

Racehorses

Racehorses are either running racehorses or harness racehorses. Running racehorses are almost exclusively Thoroughbreds or Quarter Horses. They are used for other purposes, such as for polo mounts, hunters, and calvary horses. Racing has taken place between horses from the earliest of recorded history.

Running racehorses must be extremely refined, have oblique shoulders, well-made withers, heavily muscled rear quarters, and straight hind legs. Harness racehorses need to be fast and light. The two main breeds of harness racehorses are the Morgan and the Standardbred.

Driving Horses

Driving horses are of little utility value and are mostly seen in the show ring.

SELECTION OF A HORSE

Selection of a horse depends upon many factors. Several major factors are: intended use, price range, disposition, size, gait, breed, and color.

The most important factor is the price range. Do not spend more than can be afforded. Disposition is important for everyone, especially for an amateur or child. They need a horse or pony that is quiet, gentle, and well broken. The animal needs to be controlled by the rider. Each breed has its own good points, but those are of little concern if the rider is unable to control the horse.

The size of the horse should fit the size and weight of the rider. A person that is tall will overpower a small horse. A small person will look tiny on a tall horse. An amateur will look better on a three-gaited horse, rather than on a horse doing the more complicated gaits. The breed and color of the horse are of personal preference. Lastly, the horse should be suited to the type of work that it will do.

15-6. A horse used in breakaway calf roping in a rodeo must have the capacity to perform. (Courtesy, American Paint Horse Association)

Overall, the conformation and performance of each individual horse must be evaluated. When selecting a horse, bring an experienced horseperson along for their opinion. When selecting for breeding purposes, check the horse's progeny and pedigree. Other factors to consider include style, beauty, balance, symmetry, energy level, good wind, suitable age, and freedom from disease.

COMMON BREEDING PRACTICES

Conception rates among mares are the lowest of any farm animal. Domesticated horses average a conception rate of about 50 percent. This is due primarily to lagging research and arbitrarily limiting the breeding season to late winter and early spring. Some general recommendations for breeding are discussed here.

NORMAL BREEDING HABITS OF MARES

Estrus

Mares will begin coming into heat at 12 to 15 months of age. The general recommendation is to breed a mare at three years, so that she will foal at four years of age. This way, she will not be training while heavy in foal and she will be more mature, taller, and fuller grown. Mares will continue to foal up to 14 to 15 years of age.

Heat periods are around 21-day intervals and last four to six days. Signs of estrus include relaxation of the external genitals, increased frequency in urination, teasing of other mares, desire for company, and a slight mucous discharge from the vagina.

The average gestation period of a mare is 336 days or slightly more than 11 months. Pregnant mares should be kept separate from others because they are usually more sedate. The pregnant mare should receive plenty of exercise and is best kept in an open pasture.

Parturition

The most obvious sign of nearing parturition is a distended udder. This may happen from two to six weeks before foaling. Foaling is the act of a horse giving birth. There will be a falling away of the muscular parts of the top of the buttocks, near the tailhead, and in the abdominal area about 7 to 10 days beforehand. The vulva will become loose and full. As foaling comes even closer, milk will fill the ends of the teats, she will become very restless, and may break into a sweat and urinate frequently. These signs are not one hundred percent fool proof. Always watch the mare and note changes in her behavior.

When the weather is warm, foaling may take place in a clean, open pasture away from other animals. There is less chance of injury and infection. During poor weather, the mare should be kept in a roomy, well-lighted, ventilated, quiet stall.

Just before parturition, the mare will exhibit extreme nervousness and uneasiness, biting of her sides and flanks, switching of the tail, sweating, and frequent urination. When the outer fetal membrane breaks and a large amount of fluid flows out, the water bag has broken. Foaling should take no more than 15 to 30 minutes.

The front feet with heels pointing down should be first to extend from the mare. The nose should follow as it rests on the forelegs. Then the shoulders and the rest of the body should come. If the foal is presented in any other manner, a veterinarian should be called at once.

15-7. A young colt.

Care After Parturition

Once the foal is out, check to see that it is breathing and that no membranes are covering the mouth or nostrils. Dip the navel cord in iodine to prevent infection. The mother should be lightly fed and given small amounts of lukewarm water at frequent intervals. Frequently clean the stall and observe both mother and offspring for signs of illness.

Feeding colostrum to the foal immediately after birth is very important. Colostrum contains antibodies that protect the foal against certain infections. The healthy foal should be ready to nurse within 30 minutes to two hours, but they may need assistance and coaxing.

CARE OF THE STALLION

Any stallion bought or used for breeding purposes must be a guaranteed breeder. The number of healthy foals that a stallion has sired is more important than the number of times that the stallion has serviced mares. Some farms will periodically check the semen of stallions to measure the fertility of the animal. Semen must be of good volume, have a high sperm count, the sperm must move about, and they must be morphologically correct.

Because stallions are more nervous, they need to be kept in separate quarters. The best arrangement for stallions is a roomy box stall with a 2 to 3 acre pasture. The stallion must receive daily exercise to maintain thriftiness.

BREEDING THE MARE

The mare should be in good condition at breeding time. She should not be too thin or too fat. She can be bred by hand breeding, corral breeding, or pasture breeding. The most accepted method is by hand. This will decrease the chance of injury to the mare or stallion. With corral breeding, they are turned loose into a corral. It is best to stand in a spot where you can see, but not be seen. Do not stand in the corral. Once the mare is bred, they are returned to their separate quarters.

Pasture breeding is hard to control and monitor. Valuable animals should not be pasture bred as there is an increased incidence of injury. With pasture breeding, the stallion is turned out with a band of mares. As the breeding season progresses, the stallion may become sterile.

After breeding, both the stallion and mare should be checked for injury. The mare should be pregnancy checked shortly thereafter.

FEEDING MANAGEMENT

Horses require nutrients for maintenance, growth, reproduction, lactation, and work. They require all six of the basic nutrients: carbohydrates, fats, water, protein, vitamins, and minerals, just like any other animal. They require energy, or carbohydrates and fats, in the largest amounts. Horses receive these nutrients from concentrates, forages, free choice vitamins and minerals, and water.

CONCENTRATES

Concentrates consist of various grains and supplements. They mainly supply energy and protein.

Grains

Grains are concentrates, such as oats, corn, or barley, that provide energy. Grains provide protein. Protein is necessary for the formation of bones, ligaments, hair, hooves, skin, organs, and muscles. Horses are monogastrics so they are not able to significantly synthesize their own protein. Young horses especially need high quality protein because their cecum is not fully developed. Maintenance requirements of protein are about 1 pound of protein for every 1,000 pounds (453 kg) of weight.

Oats are the most commonly used grain for horses. Oats are low in energy, but are easier on a horse's sensitive stomach. Dusty oats can cause colic, which is the number one loss of horses.

Corn is usually mixed with the oats because it is higher in energy. This is especially done in the winter. Corn should be cracked, rolled, coarsely ground, or shelled. Care should also be taken to avoid dusty or moldy corn. Moldy corn is very dangerous for any animal. If barley is fed, it must be ground or rolled because it is hard.

Supplements

The most common supplements are soybean meal and linseed meal or oil. Soybean meal is very high in protein content. Linseed meal or oil puts a shine in a horse's coat.

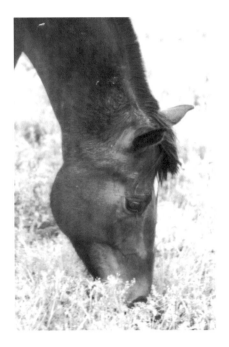

15-8. Horses benefit from good pastures.

FORAGES

Forages include pasture and fed hay. Forages should be fed at 1 to 2 pounds (0.91–1.82 kg) for each 100 pounds (45.36 kg) of body weight each day. One horse should be allowed three acres on which to graze. Hay should be of good quality and free of dust or mold.

MINERALS AND VITAMINS

Minerals are necessary for the growth and development of bones, teeth, and other tissues. The most essential and most likely to be deficient minerals are calcium and phosphorus. Steamed bone meal, dicalcium phosphate, salt, or trace minerals are the best sources of minerals.

Vitamins are needed for proper growth and development, health, and reproduction. Vitamin deficiencies are not a problem if the horse is fed a well-balanced diet. Supplemental vitamin A should be fed if the horse is not in a pasture to consume green grass. Vitamin D may be needed if the horse is kept inside.

WATER

Water is the most important nutrient that horses and all animals need. An average horse will consume 10 to 12 gallons (37.9–45.5 l) of water

15-9. Horses have gathered around the water trough.

each day. They may require more while working or in the summer. They should not drink excessive amounts of water during the summer because they will founder.

HEALTH MANAGEMENT OF HORSES

Daily management of health is imperative. Prevention is the key. Work closely with a veterinarian to devise a vaccination schedule. This will result in fewer health problems, lower veterinarian fees, and decreased animal losses. Cleanliness is necessary for the prevention of disease and the well-being of the animal.

COMMON AILMENTS

Colic

Colic is the leading cause of death in horses. It may be caused by internal parasites, improper feeding, and excessive water intake. Signs of colic include a distended abdomen, kicking or rolling, heavy perspiration, constipation, and refusal to eat or drink. Walk the horse and call the veterinarian immediately.

Encephalomyelitis or Sleeping Sickness

Encephalomyelitis is caused by a virus and is carried by mosquitos. Vaccinating and applying insecticides is the best prevention of this disease. Signs of encephalomyelitis are a sleepy attitude and localized paralysis of the lips and bladder.

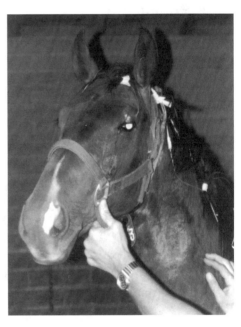

15-10. Diseased eye undergoing treatment.

Equine Infectious Anemia or Swamp Fever

Equine Infectious Anemia (EIA) is another viral disease transmitted by biting insects. It can also be spread by infected needles. There is currently no vaccine for EIA. Signs of EIA include a high, intermittent fever, stiffness, weakness, loss of weight, anemia, and swelling.

Equine Influenza

Equine Influenza is highly contagious. Symptoms include fever, loss of appetite, depression, rapid breathing, cough, weakness, and eye and nasal discharge. If influenza is suspected, consult the local veterinarian; there is a vaccine available.

Founder

Founder affects the tissue connecting the hoof wall to the foot. Severe founder cannot be cured. Founder is caused by grain or forage overeating, consuming too much cold water, overwork, rapid change in diet, or inflammation of the uterus following foaling. Signs of founder include pain in the feet, fever, and a reluctance to move.

Tetanus or Lockjaw

When an open wound becomes infected with *Colostridium tetani*, a bacterium, tetanus will result. These bacteria are found in manure, making it necessary to keep stalls and barns clean. Symptoms include stiffness about the head and slow and awkward chewing and swallowing. An annual vaccination is recommended.

Parasites

External parasites common to horses include flies, mosquitos, mites, lice, ticks, and ringworm. They are not only annoying to the animal and owner, but also carry diseases. Application of insecticides and regular cleaning of stalls will greatly decrease the external parasite problem.

Internal parasites include roundworms, bots (flies), pinworms, and strongyles (bloodsucking worms). To prevent internal parasites, do not allow the horse to eat off the ground or drink from puddles. Rotate pastures frequently. Regular worming should be performed. If internal parasites are suspected, contact the local veterinarian to examine the feces of the animal.

The extent and type of infestation will be learned and a plan of action will be recommended.

CARE OF THE TEETH

Determining Age by Teeth

The average horse lives to be 20 to 25 years old. Their best years are from three to twelve years. Age can be decided by examination of their teeth. Time of appearance, shape, and degree of wear of permanent and temporary teeth are important in aging horses.

Temporary teeth are smaller and whiter than permanent teeth. A mature male has 40 teeth, while a female has 36. As they age, the cups wear on the inside and center of the tooth. After 12 years, the teeth change from an oval to a triangular appearance and they slant forward more.

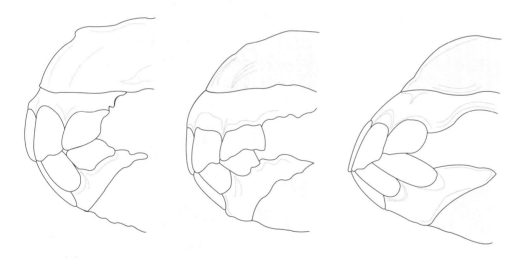

15-11. **Teeth slant forward as horses age. This side view shows the slant of teeth in horses that are 5, 7, and 20 years old.**

Floating of the Teeth

Because the upper jaw of a horse is wider than their lower jaw, teeth will wear unevenly. This causes a sharp edge to form. Teeth should be floated each year to reduce the discomfort of the sharp edges. *Floating* is using a file to smooth the sharp edges of the tooth.

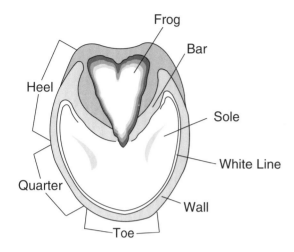

15-12. Parts of a hoof.

HOOF CARE

The ability of a horse to move will determine its worth. Proper care and maintenance of the hoof is very important. Through the domestication of horses and the change in their living environment, the care that a horse's hoof requires has significantly increased. The hoof should be kept clean, prevented from drying out, maintained at the proper length and shape, and protected from the hard surfaces they walk on.

A hoof pick should be used daily to clean out the hoof. Carefully clean around the frog. Keeping the hoof wet may be done by simply keeping the area around the water tank wet, applying dressing to the hooves, or attaching burlap to the hooves.

A *farrier* (a person who puts shoes on horses) should trim the hooves every four to six weeks. The sole of the hoof and jagged edges of the frog should be trimmed with a hoof knife. The sole of the hoof should not be trimmed unless necessary. Protection from hard surfaces is accomplished with shoes. Shoes should either be replaced or reset after the hoof is trimmed. Horses that are seldom ridden or that spend little time on hard surfaces may not need shoes.

15-13. Washing a hoof. (Courtesy, American Paint Horse Association)

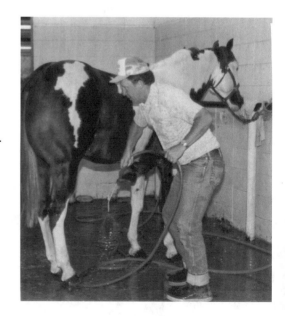

FACILITY AND EQUIPMENT REQUIREMENTS

Horse facilities do not need to be expensive or fancy. Again, no more money should be spent on facilities than can be afforded.

BARN

The facilities do need to protect the horse from the wind, sun, and other poor weather. The barn should have enough space for equipment and feed. Within the barn, the horse is given a stall. Standard recommendations for stall size are at least 10 feet by 12 feet (3 × 3.8 m) with a height of 8 feet (2.62 m). Tie stalls should be 5 feet by 12 feet (1.6 × 3.8 m). The bottom 5 feet (1.6

15-14. Small barn for two horses.

m) should be solid boards that can withstand kicking. The horse should also have access to a paddock or pasture to get plenty of exercise.

Keep the stalls clean and dry. If clay is used as flooring, remove and replace the top foot yearly. This decreases the incidence of diseases and parasites.

FEEDING EQUIPMENT

Hay or other forage should be fed in a hayrack or manger that is positioned above the ground. Grain should be fed in a box, pail, or tub. Watering is best with buckets or automatic waterers. Do not allow horses to consume large amounts of water at one time.

GROOMING

Horses should be groomed regularly. Grooming equipment needed includes a hoof pick, body brush, and a mane and tail comb. A curry comb and a sweat scraper work well when bathing the horse.

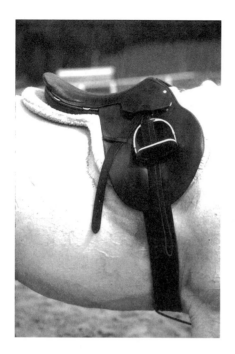

15-15. English riding saddle.

Each horse should be provided with its own halter. Make sure that the halter fits the animal and that the throatlatch is not too tight. When a horse is let out to pasture, remove the halter. They may get caught and strangle themselves.

SADDLE AND BRIDLE

There are two types of saddles; they are western and English. The type chosen depends upon the type of riding that will be done. The saddle should be comfortable for the rider and the horse. A saddle pad is placed under the saddle to comfort the horse.

A bridle is also needed when riding. A bridle is the part of the harness placed on the head that is used to direct and control a horse. There are dozens of styles and dozens of bits to choose from. A bit needs to be wide enough not to pinch the cheekbones.

EQUITATION

Equitation is riding and managing horses. It is often called *"horsemanship."* Development of good horsemanship and patience is essential for working with horses. Each animal should be treated as an individual. Horse and rider should work together as a team.

MOUNTING AND DISMOUNTING

A horse should be approached, mounted, and dismounted from the left side. The best position to stand when mounting is to place the left foot in the stirrup facing the saddle squarely. Hop and swing up and into the saddle using momentum from the right leg.

When dismounting, hold the saddle horn with the right hand and balance with the left hand on the horse's neck. Making sure the horse is steady, loosen the left foot in the stirrup and shift your body weight to the left foot. Free the right leg and gently let down.

15-16. Casually riding with friends in the open pasture is a way many people enjoy horses.

SEAT POSITION

Sit tall in the saddle. The back should be erect not slouched. The position in the saddle should allow for comfortable control of the horse and use of aids. The shoulders should be back and the arms held close to the body.

THE AIDS

Four basic aids are used to control a horse. They are hands, legs, weight, and voice.

Hands

The hands control the horse's forehead with use of the reins. Hands should be soft, light, and steady, yet firm. The arms, elbows, wrists, hands, and fingers should be relaxed. Holding the reins with a little slack will keep control, yet relieve pressure on the bit.

Legs

Legs are used to control the hindquarters and the forward movement of the horse. The horse should learn to move when the rider's legs are squeezed.

Voice

Successful use of the voice means being consistent. Words such as "back," "easy," and "whoa" are often used. The words used, as well as the tone of the voice, are important. Remain calm and do not yell at the horse.

Weight

Shifting of the rider's body weight signals the horse to shift its weight. This causes the horse to either be free to move or to be held in place. Signaling a horse by use of weight can be done by shifting your seat, moving weight from one stirrup to another, or being in rhythm with the horse by leaning the upper body forward.

15-17. This rider uses hands, legs, voice, and weight to control the Quarter Horse. (Courtesy, American Quarter Horse Association)

MOVEMENT

Movement is initiated by using the four aids. Overuse of these is a common fault of many beginning riders. There are various types of movement. The horse will become ready to move when the rider squeezes their legs. Movement is governed by the actions of the reins.

Walk

A ***walk*** is a four-beat gait. During a walk, both the rider and horse are relaxed. Slightly squeeze your legs and move the reigns to begin walking.

Jog

A ***jog*** is a two-beat gait. The rider should be sitting deep in their seat with just enough weight on their ankles to absorb some of the motion. The upper body should be leaning forward to maintain balance.

Lope

A ***lope*** is a three-beat gait. The rider should sit very deep in the saddle and again lean forward to maintain balance.

Backing

To back the horse, the rider should sit erectly in the saddle with their weight forward. Collect the horse and squeeze with the thighs. Tell the horse to "back." Alternately put pressure on each rein.

Stopping

For proper stopping, a horse should be cued ahead of time by either saying "whoa," squeezing with the thighs, tugging on the reins, or gently touching the neck. To bring the horse to a stop, squeeze with your thighs

15-18. Yearling filly class show of American Paint Horses. (Courtesy, American Paint Horse Association)

and firmly pull on the reins. The rider should be sitting erect in the saddle, grip with their thighs, and the body weight should be forward.

REVIEWING

MAIN IDEAS

Horses are used chiefly today for their recreational and personal value. They are classified as either light horses, ponies, or draft horses. Light horses are put into several types. These types include riding horses, gaited horses, stock horses, polo mounts, hunters and jumpers, racehorses, and driving horses.

Selection of a horse should be based upon the affordable price range, disposition, size, gait, breed, color, and work to be performed. Purchase a horse that suits its rider.

Conception of mares is the lowest of any farm animal. Careful and precise management of horses is a must.

Horses, like most other animals, require concentrates to provide energy and protein, minerals and vitamins, and water to supply their nutrients. Oats, corn, and barley are the most common sources of carbohydrates and fats. Supplements that are widely used include soybean meal and linseed meal. Mineral and vitamin supplements are mixed with the total feed ration.

Horses should be checked on a daily basis for signs of ill health. A health management program should be established with the help of the local veterinarian. A horse's teeth should be floated each year to reduce discomfort. Their hooves should be cleaned daily and if shoed, they should be trimmed every four to six weeks.

Facilities for horses do not have to be expensive and extravagant. A barn with a good sized stall is needed. Basic equipment needed includes a hayrack or manger that is raised off the ground, hoof pick, body brush, and a mane and tail comb.

Equitation (horsemanship) is skill in riding and managing horses. A beginner must learn how to properly mount, dismount, sit in the saddle, use the aids (hands, legs, weight, and voice), coordinate movement, and stop a horse.

QUESTIONS

Answer the following questions using correct spelling and complete sentences.

1. What is the purpose of horses?
2. How are horses classified?
3. How are light horses broken down in types?
4. What factors should be considered when selecting a horse?
5. Why is good management of utmost importance in horse reproduction?
6. Name six things that should be considered in the breeding management of horses.
7. What does an average horse's diet consist of?
8. When managing the health of a horse, what should the owner look for?
9. How should hooves be cared for?
10. What facilities and equipment are needed for horses?
11. What is horsemanship? Why is proper horsemanship important?
12. How are the four aids used in horsemanship?

EVALUATING

CHAPTER SELF-CHECK

Match the terms with the correct definitions. Write the letter by the term in the blank that is provided.

a. mare   d. lope   g. foal
b. gelding   e. farrier   h. stud horse
c. plug   f. gait

1. ____ a young horse that has not been weaned

2. ____ a male horse kept for breeding

3. ____ a person who puts shoes on horses feet

4. ____ slow, three-beat gait in which the head is carried low

5. ____ horse with poor conformation and common breeding

6. ____ sexually mature female horse

7. ____ the way a horse moves, walks, or runs

8. ____ male horse castrated before reaching sexual maturity

EXPLORING

1. Visit a horse farm. Explore the breeding, feeding, health, and facility management of the farm. Prepare a written or oral report for the class on what you learn.

2. Study common management practices used with horses and explore how this is changing. Interview authorities on horses. Use several references with your study, including *Horses and Horsemanship* (available from Interstate Publishers, Inc.).

3. Attend a horse show. Observe the breeds of horses, how they are groomed and ridden, and other areas related to horse management.

Chapter 16

DRAFT ANIMALS

Draft animals played a major role in the settlement of America and the farming of its land. They are still important in some countries. In America, when people were dependent upon themselves to grow their own food and develop their own farmstead, draft animals supplied the power necessary to farm and clear land. Draft animals in the United States were replaced by machinery and equipment which was much faster and more efficient.

Draft animals are still around today, although they are no longer widely used as the primary power source for farming in the United States. Exceptions to this are isolated communities where draft animals are still used, especially horses, in farming practices. In some areas, strong religious beliefs may not allow the use of power vehicles. Other countries still use draft animals in order to farm and do other strenuous work.

One of the first uses of the draft horse was in warfare. It took an extremely large, well-muscled horse to carry a person in full armor as well as the armor that was on the horse to protect it. These horses could carry the load and run over obstacles in their paths as well. When military tactics were changed and gun powder was developed, the horse was no longer vital in battle. The horses' role evolved into one in agriculture because they were strong and could be very useful.

16-1. These Percheron horses are plowing a field with a walking plow.

349

OBJECTIVES

This chapter provides basic information on draft animals. It has the following objectives:

1. Describe draft animals and list major kinds and breeds
2. List the possibilities in draft animal production
3. Explain important management practices with draft animals
4. List the nutritional requirements of draft animals
5. Describe the facility and equipment needs with draft animals.

TERMS

automated-feeding	gregarious behavior	roan
breeding period	hand-feeding	ruminate
by-product feeds	harness	self-feeding
curry comb	mule	social ranking
draft	ox	tri-purpose animal
draft animal	power	yoke
feather	regurgitate	

DRAFT ANIMALS

Several different animals are used for draft. A ***draft animal*** is an animal which has been trained and is used for pulling heavy loads. Most often, draft animals are horses or oxen. However, donkeys, mules, camels, and buffalo have been used as draft animals and still are used today in some countries.

When a draft animal is selected, consider the climate, the availability of the animal, the cost of the animal, the type of work to be done, and the social and religious traditions which might limit the ownership or use of some animals.

OXEN

The most popular draft animal, the *ox* is any animal of the bovine (cattle) family. They are primarily castrated males or bulls, but sometimes females are used. Oxen are preferred over other draft animals because they are ruminants and use their feed better, but they are much slower than the other animals. Chapter 9 gives you more information on

16-2. **A team of oxen being used to plow a field. (This team is made up of Shorthorn steers.)**

beef cattle and Chapter 12 gives you more information on dairy cattle. Oxen are classified as a ***tri-purpose animal*** because they served three purposes: work, milk, and meat.

DRAFT HORSES

Draft horses are common draft animals in the United States. Some breeds of horses are defined as draft horses. They are larger and heavier than the horses used for riding and pleasure. All horses and horse-like animals are from the equine family. Draft horses were developed to be 1,500 to 2,500 pounds (680–1,134 kg); have a low center of gravity; and be wide, deep, compact, strong, and large-boned. Other members of the equine family, such as mules, are used because they are cheaper to buy than horses.

Five breeds of draft horses are found in North America: Belgian, Percheron, Shire, Clydesdale, and Suffolk.

Belgians, which are a Flemish breed, are brown to light blonde and sometimes even **roan,** which is a mixture of white and colored hairs, and became popular because of their size, endurance, tremendous power, excellent muscling, and style. Belgians are the number one breed of draft horses in the United States today, replacing the Percheron that was previously the number one.

Percherons, which are from France, are more refined and balanced than some of the other draft horse breeds. Fifty percent of Percherons are black and the other 50 percent are black and gray.

Clydesdales, which are the third most popular breed in the United States, are brown or black with white faces; have long white hair on their legs, known as **feathers,** which sometimes extend all the way up to their bellies; and black manes. Clydesdales originated in Scotland and are well-known for pulling heavy wagons.

Shires are black in color and are like Clydesdales with the feathers on their feet, but they have greater bulk than the Clydesdale. Shires were developed in England.

Suffolks were also developed in England. Unlike the other four breeds, their development was solely for agricultural reasons. They are chestnut color (sometimes referred to as sorrel) and have very little white on them. They have very little or no feathers. Suffolks are easy keepers, which means they stay in good condition with less feed than other horses require for the same condition. Suffolks have a very fast walk, great stamina, longevity, and a willingness to work.

MULES

A **mule** is the hybrid offspring of a male ass (donkey, jack, or jackass) and a mare (female horse). Mules resemble both donkeys and horses. They have the large ears of donkeys, are strong like a horse, and tend to pace themselves for long-term power like a donkey. Horses tend to do work at a fast pace and not hold up for the same long hours as mules. Mules are larger in size than donkeys, but smaller than draft horses. As hybrids, mules do not reproduce, though they are born as males and females. Male mules are gelded (castrated) as young animals.

Mules have advantages over other draft animals. Mules endure heat better than horses. They are less sensitive to their feed than horses and have fewer feed-related diseases. Mules eat lower-grade feed, which saves money.

Mules are physically sound, have fewer defects than horses, are surefooted and careful, and live a long time. Mules tend to have stubborn dispositions somewhat like a donkey but not to the same extent. Many farmers in the United States relied on mules to provide power for long hours of the day doing tedious crop cultivation. In the late 1800s and early 1900s, the large cotton plantations of the South often used hundreds of mules for power.

Today, some people view mules much as horses. They train them for show and dress them in fancy ways formerly reserved only for horses. The mules are pets!

16-3. Mule being used to drag a small log. (Courtesy, Lucky Three Ranch, Loveland, Colorado)

16-4. Team of mules pulling a covered wagon. (Courtesy, American Donkey and Mule Society, Denton, Texas)

16-5. Mules may be trained for show and competitive events. (Courtesy, Lucky Three Ranch, Loveland, Colorado)

DRAFT ANIMAL PRODUCTION

After a period of decline, draft animals are increasing in numbers. Special events feature draft animals performing power events as well as skill events.

Many enthusiasts show their draft horses in halter classes where the animal is judged on the way it is put together and hitch classes where they have classes for hitches of anywhere from one to six horses. There are also clubs of draft horse owners who put on old-fashioned farming demonstrations for the public. Loggers in the north are starting to use the draft horse again because they cause much less damage to the environment than their mechanical counterparts. Horses are also teamed up with another horse and compete in horse pulls where the team is hooked either on a boat or on a machine and they pull the weight. The team which pulls the most is the winner.

16-6. Three teams competing in a plowing competition at an old fashioned farm days weekend event. The participants are judged on how deep and straight they plow.

DRAFT ANIMAL REPRODUCTION

Reproduction is a key component in the draft animal industry. Strong, well-bred animals are needed.

Estrus is the time when the female animal is most receptive to be bred by the male animal or artificially. The time of the year when the female has her estrus periods and estrous cycle is the **breeding period.**

Oxen

In oxen, the females reach puberty at 8 to 15 months of age. The duration of their cycle is 12 to 20 hours every 21 days year round. Breeding must occur during this time.

Even though cows cycle year round, most producers prefer to have the females bred about the same

16-7. **Horse mare with mule colt. (Courtesy, Lucky Three Ranch, Loveland, Colorado)**

time so they calve about the same time of the year and all of the young are close to the same age. When it comes time to make management decisions such as vaccinating or castrating of the male calves, it is easier when the whole group can be done at once. Oxen in estrus stand to be mounted by herdmates, lack an appetite, act nervous, restless, and are very vocal. After the cow has been bred, it takes about 9 months or 270 to 295 days for her to calve.

Draft Horses

Draft horse females reach puberty at 12 to 15 months of age. Unlike oxen, females have a cycle that lasts four to six days and comes every 21 days during the estrous cycle, which begins in the spring and stops in midsummer. Females must be bred during this time of the year.

When females are experiencing estrus they are very vocal, get nervous,

16-8. **A five-month-old Belgian filly (female) is being broke to lead.**

restless and irritable, squat and urinate frequently, and seek out the stallion. Mares should be bred on the second day of estrus and every other day for two to four days. After the mare has been bred, either artificially or by the stallion it takes 11 to 11½ months, or 330 to 340 days, for her to foal.

YOUNG ANIMAL CARE

Both female horses and oxen clean their young with their tongues after giving birth. This gets blood circulating in the newborn and encourages the newborn to stand and nurse. Instinct makes mothers become very aggressive in protecting their young and even a calm pet may become ornery.

DRAFT ANIMAL MANAGEMENT PRACTICES

Understanding animal behavior is essential for good management of draft animals. First of all, **draft** is the pulling of an animal to move any object. In order for an animal to move an object it must exert a force equal to or greater than the weight of the object.

DRAFT CAPACITY

The draft capacity of an animal increases with the weight of the animal. The rule of thumb is that an animal can exert a constant pull at its normal pace of one-tenth its body weight. If an animal is required to exert a concentrated effort, the time the animal can work will greatly diminish. Horses, mules, and oxen are preferred because they can pull loads at a normal speed over long time periods and exert extra energy when needed.

16-9. A Belgian draft horse team is speeding up and exerting extra energy to pull a heavy load.

Power is the combination of pulling capacity and speed. A draft horse will pull a 150 pound (68 kg) load at a steady rate of $2^{1}/_{2}$ miles per hour, which is defined as one horse power. One horse power may also be defined as pulling 150 pounds (68 kg) out of a 220 feet (67 m) deep hole in one minute. An oxen pulling the same weight will travel at $1^{1}/_{2}$ miles per hour, so at the end of the day the horse has travelled further. Also, shorter animals have a lower center of gravity, so they must be hitched lower so that less work is required for them to get the task accomplished.

All animals have certain behavior patterns which must be recognized for proper health and overall well-being of the animal. Some animals pull better and reach maximum capacity when worked alone but not when worked with other animals. Animals work well only when they have been properly trained and properly outfitted with a correct fitting harness (for more information on the harness see the equipment section of this chapter).

An ox can deliver 25 to 50 percent more horsepower when harnessed with a collar instead of a yoke. This is because of the ox's lower center of gravity and the wider pushing area of the collar.

DRAFT ANIMAL SELECTION

The type of draft animal selected depends on the work to be done and the time allotment for it to get done. For example, horses work much faster than oxen, but oxen are able to eat more efficiently because they are ruminant animals. If horses are considered, there are other animals in the equine family which may be cheaper to purchase than a horse, such as a mule. All of these factors need to be taken into account by the owner when a draft animal is selected.

When selecting animals, it is also important to note that horses and oxen in herds develop social rankings. *Social ranking* is the order the animal falls within the herd. Social rank is determined by the age, size, strength, genetics, if they have horns or are polled (which means they are bred to be born without horns), and previous experience of the animal.

There are many sources available to assist producers of draft animal owners and producers. Veterinarians, the Cooperative Extension Service, state universities, state experiment stations, feed suppliers, other animal owners, animal associations, and libraries are all important sources for people to gather information pertinent to their animal and situation. Since animals and situations are different, it is impossible to describe an answer to every and all problems that may occur.

NUTRITIONAL REQUIREMENTS OF DRAFT ANIMALS

Animals vary in their feed needs, just as they vary from breed to breed and horse to oxen. A good feeding program is vital in maintaining an animal's health and strength. In determining the feed ration for a draft animal, first the species, development stage of the animal, weight and size of the animal, if they are or are not pregnant, and how often and hard they are worked must be decided.

16-10. Most draft animals are outside for at least part of the day, so it is easier to water them out of a tank.

Water is the most important for all animals and should be available to them at all times, except during and right after working. Draft horses and oxen both require approximately 8 gallons (30.3 l) of water per head per day.

All draft animals need protein. It may be fed in the form of alfalfa hay to both horses and oxen and/or cottonseed or cottonseed meal to oxen. The feed ration needs to be complete, but economics must also be taken into account.

High fiber and low fiber feeds are also important to draft animals. In the high fiber category, horses and oxen both benefit from oat hay. Oxen also benefit from corn silage and haylage. In the low fiber category, both oxen and horses eat corn, grain, barley and oats. There are also **by-product feeds,** such as pulp from citrus or beets, barley malt, and others. The availability and types of by-products depends upon the area in which the producer lives.

FEEDING OXEN

Oxen are ruminant animals, which means they have four compartments in their stomachs. They swallow their food as soon as they bite and salivate over it. Later they **regurgitate** the food. This is returning eaten food to the mouth for chewing. The food in the mouth is a cud. Oxen graze from four to nine hours per day and ***ruminate*** (chew their cud) for four to nine hours per day.

Extreme heat or cold lessens the amount of food consumed by the animal. Oxen are similar to beef cattle in many nutritional needs. An exception is that they need feed with higher energy when they are being worked.

FEEDING DRAFT HORSES

A horse is a single-stomached (monogastric) animal. It eats and chews its food immediately and then the food is used to give the animal energy. Ruminants are sometimes preferred because it has been said they use their food more efficiently.

Vitamins and minerals are important to assure the draft animal remains healthy and strong. The three minerals often added as supplements to the feed ration are salt (sodium chloride), calcium, and phosphorus.

Draft horses need the following seven major elements: calcium, phosphorus, magnesium, sodium, chloride, potassium, and sulfur. There are also some trace minerals which are required: chromium, nickel, silicon, vanadium, tin, copper, cobalt, and zinc. The vitamins needed by horses are A, D, E, and C. In most cases, good quality hay will eliminate the need for vitamin supplements.

Horses should be allowed to cool down after working before they are fed. They may be given some water after they have cooled, but horses will get lame or foundered if they eat or are watered when they are too hot. Give a horse an hour or so to cool down before feeding or watering it. Also, give a horse 1/2 hour to 45 minutes to digest food before harnessing the horse and making it do hard work. When feeding horses, make sure the hay is dust- and mold-free to prevent illness. The harder the horse works, the more feed the horse should be fed.

FEEDING SCHEDULES

In addition to feeding the correct ration, it is important to develop a feed schedule so animals are fed at the same times each day. The animals may be fed by ***hand-feeding***, which is giving animals the same amount of feed either one or two times daily. This would entail their grain and hay. For example, hay may be given two or three times daily and grain only once or twice.

Self-feeding is when animals have feed in front of them at all times. In this system, the producer needs to ensure some way of having feed available at all times. Many times, hay is available in the self-feeding system and grain is given through the hand-feeding system. This tends to work well since draft horses will overeat and make themselves sick on sweetfeeds, such as grain.

The third option of feeding is **automated-feeding.** Animals are fed mechanically so no one needs to haul feed at feeding time. Usually, an auger is used to bring the feed to the animals. This is a common way of feeding silage to cattle.

Draft animals ready for weaning from their mother should have their mother's milk supplemented with hay and grain. When the weaning takes place, the young animal will do better if it is used to eating food along with its mother's milk. It will cause less stress to the young animals if two or more of them are weaned at a time. The companionship is important to these animals and they will do better if they are not alone.

DRAFT ANIMAL HEALTH PRACTICES

Animals respond differently to their environment. Some things may cause stress in one animal while not in another. It is vital that owners understand their animals' attitudes and behaviors. If an animal is behaving strangely, it may be ill, under stress, ready to be bred, or have other problems which may require attention.

There is no one who knows an animal better than the person who cares and works with it everyday. This person must always pay attention to the actions and attitudes of the animal.

CONDITIONING

A very important health practice is to keep the draft animal well-conditioned. If the animal does not work hard over the winter, start the spring with a smaller work load and shorter days until the animal has gotten back into condition.

Just as any good athlete needs to stay in condition, so does a good draft animal. Since they do not have the power to reason and decide to

16-11. These Belgians are being used to rake hay. They were kept in condition all winter so they are well-conditioned to work all day.

stay in shape on their own, it is the responsibility of the owner to ensure that an animal is in shape before hard work. Animals may pull a muscle or physically hurt themselves if they are not conditioned. The least that will happen will be a break in their spirit because they will experience defeat.

CONSIDER HEALTH

In order to have a good, strong, sound draft animal the producer needs to begin by buying that type of animal. The stock purchased will make a big difference in the amount of work actually done. It is an accepted practice to take a veterinarian to inspect an animal before it is bought. The buyer is usually the one who has to pay for the veterinarian's visit, but the buyer then knows the soundness of the animal.

General Health

A draft horse has a normal heart rate of 28 to 50 beats per minute and an average temperature of 100.5°F (38°C) with ranges of 99.0 to 100.8°F (37.2–38.2°C). An ox has a normal heart rate of 40 to 70 beats per minute and an average temperature of 101.5°F with ranges of 100.4 to 102.8°F (38–39.3°C) temperature. When the owner has a feeling the animal is not feeling well, a higher temperature is a good indication that something is wrong. There are many possibilities of what can go wrong with an animal, and a veterinarian is a good person to call for assistance.

Nutritional diseases may affect draft animals. If animals are not fed the proper ration or supplements, they may have deficiencies. Too much feed can also cause problems. Metabolic disease may result from erratic feeding times and feeds themselves.

Parasites

Internal parasites may cause major problems in animals. They live at the expense of the animal they are living inside. External parasites get their nourishment from the outer surface of the animal's body.

Worming and vaccinations of draft animals are important to stop or prevent parasites. Vaccinations are the injection of an agent into the animal for the purpose of developing resistance to what the animal is treated for. Animals pick up worms through their feed and pasture. As long as the animals are treated for parasites periodically, they will be okay.

Small precautions, such as brushing and grooming draft animals, will cut down on any manure on their coats, which attracts flies and other insects which may carry disease. A curry comb and a brush are both needed to groom the draft animal. The **curry comb** is an oval-shaped plastic or metal device used to loosen sweat, manure, and other foreign materials from the animal's coat. The brush is used to remove the materials from the animals coat. If an animal is not bothered by flies and insects, it will be happier and healthier.

Foot Care

Another important area of health care is the feet of the draft animal. Since the animals are used to do heavy work, their feet must be maintained in good condition. If an animal has sore feet it will work much less and put forth minimal effort. Feet must be trimmed about every six months, depending upon the growth of the animal's hoof.

16-12. Examining the foot of a mule. (Courtesy, Lucky Three Ranch, Loveland, Colorado)

Horses have metal shoes formed to fit their feet. The shoes are nailed on to give them better traction for pulling loads. Proper shoes assist the horse in digging its feet into the ground to pull. Proper shoes and trimming lessen the possibility of the animal falling or pulling a muscle or developing a foot disease.

DRAFT ANIMAL FACILITY AND EQUIPMENT NEEDS

In order to develop proper housing for a draft animal, the personality and characteristics of the animal must be understood.

Gregarious behavior is the instinct of animals to flock or herd. When an animal that normally is in a flock or herd is not able to be with other animals, this causes stress in the animal and makes them difficult to deal with. Although draft animals prefer to be in herds, they also need to have enough space to live comfortably, because overcrowding of animals is also a source of stress for the animal.

When constructing windbreaks or barns, make sure the builder takes into account that the area needs to be easily cleaned and well-drained so animals can stay dry and out of the mud.

Draft horses need shelter or a windbreak, especially in the winter from the wind. Mares who are going to foal need to be kept separate from barren mares and be in an area where they can get plenty of exercise to stay in peak physical condition. If they are bred to foal in the winter, they need a warm, dry place to give birth. Some horses do foal outside, but the survival rate of the newborn is better and there are less complications if they are born inside.

Draft horses prefer to be in the company of other horses. If a person plans to work a horse—one that is usually worked as part of a team—alone, they may have trouble. Many times, when horses are worked together as a team, it makes it difficult to then separate the team without the horses getting unruly or upset. Horses tend to be naturally curious, so owners may put toys in their stalls to keep them entertained or leave them out in the pasture longer so they do not get so bored in a stall. Remember, oxen are ruminants so they spend many hours per day laying around and chewing their cuds, while horses have to search for pasture longer since they only have one stomach.

Horses may be kept out on pasture where they have access to a windbreak, or kept in box stalls or tie stalls. Tie stalls are okay, but not for long periods of time, because the horses will get bored and start causing trouble, such as chewing on the boards they can reach, pawing or whinnying, and being vocal. These are all signs of boredom in the horse. If horses are pastured in lots next to each other, they may run the fence together and be very vocal with one another.

STRONG FENCES AND STALLS

Oxen, on the other hand, because they are so big, tend to break through fences to get with other cattle if they are separated and penned alone. Oxen prefer shady areas in the warm weather and gather together under shelter or windbreaks in the cold weather.

16-13. Strong fences are needed in training facilities. (Courtesy, Lucky Three Ranch, Loveland, Colorado)

In very cold weather, a cow ready to calve should be kept inside because the newborn has a better survival rate if it is born inside, out of the elements. A cow and calf require about 300 square feet (27.9 m²) of space and a steer needs 150 to 200 square feet (13.9–18.6 m²) of space. The fences should be 50 to 60 inches (1.3–1.52 m) high; however, this depends on the size of the cattle.

Oxen are followers and they will tend to follow other cattle in the herd, so when animals must be moved, an easy way is to get the leader moving and the rest will follow. When moving either horses or oxen, it is necessary to move slowly and quietly so the animals do not get any more nervous and have extra stress. If they are to be loaded, make sure they only see one way to go. Draft animals have a keen sense of smell and notice loud, unfamiliar noises and strange objects they may see. It is best to avoid exposing them to as many strange, unfamiliar things as possible when moving them. Many animals get scared and even unruly in strange surroundings with other animals and people they do not know. Working calmly and slowly will quiet the animal and eventually pay off because they will do what is expected sooner than if they are scared and try to run through things and over people.

EQUIPMENT

The most important equipment in using draft animals is the harness and bridle. The *harness* includes the attachments that make it possible for a draft animal to pull. The harness is the key to the amount and efficiency of the power of the animal. There are different types of harnesses, from fancy to plain and from leather to nylon. It all depends upon the owner. It is vital that the harness fits properly. A good-fitting harness will allow the animal to maximize its power. In actuality, an animal is pushing instead of pulling. The horse pushes on the collar and this pulls on the hames and the tugs which causes the load to move.

16-14. A side view of a harnessed team of Percherons. The bridle is over their head, the collar is the red piece around their necks, and the hames are attached over the collar.

Each animal usually has its own collar, harness, and bridle, which are adjusted to fit perfectly. The collar goes over the horse's head and rests around the neck and on the shoulders. The collar should fit snugly, but not bind on the horse. If the horse is worked, and a sore results, the collar is not properly fitting the horse.

16-15. **Holiday season parade entry. (Courtesy, Lucky Three Ranch, Loveland, Colorado)**

The harness is put over the top of the horse and the hames are attached around the collar. It is vital the hames are a good fit with the collar. If the fit is improper, a sore will once again result. All of the power comes from the collar, hames, and to the tugs which run the length of the horse and are attached to the load at the back end of the horse. The bridle goes over the head of the horse and is hooked to the lines which are used to steer and stop the horse.

After the horse is harnessed, it is hooked onto a pair of whipple trees. The tugs hook onto the whipple tree, which is then hooked onto the load. It is imperative to hook the animals the proper distance from the load; otherwise, they may get their heals scrubbed or be so far out and away from the load that they lose the advantage of being hooked low and close.

If one has never harnessed an animal before, it is best to find someone who knows how and ask for assistance. Many times, the people the animal is bought from can give good advice as to the collar size and the way it was hitched before.

A harness can be a very expensive piece of equipment and must be cared for. A properly cared for harness will last virtually forever. All that is required is to keep the harness clean and well-oiled with a good harness oil. Amish communities have harness shops that can dip harnesses in oil, eliminating the owner's mess and problems of what to do with all of the old oil.

Just as horses push on their collars, oxen push through their yoke to move their load. The *yoke* is a wooden bar which hoods two animals together with a bar between them. The oxen are hooked around their necks or horns with a bar up top between their heads. They hold their heads low and since that is where their strength is, they are able to move more hooked this way. Oxen do not have bridles like horses do. Some oxen are steered with lines hooked to a halter and some have nothing on their faces.

REVIEWING

MAIN IDEAS

Draft animals have been trained to pull heavy loads. Once used for power in agriculture, they have been replaced in North America by machinery. Today, people enjoy having draft animals as pets and for competitive events.

Draft horses, mules, and oxen are the most common draft animals. The horses and mules are often trained for showing and groomed in fancy ways.

In order to effectively work with and raise draft animals there are many considerations to be taken into account. Breeds, the type needed for the work to be accomplished, good management practices, a nutritionally complete diet, proper health program, and the correct facilities and equipment are all important to healthy draft animals.

A healthy draft animal will work hard to accomplish the tasks needed to be done. Diseases, parasites, and nutritional problems affect most draft animals in one form or another.

Appropriate facilities and equipment are needed. Since draft animals are very strong, all pastures, pens, and other structures around them should be built to withstand their power.

QUESTIONS

Answer the following questions using correct spelling and complete sentences.

1. What is a draft animal? Name three and compare their performance.

2. What are the major differences between reproducing oxen and draft horses?

3. What are the power and drafting potential of animals?

4. What are the nutritional needs of draft animals?

5. What are the major health practices with draft animals?

6. What facilities and equipment are needed with draft animals?

7. In constructing facilities, why is the power of the animal important?

8. How is an animal or team harnessed to its load?

EVALUATING

CHAPTER SELF-CHECK

Match the terms with the correct definitions. Write the letter by the term in the blank that is provided.

a. mule   d. harness   g. yoke
b. power   e. social ranking   h. gregarious behavior
c. draft   f. regurgitate

1. ____ the pulling of an animal

2. ____ hybrid of ass and horse

3. ____ combination of pulling capacity and speed

4. ____ instinct of animals to flock or group together

5. ____ wooden bar that holds two harnessed animals together

6. ____ returning undigested food to the mouth for chewing

7. ____ the order an animal has in a herd

8. ____ attachments put on an animal so that it can pull

EXPLORING

1. Attend a draft horse, mule, or oxen pulling event. Observe how the animals are handled and the kind of work they do. Give a report in class on your observations.

2. Assist in harnessing a team of animals. Study the different parts of the harness and how they are connected. Be very careful; draft animals can be dangerous.

Chapter 17

COMPANION ANIMALS

Companion animals are important to people. They provide fun as well as help make life better. People relax when they have an enjoyable companion animal. Medical experts say that the heart rate of people decreases when people pet a dog. Some companion animals help people, such as leader dogs for the blind.

Nearly everyone has a favorite companion animal. Cuddly kittens, perky puppies, and fluffy bunnies get the special attention of most people. Having an animal is more than adoring a young kitten, puppy, or rabbit. It involves care, training, and attention on a daily basis.

Some of the work with companion animals is not fun. Who likes to clean out a cat's litter box or wash a fish bowl? But, these are important to the well-being of animals. If they are not done, the animals are not well cared for.

In addition to companion animals at our homes, many people are involved in the business of providing what is needed for the animals. Pet stores, kennels, animal shelters, and other places need good employees who know how to care for animals. Some people are entrepreneurs and start their own business.

17-1. Dogs are preferred as companion animals by some people. (This dog is a young Chocolate Labrador Retriever.)

OBJECTIVES

This chapter presents basic information on companion animals. It has the following objectives:

1. Describe companion animals
2. Explain the possibilities of companion animals
3. Explain the management practices with companion animals
4. List the nutritional requirements of companion animals
5. Explain important health practices with companion animals
6. Describe facility and equipment needs with companion animals

TERMS

bitch	hound	reptile
carnivore	kitten	spaying
companion animal	litter pan	sporting dog
declawing	nocturnal	tomcat
diurnal	non-sporting dog	toy breed
hairball	pet carrier	whelp
herding dog	queen	working dog

COMPANION ANIMAL POSSIBILITIES

A ***companion animal*** is a domesticated animal kept by humans for enjoyment in a long-term relationship. Many times companion animals are called pets. In some cases, they do valuable services, such as guarding property and assisting people with disabilities. Regardless, people and animals may develop very close relationships.

Companion animals are used for company and friendship. They keep people from being lonely and they provide entertainment. Some are taught to do tricks and perform useful duties, such as retrieving the newspaper from the driveway. Not only do people enjoy their companion animals, but companion animals enjoy their people. Of course, the well-being of the animal must be considered.

Many animals are used as companions. A few common companion animals are dogs, cats, rabbits, gerbils, hamsters, guinea pigs, rats, and mice. In addition to these small animals, some people use horses, cattle, pigs, llamas, goats, and others as companion animals. Selecting a companion is important. People need to know the background information on these animals and if local laws allow them.

DOGS

Dogs are versatile animals and may be used for a variety of purposes. The selection of the dog is based on the reason the owner wants to have a dog. There are several items to be taken into account when looking for a dog.

Biology

Dogs are ***carnivores,*** meaning that they are flesh-eaters. Dogs are also mammals and monogastric. The scientific name for the domestic dog is *Canis familiaria*. They are often called canines because of their family (*Canidae*) and genus (*Canis*).

The body structure of dogs varies with the size and breed of the dog. The weight of dogs varies. A mature Chihuahua may weigh less than 6 pounds (2.7 kg). A mature Saint Bernard may weigh as much as 200 pounds (90 kg). Most dogs have similar skeletons of about 320 bones. The major difference is in the size of the bones.

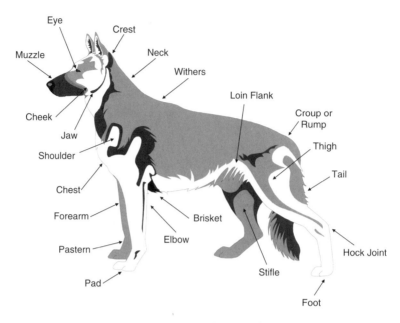

17-2. Major external parts of a dog.

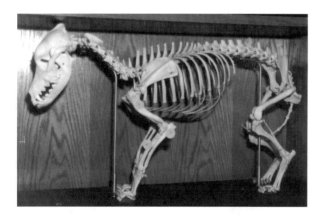

17-3. Skeleton of a dog. (This skeleton is used by students of veterinary technology at Hinds Community College, Raymond, Mississippi.)

Dogs may live several years. The smaller breeds may live 15 to 18 years. The larger breeds, such as the Saint Bernard, have a shorter life span, often only 8 to 10 years. Babies are born in groups of 1 to 10 puppies, known as litters. The female, known as a *bitch* or dam, has a gestation period of about 63 days, depending on the breed. The male is the sire or stud dog. (More information on dog care is given later in the chapter.)

Classes and Breeds

About 300 breeds of dogs are found throughout the world. In the United States, the American Kennel Club (AKC) classifies and maintains records on dog breeds. Seven classes have been identified, based on the use and characteristics of the breeds. This information is helpful in selecting a breed.

Herding Dogs. The *herding dogs* are popular with many people who have sheep and cattle and as pets. The dogs are easily trained to assist in herding animals into pastures and barns. Common herding species include the Border Collie, Australian Cattle Dog, Shetland Sheepdog, and German Shepherd. Select these dogs as pets only if space is available for them to run and enjoy. A herding dog does not usually like to live in confinement.

17-4. A dog herding sheep.

Sporting Dogs. *Sporting dogs* are those such as Pointers, Setters, Spaniels, and Retrievers. The Pointers and Setters are hunters, which run and then stop when they find their prey. Retrievers swim and retrieve game in both the water and on the land. Spaniels run and scare their game out of the cover.

17-5. This Australian Cattle Dog works in a bull test station and helps in penning and moving the bulls.

Tracking Dogs (Hounds). *Hounds* are used for tracking. They have good abilities to follow the scent left by animals or people. Hunters may use them to locate game. Law officials use them to find escaped prisoners or lost people. The most common breeds are the Beagle, Basset, Dachshund, and Greyhound. Some breeds run quickly to track and others work slowly and thoroughly to find their catch.

Working Dogs. *Working dogs* are used by people to get something done. Examples of the work done by these breeds is pulling sleds, protecting property and other animals, and sniffing bombs and drugs. Some of these dogs are used to guard property and for human protection. Breeds of

17-6. A Samoyed patiently waits outside a store at Tahoe, California, while its owner shops.

17-7. A female Alaskan Malamute (also known as a Husky) trained as a sled leader dog in a Tok, Alaska, kennel.

17-8. The Rottweiler is a massive, powerful dog used for human protection. (Rottweilers are very protective of their owners and can attack strangers, causing serious injury. This Rottweiler is kept in a specially-fenced back yard.)

working dogs include the Alaskan Malamute, Collie, Rottweiler, Saint Bernard, and Samoyed.

Terrier Dogs. Twenty-five different terrier breeds are recognized by the AKC. The name, terrier, is from Latin, meaning earth. The breeds are noted for following animals they are chasing down into the earth. They dig in the earth to capture their prey. Terriers fall under this category, as does the Schnauzer.

Toy Breeds. The *toy breeds* are small and only weigh between 4 and 16 pounds. These animals are known for their companionship and their long lives. Examples of toy dogs are Yorkshire Terriers, Toy Poodles, English Toy Spaniels, and Chihuahuas.

Non-Sporting Breeds. *Non-sporting breeds* of dogs were developed for specific purposes, but are now primarily pets. Examples of these are Dalmatians, Bulldogs, and Poodles.

Purebred or Mixed Breed

After deciding on the type of dog, it is necessary to decide if the owner wants a purebred or mixed-breed dog. There are positives and negatives with either choice. The biggest problem with a mixed-breed dog is the owner does not know what the mature dog will end up like. A young puppy can change significantly as it grows and matures. The mixed-breed dog may also never be shown in breed shows or raise pups to be sold as purebreds. Of course, keeping records and making application for breed registration requires effort with purebred dogs.

Space Requirements

It should also be decided if the dog will live inside or outside before selecting and bringing home a dog or puppy. Most breeds can live healthy lives outside as long as they have shelter from the elements. Dogs should not be shuffled inside to sleep and then outside to sleep. It is difficult for the dog to adjust to the temperature changes and get any rest.

Inside dogs require much more training than outside dogs. This is because dogs have to learn what they can and cannot do. Owners must be involved in the training.

17-9. Dalmatians are easy to identify and noted for riding on fire trucks.

17-10. A mixed-breed dog selected to work on the farm because of the breed of each of his parents.

Hair Length

Short-haired and long-haired breeds of dogs are another decision. The owner must note the extra time required to properly care for and groom a long-haired dog. A short-haired dog does not need to be brushed as often and it does not take as long when they are brushed.

Gender

The sex of a dog should be considered in the selection process. Most dog breeds reach sexual maturity between 9 and 10 months of age. There are positives and negatives with both males and females.

If the puppy or dog is a female, precautions must be taken so it does not breed. If a mature female is allowed outside, the male dogs may pick up her scent and come over to breed her. If the owner does not intend to enter the female dog in breed shows, then she may be spayed. *Spaying* is a surgical operation in which the ovaries and uterus are removed, thus eliminating reproduction.

17-11. Cats make interesting pets.

Male dogs or pups will try to break out of pens or off of leashes if a female in heat is nearby. In some cases, the males may run off for several days looking for females ready to breed. Male dogs also fight, so the dog may come home injured. Male dogs may also be neutered or castrated. Just as in the case of the female, if the owner intends to enter the animal in breed shows, it may not be shown if it is castrated. Neutered or castrated dogs may be shown in obedience classes.

CATS

Cats make loving playful pets once they are used to their home and owners. They have not been used for companion animals as long as dogs. The number of cats used as companion animals is less than dogs. Most cats are smaller than dogs and are often well suited to living in an apartment or small home.

Biology

Though cats and dogs often do not get along with each other, they descended from the same ancient mammal, *Miacis*. Cats are carnivores and often will catch birds, mice, and other small animals as prey. The scientific name of the domestic cat is *Felis catus*. Many breeds of cats with unique features are used as companion animals.

Typical domestic cats weigh 6 to 15 pounds (2.7–7 kg), though some are larger and some are smaller. Cats have about 250 bones. The exact number depends on the size of the cat and length of its tail. Their muscles, bones, and structure are designed for speed and quickness. Cats' feet have claws and other structures that allow them to climb trees. Pads on the bottoms of feet serve as cushions and help the cat walk quietly in stalking prey. The eyes of cats have interesting color combinations. The whiskers of cats are tactile hairs. They are connected to sensitive nerves that help the cat protect itself and find its way in darkness. The claws on cats' feet are dangerous to people and can damage furniture. Some cats are declawed. **Declawing** is cutting the claws from the feet of the cat.

Cats may live 12 to 18 years, with some living as long as 30 years. The female cat is known as a **queen,** while the male is a **tomcat.** Baby cats are **kittens.** Gestation is about 63 days. Queens often have litters of

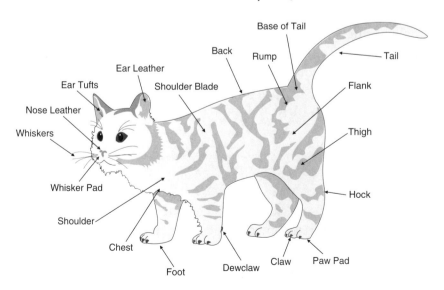

17-12. Major external parts of a cat.

three to five kittens, though some have larger litters. The eyes and ears of newborn kittens are sealed; therefore, they cannot see or hear. Both eyes and ears are open at about two weeks of age. The kittens nurse their mother for about six weeks.

Classes and Breeds

The breed of cat selected by a person largely depends on personal choice. Basically, there are two choices: either the common house cat or a purebred cat.

The common house cat is not a specific breed, but a mix of various breeds. It is produced by chance and not by planning for a specific breed.

The purebred cat is bred by design. Careful planning and selection of bloodlines is used with purebred cats. The purebred cat falls into the two categories of short-haired and long-haired. Thirty-six breeds of cats are recognized in the United States.

Short-haired Breeds. A short-haired cat has low maintenance and requires little or no brushing to look good compared to a long-haired cat. Some of the most common short-haired breeds are Abyssinian, American Shorthair, Burmese, Colorpoint Shorthair, Exotic Shorthair, Havana Brown, Japanese Bobtail, Korat, Manx, Rex, Russian Blue, and Siamese. In all, 26 breeds of short-haired cats are recognized.

Long-haired Breeds. Long-haired cats require more attention to look their best, but they are very pretty when groomed. Long-haired cats also get hairballs, which they must have removed. A ***hairball*** is a wad of hair that collects in their digestive tract. It can clog the movement of food in digestion and cause serious problems for the cat. Some of the most popular long-haired breeds are Persian, Balinese, Birman, Himalayan, Maine Coon, and Turkish Angora. Thirteen long-haired breeds are recognized in the United States.

Age

Kittens are usually cute and adorable, but they are also a lot of work. They require special attention, care, feeding, training, and time to learn. Kittens are more active and playful than grown cats. However, they do adapt to a new home and other animals much quicker and with less stress than cats. It usually takes less than three weeks for the kitten to adapt because it does not have a very long memory of living in another place.

Gender

Male and female cats vary little as kittens, but major behavior differences develop at maturity. If the owner wants to raise kittens, then an obvious choice is a female; but if they do not want kittens, they could have either sex cat. When the owner chooses a male or female, the option of having the cat spayed or neutered at a young age should be considered. Female cats are spayed, which means that they no longer have kittens. This not only eliminates unwanted pregnancy, but also eliminates the period of heat. Female cats have different behavior during heat. Male cats that are not neutered urinate frequently to mark their territory. This is annoying in a home. When let outside, tomcats may stray from home to look for a female to breed or find another male cat to fight.

Purebred or Mixed Breed

Most people enjoy mixed-breed cats. Some prefer purebred or pedigreed cats. If the owner plans on raising kittens or showing the cat, they may want to consider a pedigreed cat. However, it is more expensive to buy pedigreed cats. A pedigree will allow the owner to trace the animal's family history.

RODENTS

Many rodents, such as gerbils, hamsters, mice, and rats are kept as pets. They require a clean cage, a well-balanced diet, clean water always available, and attention from their owner. The owner is repaid with an attentive, loving pet.

Gerbils

Gerbils make fascinating pets because they are very quick and curious. The gerbil came to America in the 1950s for medical research. Unlike most rodents, gerbils sleep during the night and are awake during the day, they are **diurnal.**

17-13. Hamsters in a clean environment that provides for their well-being.

The adult gerbil's body is four inches long, its tail is another four inches long, and it weighs about three ounces. Gerbils are dark brown on their backs, light brown on their sides, and light gray underneath. They have black and white whiskers.

Hamsters

Hamsters are fun because they can ride around in your pocket, and can be trained to do tricks. The hamster came to America in 1938. It is *nocturnal* and sleeps during the day and is awake at night.

Three types of hamsters are used as pets. The Syrian or Golden hamster has a golden body with white patches on its cheeks, front legs and hind feet. The Angora hamster has long, fuzzy hair, which comes in many colors, but is most common in tan, gray, or white. The Chinese hamster is solid black or gray, but has short hair so it can be distinguished from the Angora hamster.

Mice and Rats

Mice and rats can be good company because they are friendly and curious. They can be trained to climb ladders, beg for food, do tricks, and ride on people's shoulders and in their pockets. Tamed mice and rats have gotten a bad reputation because of their wild relatives, which do cause damage and are not loving and fun like tame mice and rats. The history of mice and rats dates back to about 4,000 B.C. where they appear in poems and stories. They are very useful in psychological, biological, medical, and nutritional studies.

Mice and rats come in various colors. Rats come in either all white, black, brown, or hooded, which is white with brown or black around the head and shoulders. White rats are easier to tame and make nicer pets. Mice come in a variety of colors, such as white, black, tan, brown, or

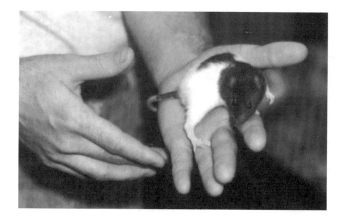

17-14. Hooded rats make colorful pets.

spotted, and there seems to be no difference in their personality based on their color.

Guinea Pigs

Guinea pigs make wonderful friends and greet their owners with whistles. Before 1870, guinea pigs were used as pets. Between 1870 and 1890, they were primarily used for research, and, in the 1900s, they were used as show animals. They were used for research because their bodies are much like humans so they work well in the laboratory.

17-15. Silky guinea pigs are appealing animals.

The guinea pig has a short, heavy body, short legs, and no tail. There are three strains of guinea pigs—the English, Peruvian, and Abyssinian. The English and Peruvian strains of guinea pigs both have short hair and are found in a wide variety of colors, but the Peruvian has rough hair that stands up in all directions, which distinguishes it from the English strain. The Abyssinian strain has long hair, but is also in either solid or mixed colors.

REPTILES

Reptiles are popular with some people. A **reptile** is an ectothermic animal with dry scaly skin and lungs for breathing. Ectothermic means

17-16. The iguana is a popular reptile pet.

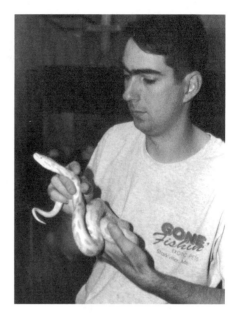

17-17. Snakes are sometimes kept as pets. This shows an Albino Burmese Python.

17-18. The Lesser Sulfur Cockatiel is an interesting bird.

that the animal's body temperature adjusts to its environment. Some people refer to reptiles as cold-blooded animals because they feel cold to the touch.

Two groups of reptiles are sometimes kept as pets: lizards and snakes and turtles. More species of lizards and snakes are kept than of turtles. People planning to use reptiles as pets need to carefully study the animal they propose to use. They need to know how to safely keep it and the requirements for its well-being. Since many people are afraid of reptiles, owners must handle them carefully and keep the animal secure so that it does not escape.

BIRDS

Several kinds of birds make excellent pets in homes and apartments. Special cages will be needed to care for them. Cages must be kept clean. The birds must be provided food, water, and a good environment. Canaries, parakeets, parrots, cockatiels, and macaws, a kind of parrot, are a few examples.

FISH

Fish are quite popular as pets. These fish are known as ornamental or tropical fish. Some live in freshwater, while others live in saltwater. Having fish as pets requires the maintenance of an aquarium. This requires knowledge and skill to keep it in good condition. More details on ornamental fish are presented in Chapter 18.

17-19. The Mexican Redhead Amazon Parrot is a colorful bird.

OTHERS

Many other animals are sometimes used as pets. These include insects, ferrets, hedgehogs, deer, rabbits, raccoons, and farm animals, such as horses, goats, pigs, and cattle. People who have wild animals near their homes often put out food and become friendly with them. Squirrels commonly live in trees in urban areas and become friendly to people. Some of these are wild animals that have not been domesticated. They can suddenly change behavior and attack people.

17-20. A Rose Hair Tarantula kept as a pet.

17-21. Ferrets are active pets and members of the weasel family.

17-23. The sugarglidder is a marsupial, which means that its young develop in a pouch similar to a kangaroo. This female has two babies in its pouch.

17-22. This African Hedgehog is being fed its favorite food—a worm.

MANAGEMENT PRACTICES AND UNIQUE NEEDS

All companion animals require more than just being fed. They require a commitment of time and money from their owner. They need attention, feeding, exercise, pen cleaning, and caring every day.

DOGS

Properly selected dogs can be an asset to the home. Since they often live to be 10 or more years old, this decision is a lasting one. It is important to teach your dog to bark when strangers approach, but to train it not to bark for no reason at all. A dog should be trained to stay at home and not roam on other's property when it is exercised.

If the owner decides to purchase a puppy, it should be bought between eight and 12 weeks of age. This is when it is old enough to be removed from its mother and is still forming its personality. If the owner wants to have a dog which is an average size for the breed, then an average size puppy should be selected.

CATS

Cats enjoy being handled, as long as they are properly handled. When picking up a cat, slip one hand under the cat's chest and hold its front legs gently but firmly with your fingers, and put the other hand under the cat's hindquarters. Treat your cat carefully and make sure it feels secure.

Owners should remember that every cat is different, so give the animal time to get used to its new home and owners. It is wise to keep the cat locked up so that it cannot run away until after it is used to the new owners and home. Give the cat time and space to get used to its new owners and do not ever force attention on the cat. While working with the cat, repeat its name often so the animal learns

17-24. Cat condominiums provide a place for cats to play and exercise.

its name. If the cat is frightened, it will fight and try to run away. Patience and time will allow the cat to feel loved and comfortable in its new surroundings.

RODENTS

In the wild, gerbils, hamsters, guinea pigs, and mice and rats can cause heavy damage to crops since they are rodents. It is important to make sure the animals do not escape. If they escape and multiply, they will cause serious crop damage. Gerbils are illegal to have as pets in California; and the U.S. Department of Agriculture has issued a warning about taking them into desert regions of the west, such as Texas, Arizona, and New Mexico.

Gerbils

Gerbils are shy animals, so be gentle when handling them. Scratching the ears and back of the gerbil will relax it. When holding a gerbil, keep your hand firmly around its entire body to make sure it does not get dropped.

Hamsters

Hamsters must be handled frequently to remain tame. The best way to do this is to stroke its back until it gets used to your touch, then carefully pick it up, place your hand underneath it to support it. After your hamster learns to trust you, it will crawl all over you and ride in your pocket. Always protect your hamster from dogs, cats, and anything else that may scare it, or it may try to escape. A good way to train hamsters is to give them treats for doing things you like. Peanuts, cashews, and other nuts make good treats for the animals.

Mice and Rats

Mice and rats are usually very tame animals, especially when the animal is acquired at a very young age. The younger the animal, the easier it is to tame, and the fastest way to tame it is to hand-feed it. When picking up a mouse or rat, grab it by the tail near its body, place it in your other hand and then gently stroke its head and back. After the animal gets to know its owner, it will be at the door waiting to get out of its cage to crawl around on its owner. Treats may be used to encourage the animals. Since the animals are nocturnal, it is best to play with them in the early evening.

Guinea Pigs

Guinea pigs require attention and enjoy being talked to, especially when eating. When handling the guinea pig, even when it is small enough to fit into your hand, hold the animal with your thumb and forefinger just behind the animal's head and in front of its front legs. As the animal is lifted, use your other hand to support its back end. If the animal remains comfortable, it will remain calm, but if it feels uncomfortable, it may scratch you.

REPRODUCTION

Companion animals should be reproduced if there is a need for the animal. Demand is greater for some kinds of animals than others. People need to help control animal breeding so that the number of pets does not become greater than demand.

DOGS

Dogs should breed and raise puppies only if there is a demand for the puppies. Many dogs are put to sleep (humanely killed) each year at shelters because they are not wanted.

Breeding

The female dog comes into heat the first time when it is between six to 12 months for small breeds and eight to 18 months in larger breeds—they mature slower. The owner can tell when the bitch is in heat because blood is discharged. After the discharge, the bitch should be exposed to a male in 10 or 11 days. If puppies are not wanted, the female should be confined, because male dogs from the neighborhood will approach her when she is outside.

When selecting a male to breed the female to, it is important to have parents whose weak points compliment each other. Although registered dogs have backgrounds on their bloodlines on their papers, know that about 50 percent of the genetic make-up of a puppy comes from one parent.

Whelping

The bitch will give birth, or **whelp,** 63 to 67 days after she is bred by the male, and take care of her puppies with little or no assistance. A whelping box should be constructed for the bitch about two to three weeks before she gives birth so she can get used to it.

The box should be large enough for the female to lay down and stretch out comfortably. Bedding from the female's bed should be put into the whelping box, along with newspapers, shredded paper, or carpet, for her to use at whelping time. The box should be placed in a warm, quiet, secluded place where there is very little, if any, traffic.

17-25. Five-day-old Bulldog puppy with eyelids sealed.

Caring for Puppies

The puppies should be handled as little as possible for the first 14 to 21 days. For the first few days, they need a temperature of about 85°F (29°C).

Newborn puppies cannot see at all nor hear very well for 10 to 15 days following birth. Their ears and eyelids are sealed. Puppies begin walking at about two weeks of age when their ears and eyes open.

Newborn puppies cannot urinate or defecate. They do so when their mother licks under the tail. Puppies begin to wag their tails and bark at about three weeks of age. When they are so young, they need warmth, their mother's milk, and plenty of sleep.

CATS

The female cat's heat period is usually easy to detect. She will growl, chew, maul, and call for the male. After she is bred, it will take between 60 and 65 days for her to give birth. The female may continue her normal routine until the last week of her pregnancy, when she should no longer be running up and down stairs and jumping on furniture.

During the last week of her pregnancy, the female will be restless as she looks for a safe place to have her young. A maternity box should be prepared for the cat and placed in a dark, quiet, warm, draft-free, traffic-free area of the house. All that is needed for the maternity box is a cardboard carton large enough for the mother to stretch out at full length on her side and have room left over. The top should be left attached to keep the area dark. A door should be cut with 3 to 5 inches of cardboard left at the bottom of the door so the kittens do not escape. The bottom should be lined with shredded paper and a towel.

The mother will give birth to a litter, which is usually three to four kittens. After the mother delivers the kittens, she should be left alone. No strangers should be allowed to see the mother or her kittens for at least three days.

At three to four weeks of age, the kittens are ready to begin eating food. Feed the kittens the same food as the mother. If it is dry food, it is a good idea to moisten it with water to make it easier for the kittens to eat. At six to eight weeks of age, the kittens are eating food and are old enough to be weaned. After weaning the kittens, do not overfeed the mother because it will cause her to gain weight and overproduce milk.

RODENTS

Gerbils

Gerbils are ready to mate at three months of age. A gerbil desires only one mate. Potential mates should be kept separated from each other until they grow accustomed to each other's scent. After the female is bred, it takes 24 to 25 days for her to have her litter of about five babies. Both parents stay with the litter until they are weaned at six weeks. However, no contact is encouraged with the young until their eyes are open, or they could be harmed by their mother.

Hamsters

Hamsters together in a cage will breed if there is a male and a female. If you do not want to raise hamsters, it is best to have only one since they do not get lonely nor require the attention of another hamster.

Hamsters should not be bred until they are eight weeks old, and many owners do not recommend breeding until they are three months old. After the female is one year old, it is very rare for her to produce offspring. Only well-developed, healthy animals should be used for breeding. It only takes a hamster 16 days to give birth from the time she is bred. You can have too many hamsters if you do not plan your program carefully.

Start out the breeding process with hamsters by putting the cages of the male and female close to each other so they get used to one another. Place the female in the male's cage. If the male is put into the female's cage, she may kill him because she feels threatened. Wear gloves when handling the hamsters at this time because if she is not in heat they may fight and you may have to remove her. Put the female into the male's cage at night about 9 o'clock because most breeding takes place between 9:00 and 10:00 p.m.

Leave the female alone for nine days after she gives birth, which includes not cleaning her cage. A frightened mother may kill and eat her young. When the young reach two weeks of age, they are old enough to be weaned from their mother.

Mice and Rats

A female mouse may be bred at 8 to 10 weeks of age and a female rat at 3 months. It takes both animals about 21 days to give birth after they are bred. The male and female may be left together until the female gives birth and then the male must be removed until the babies are weaned at about three weeks old. Baby mice have a habit of jumping, so anyone holding the animals should be aware. Also, since the babies reach maturity so quickly, it is important to separate the males and females or more babies will be born. Since a litter consists of six to eight babies, it does not take long to be overwhelmed with young.

Guinea Pigs

Female guinea pigs are called sows and male guinea pigs are called boars. Sows should not be exposed to the boar until they are three to five months old. The boar is always put into the sow's cage and she is never put into his cage. They are left together for three weeks. The sow will give birth between 63 and 72 days after she is bred. The sow will breed again within a few hours of giving birth to her litter and if she is not bred then she will wait until the litter has been weaned. The litter should be weaned at three weeks old and the males and females in the litter should be separated because it is possible for the females to get bred.

NUTRITIONAL REQUIREMENTS

All companion animals require proper nutrition. If not, they will become diseased, will not grow well, and make poor pets. Proper feeding is part of animal well-being.

DOGS AND PUPPIES

Dogs and puppies are fed differently. For a new owner, the key to feeding a puppy is to get food from the person you are getting the puppy from. The same feeding schedule and food should be used the first days your puppy is in its new home, because it has enough other new things to get used to. Starting the second day, gradually switch your puppy over to the food you will be feeding it. After about four days, the puppy should be completely switched over.

Feed the puppy three times a day. The amount of food should not exceed what a puppy can consume in 15 minutes. A high-quality, balanced puppy ration is the best food for puppies and there are many commercial types on the market. If the puppy is eating dry food, moisten it with water so it is easier for the animal to eat. It is important to get the puppy on dry food now, so he will be content eating it the rest of his life.

Dry food for dogs and puppies is much more economical than the other foods. The dry food contains less water, so the animal is getting more nutrition in dry food and less water. The feeding of dogs and puppies is a major factor in keeping the animal healthy from puppyhood into adulthood.

Commercial dog food has been tested and is a complete ration, so it is acceptable to feed it to the dog. The bag of food will indicate how much food the dog should be fed, but it should be what the dog will consume in 15 to 20 minutes. The dog should be fed once per day. The exception is a very large dog or a hard working dog, then twice a day is acceptable. The dog should be fed on the same schedule each day as well as the same feed. Table scraps may be added to get the dog to eat its food better, but do not rely on them to add anything to the dog's diet except fat and energy.

17-26. Dry dog food often comes in large, economical bags.

CATS AND KITTENS

Cats will remain healthy and happy if they eat a well-balanced diet and always have water available. Many owners give in to the idea of feeding their cat from the table. This may sound like a good idea, but it does not provide the cat with a well-balanced diet.

Cats require a high protein, high fat diet. The surest way to feed a cat a well-balanced diet is to feed a high-quality commercial feed. The food is available in either dry, semi-dry, canned specialty, and canned maintenance. Cats have different needs and different tastes so it is up to the owner to decide which food works best for their cat. Labels list the information needed to select the proper food for the cat.

17-27. Clean bowls for a cat's food and water.

If the cat is pregnant, she should continue to receive her same daily feed until the last 20 days of her pregnancy, then she should be allowed to eat all she wants. After the kittens have arrived, the first few days the mother may not eat, but make sure she has both food and water available. During the time the mother is nursing the kittens, feed her all of the food she wants. Producing milk is hard work for the mother's body and it requires lots of food.

The day the kittens are weaned, take all food from the mother. On the second day, increase feed to one-fourth of what the mother used to eat. On the third day, feed one-half the amount. On the fourth day, feed three-fourths, and on the fifth day, feed the normal amount. Limiting the amount of food the mother intakes will help decrease her milk production.

RODENTS

Gerbils

Gerbils only eat one tablespoon of food per day, only need to be fed once per day, and also drink very little water. Gerbils eat grain, seeds, roots, grasses, sunflower seeds, corn, oats, wheat, watermelon seeds, bits of apple and lettuce, and may enjoy fresh grass as a treat. Different gerbils enjoy different foods. So, experiment to see what the animal enjoys the most.

Hamsters

Hamsters eat about one-half ounce of food each day and should be fed at night when they are active. The natural food of hamsters is grains, seeds, and vegetables. Another option for a hamster is to feed corn, oats, or wheat and mix them with prepared dog food.

As hamsters eat, they stuff their cheek pouches, hide the food, and return to do the same. In the wild, hamsters burrow into the ground and make one main tunnel with many side chambers in which they store food. As you clean the cage, you may find pockets of food; it is okay to put the food back where you found it, as long as it has not spoiled.

Mice and Rats

Mice and rats have the same water requirements as all other animals. So, make sure they always have plenty of clean, fresh water. They eat dry dog food and that is enough to keep them healthy, but you may want to surprise them with foods, such as seeds, nuts, rabbit food pellets, hard-boiled eggs, bread, breakfast cereals, rice, leafy foods, or raw potatoes. Feed dishes should be small enough so the animals cannot sleep in them and attached to the side of the cage above the floor. The food will stay clean from the animal's droppings.

Guinea Pigs

Guinea pigs drink very little water because they eat so many greens. However, make sure they always have some clean water. Feed should be available at all times. At the end of the day, old food should be removed because the animal can get sick from rotten food. Favorite foods are alfalfa, apples, carrots, corn, dandelions, lettuce, cauliflower, clover, celery, lawn clippings, spinach, and tomatoes. Dry foods, such as rabbit food pellets, and a salt spool must also be available.

Guinea pigs, like humans, like different foods, so experiment to see what the animal prefers. Guinea pigs, monkeys, and humans are the only

17-28. Carefully select the commercial food that is best for a companion animal.

three animals that cannot make their own vitamin C, so make sure the guinea pig gets some of the foods mentioned above because they all contain some vitamin C.

HEALTH PRACTICES

Owners and producers of companion animals want to keep the animals in good health. Sanitation goes a long way in keeping all animals healthy.

DOGS AND PUPPIES

Dogs and puppies require routine visits to the veterinarian to make sure they are healthy and to receive all of their vaccinations against diseases and wormings against parasites. Many times, a breeder will take the bitch after she is bred for a re-vaccination. This allows the puppies to be born with antibodies to those diseases the mother was vaccinated for.

The external parasites dogs can get are fleas, lice, ticks, mange, and ringworm. There are also some diseases dogs can transmit to humans, such as skin diseases, rabies, and roundworm; however, the incidence of this is very small. The best way to keep a dog or pups healthy is to keep its area clean, remove droppings, and to take it to the veterinarian for routine visits.

17-29. This puppy has a clean environment in an animal shelter.

CATS AND KITTENS

Cats are generally healthy animals, especially if proper precautions are taken. First of all, the new owner of the cat should take it to the veterinarian to have it completely checked out. Also, vaccinations will be given to the animal to prevent it from getting some diseases. The veterinarian will inform the owner about other shots, wormings, and office visits the animal needs in order to stay healthy.

If the cat does not feel well, it may show some of the following signs: coughing and sneezing; vomiting; runny nose; red, watery eyes; severe, prolonged diarrhea; loss of appetite for several days; dull coat; and a non-caring attitude. If the cat shows any of these signs, a visit with the veterinarian is in order.

RODENTS

Gerbils

Gerbils are fairly disease-free. Gerbils will use one corner of their cage as their toilet, so it is important to clean it every one to two days. Gerbils do not like cold temperatures or drafty areas. Keep the temperature between 65 and 80°F (18.3 to 26.6°C) and, if the gerbil is in the sunlight provide it with an area for shade. Gerbils also need hard, dry food or even a block of wood to chew on. This will keep their teeth filed down, so they can chew properly.

17-30. **Meeting the mineral needs of companion animals helps keep them healthy. Specially designed products often help and entertain animals.**

Hamsters

Hamsters that are healthy exhibit the following signs: soft, silken fur; a plump body; and prominent, bright eyes. A hamster that is long or skinny, has watery eyes, a runny nose, or a wet tail may be ill and if clumps of fur are missing, the animal may have a disease called mange.

Hamsters and guinea pigs may get diseases or parasites from wild mice or birds. If they do get fleas, lice, or other insects, it is okay to treat them with an insecticide for cats. Never use an insecticide for dogs, but one for cats is fine.

Mice and Rats

Mice and rats will live a healthy life if they are kept in good health. They need a block of wood to chew on to keep their teeth in good condition and a clean, dry cage. Like hamsters, it is okay to dust them with cat flea powder and disinfect their cage. However, never use a flea powder designed for dogs.

Guinea Pigs

Guinea pigs are very healthy animals. If they are not feeling well, the owner will be able to tell because the animal will sit perfectly still, all hunched up, its coat will be ruffled and messy, it will not eat and will lose weight, and its droppings will be loose and watery. If any of these symptoms appear, the animal should be separated from other animals and taken to the veterinarian. If the animal picks up parasites, it may be treated the same way as the hamster.

FACILITY AND EQUIPMENT NEEDS

Companion animals need certain facilities and equipment. Owners have the responsibility of caring for the well-being of animals. The owner must also keep the animals from disturbing other people. People who do not have animals may not appreciate your animal.

DOGS AND PUPPIES

Dogs and puppies kept outside need shelter from the heat and cold. An outside dog requires a house that is warm in the winter and cool in the summer. At all times, it should be draft free, dry, easy to clean, and with an area for exercise. The house should be big enough that the dog can lay in the back and warm it with its body heat at night and lay in front during the day with its head out to protect the area. Some owners even put a partition between the two areas.

17-31. A spacious pen with house that provides for the well-being of the Alaska Husky.

The biggest problem is that houses are often made too large and the animal's body heat cannot warm the area. A dog the size of a beagle needs about 2 × 2 feet (0.7 × 0.7 m) of floor space and a roof height no taller than 18 inches. A medium-sized dog of about 40 to 75 pounds (18 to 34 kg) requires floor space of 4 × 4 feet (1.4 × 1.4 m).

An owner should also provide a collar, warm bedding, water and feed dishes, and a brush for the animal. If the dog gets away, it is important to have a tag on the collar identifying the owner and the address where the dog belongs.

CATS

Cats, if they are kept inside, require a bed, carrier, litter pans, a scratching post, and possibly some toys. The equipment does not have to be expensive, but in order for your cat to be happy and healthy the items are needed. If your cat stays outside, it does need protection from the elements and a safe place to eat where other animals do not eat its food and drink its water.

A *pet carrier* is a carrying case for a cat or other animal. Since all cats must be taken to the veterinarian, they will, at some time, travel. The case should be large enough for the animal to turn around in, well ventilated, easy to clean, and lined with a towel in the bottom.

17-32. Carrier for a cat.

17-33. Covered litter pan that reduces odor.

The bed for the cat does not have to be a fancy bed purchased from a pet store. It may be something as simple as a cardboard box with fairly high sides and the bottom lined with a cushion. Cats prefer quiet places to sleep that are secure, warm, and draft-free. To avoid drafts, it is a good idea to place the bed several inches off the floor.

A *litter pan* is a container designed for the cat to use in urinating and defecating. Most are made of plastic or baked enamel on metal, which will not rust, so they work best. The pan should be washed with soapy water weekly, when the litter is changed. As far as litter, all types are available, such as prepackaged sawdust, shredded newspaper, sand, clean soil, peat moss, or wood shavings, which are often used with a small amount of baking soda to help eliminate odor. If droppings are removed daily, changing the litter once a week should be sufficient.

Cats are curious and energetic animals, so they need a scratching post for exercise. Contrary to popular belief they do not need it to sharpen their claws. If the owner makes the post, they need to keep in mind that it needs a wide base and do not cover it in carpet. It is hard for a cat to know the difference between carpet on the post and carpet on the floor.

Cats also enjoy toys. However, avoid toys that are rubber, furry, or woolen, and can be torn apart and eaten. Watch the cat to see the types of toys it enjoys. Many toys can be made from things already around the house, such as a paper bag, shoe box, or a spool attached to a strong string and hung from a door.

RODENTS

Gerbils

Gerbils, because of their curiosity, prefer interesting homes, which may be purchased or made by the owner. If a cage is made or purchased, the best choice would be one with two floors so the animal can get exercise

17-34. Snakes prefer gravel and small stones in their containers. This shows two Red Tail Boa snakes in a covered aquarium.

by running up and down and all around. To create fun and excitement for the animal, the owner may want to create stairs, holes, and other play things for the gerbil. Make sure the cage has a top and keep it covered so the gerbil does not escape.

Gerbils need bedding in their cage, which is absorbent and clean. Cedar chips, sawdust, and small animal litter work well when put into the cage an inch thick. Gerbils build nests, so put a small piece of burlap, paper, or cardboard in the cage for the animal to use to build its nest.

Hamsters

Many hamster owners buy bird cages or use old aquariums to house their animals. If a wooden cage is used, be sure to cover the wood on the inside so the hamster does not gnaw its way out.

Hamsters need a clean, dry cage away from drafts—55 to 80°F (12.8 to 26.7°C). If the temperature falls below 45°F (7.2°C), the hamster will go into hibernation. It may appear as though the animal is dead, but it is hibernating. Hold the animal to slowly warm it and feed it warm milk, one drop at a time, until it revives. When returning the animal to its cage, cover the cage with a heavy cloth. Hamsters also need a hard, dry food or even a block of wood. Chewing on a hard material will keep their teeth filed down so they can chew properly. Give the female scraps of cloth, tissue, paper, or cotton to line her nest. Hamsters use one corner of their cages for a toilet, much like gerbils.

Mice and Rats

Mice and rats must be kept in wire or metal cages because they will gnaw through wooden cages. They need a hard, dry food or even a block of wood to chew on to keep their teeth in top condition. A solid upper platform should be provided for the animal to rest on and platforms and ramps should connect the floor and platform.

Since mice and rats are curious animals, toys, such as swings, perches, and exercise wheels should be included. Sawdust, cedar shavings, or cat litter should be used as bedding to keep the cage clean and dry. For nesting, cotton or shredded paper should be supplied.

Guinea Pigs

Guinea pigs are not fighters and they do not jump or climb, so their cages may be small. Their cages should have some type of cover, even if it is just wire, to keep other animals away. A guinea pig needs at least one square foot of floor space, but many animals may be housed in one cage, as long as the cage is large enough. For example, a cage large enough for one male and three females would be 36 by 24 inches (0.9 by 0.6 m).

Guinea pigs need a dry floor of which at least half of it is solid and not made of wire. To keep the cage dry, a thin layer of shavings, sawdust, or straw may be put down, but make sure it is not dusty material or the animal may get sick. Guinea pigs are also temperature sensitive and will not grow well if the temperature gets below 65°F (18.3°C) in the winter or if they are not protected from the extreme heat of summer.

17-35. **An easy-to-keep bird cage that provides a swing, feeder, waterer, and other features to entertain birds.**

REVIEWING

MAIN IDEAS

Dogs, cats, rodents, guinea pigs, and many other animals are used as companion animals. They provide entertainment, relaxation, and friendship to their owners. Regardless of the type of animal used, it is important to care for their well-being.

It is important to know the use for the animal, other than companionship. If the owner wants an animal to do certain tricks, or be easily trained, the best thing to do is study the types of animals that fall into the proper category. After studying, talk to people that own the animals and find out the animals' strengths and weaknesses.

Once a pet is purchased, take good care of it, from feeding to regular visits with the veterinarian, to ensure the animal of a healthy, long life. A pet could be a part of the family for up to 10 years, so make the decision a family decision.

QUESTIONS

Answer the following questions using correct spelling and complete sentences.

1. Why are companion animals important to people?

2. What are the classes of dogs? Briefly describe each class.

3. What are the two classes of cats? What are the advantages and disadvantages of each class?

4. What rodents are commonly used as companion animals? Briefly describe each.

5. What are reptiles? Name examples that are used as companion animals.

6. What important management practices are followed with companion animals?

7. What important production practices are followed with companion animals?

8. What are the nutritional needs of companion animals? How are these needs met?

9. What are the important health practices with companion animals?

10. What equipment and facilities are commonly needed with companion animals?

EVALUATING

CHAPTER SELF-CHECK

Match the terms with the correct definitions. Write the letter by the term in the blank that is provided.

a. queen
b. whelping
c. diurnal
d. nocturnal

e. pet carrier
f. companion animal
g. spaying
h. carnivore

i. kitten
j. tomcat

1. ____ baby cat

2. ____ meat-eating animal

3. ____ neutering a female dog

4. ____ container for carrying pets

5. ____ mature female cat

6. ____ bitch giving birth

7. ____ animal that sleeps at night and is active during the day

8. ____ mature male cat

9. ____ animal that sleeps during the day and is active at night

10. ____ domesticated animal kept by humans for enjoyment in a long-term relationship

EXPLORING

1. Visit a nearby store that sells companion animals. Make a list of the species in the store. For each species, list the equipment and facilities needed to care for it. Also, list the food, health care, and other items needed to care for the well-being of the animal.

2. Volunteer to pet-sit or care for animals for other people. This can become a profit-making enterprise by looking after pets while their owners are away. Be sure to visit with the owner about the needs of the animal and how it is managed. Develop a schedule of activities for the time you are responsible for the animal.

3. Tour a veterinary medical facility that treats small animals. Interview the veterinarian about the work. Determine the kinds of animals and the common veterinary medical procedures that are performed. Study how animals are restrained and moved. Draw a floor plan of the arrangement of the facility, including the animal boarding area. Note how the pens or cages are constructed and maintained.

Chapter 18

ORNAMENTAL FISH

People enjoy watching fish! They like to see them swim and go about life in an aquarium or open pool in a courtyard. Watching fish even relieves stress and helps people lead healthier lives, according to research.

Setting up and keeping ornamental species is an interesting hobby. It allows people to express themselves in the design of the aquatic environment. The fish or other species may be kept much as people keep dogs and cats as pets. They must have proper care.

Ornamental fish form a large industry in the United States. These fish are more valuable than food fish based on weight. People want to buy new fish and all the supplies. This means that ornamental fish must be raised. Some people find ornamental fish a profitable crop to grow.

18-1. Caring for a goldfish can be fun.

OBJECTIVES

This chapter provides an introduction to raising and keeping ornamental fish. It has the following objectives:

1. Explain ornamental species and list examples
2. Explain the equipment and facilities needed for ornamental species
3. Describe the water environments for ornamental species
4. Describe how to care for ornamental species

TERMS

aeration	egg-laying fish	ornamental fish
aquarium	filtration	oxygenation
aquarium maintenance schedule (AMS)	freshwater ornamental fish	salinity
biological filtration	habitat	saltwater ornamental fish
biological oxygenation	livebearing fish	shoal
chemical filtration	marine aquarium	synthetic seawater
companion fish	mechanical filtration	tapwater
earthen pond	mechanical oxygenation	tropical fish
	nitrogen cycle	vat

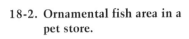
18-2. Ornamental fish area in a pet store.

ORNAMENTAL SPECIES

Ornamental fish are fish kept for their appearance and personal appeal to people. Some ornamental fish have bright colors and large, fancy fins. They have high personal appeal to people. These fish are not typically used for food.

HOW ORNAMENTAL SPECIES CAN BE GROUPED

Ornamental species are grouped as tropical, companion, and other species.

Tropical Fish

Tropical fish form a subgroup of the ornamental species. *Tropical fish* refers to small, brightly-colored fish that are popular in home and office aquaria. These fish typically thrive in warm water and breed rapidly. They are usually tolerant of the conditions in home aquaria. Tropical fish range from 1 to 12 inches in length.

Companion Fish

Companion fish are kept in homes for human companionship or as pets. An aquarium with fish may provide an important use of leisure time. Companion fish amuse, entertain, and provide enjoyment. Have you ever stood and watched fish swim about in the water? Companion fish may be tropical fish or they may be other species of fish, as well as other aquatic animals, such as snails or turtles, and aquatic plants.

Other Species

Some species of ornamental fish are neither tropical nor companion fish. They are kept for their beauty and appeal to people. These are often the larger species in indoor or outdoor pools or large aquaria. The koi is a good example.

KIND OF WATER

Tropical fish usually live in freshwater and saltwater. A few live in brackish water. The freshwater species are usually considered easier to keep and raise. Setting up and keeping a saltwater aquarium is more demanding because of the water. Materials may be added to freshwater so that it has the qualities of saltwater. Providing the right environment can be a tremendous challenge! (Refer to Chapter 14 for more information on water.)

SOURCES OF ORNAMENTAL FISH

Ornamental fish are captured from the wild or raised in confinement. In the past, most ornamental fish were caught from streams and lakes. They brought with them any diseases or problems that existed. Many of them often died. People have found cultured ornamental fish to survive better in aquaria. A few people raise ornamental fish for stores and pet shops. Florida is the leading state in the production of ornamental fish.

Wild ornamental fish are typically caught in the tropical waters of the Pacific Islands, Asia, and South America. Proper treatment of captured fish can rid them of disease and help them survive in captivity. Providing good conditions to transport the fish will help them survive.

EXAMPLES OF ORNAMENTAL FISH

Hundreds of different fish may be used as ornamental fish. Some are much easier to keep than others. People who are just beginning to raise ornamental fish should start with those that are easier to keep. With experience, they can begin to raise those that require increased care.

Most ornamental fish are not native species. This means that they have been brought into North America from another place. In some cases, releasing these fish into streams is illegal. In other cases, imported fish will not survive if released because they are not adapted to the climate or water.

Freshwater Ornamental Fish

The *freshwater ornamental fish* are most popular. These are species that grow in freshwater. They will tolerate only a very small amount of salt in the water.

The freshwater ornamental fish reproduce in two ways: laying eggs and giving birth to live young.

Egg-Laying Fish. *Egg-laying fish* reproduce by the female fish producing eggs that are fertilized by sperm from the male fish. Fertile eggs hatch after a period of incubation. Incubation usually lasts only a few days.

Popular egg-laying ornamental fish include koi, goldfish, gouramis, tetras, barbs, and catfish.

Koi—The koi is a variety of common carp (*Cyprinus carpio*) that was developed by selecting fish with distinctive coloring. The colors may be bright red, black, gold, or white. Koi, a freshwater fish covered with scales, may reach a length of 37.5 inches (95 cm) and weigh 3 to 10 pounds or more. Most koi grow in pools or large tanks. Koi clubs are found in some places. The members raise and enter koi in various competitive events. Koi-keeping is popular in Japan.

18-3. Large koi.

Goldfish—People have been keeping and raising goldfish (*Carassius auratus*) for over 2,000 years. In the same family as the koi, and sometimes known as golden carp, goldfish are smaller than koi. They are adapted to a wide range of environments. New varieties of goldfish with bright colors, fancy fins, and interesting eyes have been developed. Goldfish are hardy and easy to keep. Their size depends on the size of the container they are grown in and the nutrition they have. They range from a couple

18-4. Koi in a decorative interior water garden.

18-5. Gouramis being watched in a tank.

18-6. Small aquarium used for tetras.

of inches long to over 2 feet long. A goldfish must have at least 2 gallons of water to grow to 2 inches (5 cm) in length. The goldfish is a good species for the person who is just beginning with ornamental fish. However, their water must be cleaned often because goldfish are messy. Goldfish will live up to 15 years.

Gouramis—Several different species are known as gouramis. The most interesting is the kissing gourami (*Helostoma temmincki*) because of the unique kissing behavior of pairs of the species. Other popular gouramis include the three-spot gourami (*Trichogaster trichopterus trichopterus*), blue gourami (*Trichogaster trichopterus*), and the pearl gourami (*Trichogaster leeri*). Gouramis are freshwater fish that may reach lengths of 12 inches, depending on the species. Most gouramis prefer water that is in the range of 75°F (24°C).

Tetras—The tetras include several species that are easy- to medium-care ornamental fish. The neon tetra (*Paracheirodon innesi*) has a distinctive electric-blue stripe the entire length of its body. The cardinal tetra (*Cheirodon axelrodi*) has a brilliant reddish color. Several other tetras are often raised. Most tetras grow to a maximum length of 1.5 to 3 inches (3.75 to 7.5 cm). Tetras prefer water that is 72 to 85°F (22 to 28°C).

Barbs—Barbs are popular. They are easy to raise and are in the same family as the goldfish. Barbs will reach a mature size of 2 to 4 inches (5 to 10 cm) in length. They prefer freshwater that is 70 to 80°F (21 to 27°C). Most barbs like aquaria with plenty of light. Three common barbs are the spotted barb (*Barbus binotatus*), rosy barb (*Barbus conchonius*), and tinfoil barb (*Barbus schwanenfeldi*). Some barbs eat their eggs after spawning.

Catfish—Some species of America's most popular food-fish are grown as ornamentals. The upside-down catfish (*Synodontis angelicus*) is interesting to watch in an aquarium because of its behavior in swimming upside-down. The glass catfish (*Kryptoptereus bicirrhis*) and electric catfish (*Malapterurus electricus*) are also interesting species. The glass catfish has a clear-like body that may appear as a rainbow in the right light. The electric catfish can give off an electric shock that will kill smaller fish. Catfish prefer water that is 70 to 80°F (21 to 27°C). Most catfish prefer the darker areas of an aquarium.

18-7. Six common aquarium species: comet goldfish (top left), cardinal tetra (top right), tiger barb (middle left), catfish (middle right), guppy (bottom left), and male swordtail (bottom right).

Livebearing Fish. The *livebearing fish* give birth to live young. They like to live in a group of five or more, known as a **shoal.** Rather than releasing sperm over the eggs, as with egg-laying fish, the males deposit sperm inside the female fish where the eggs are fertilized. The fertilized eggs develop into nearly fully-formed fish. The females of some species can store sperm from the male until needed for reproduction. Four kinds of livebearers are most common: guppies, swordtails, mollies, and platys.

Guppies—The guppy (*Poecilia reticulata*) is the most popular livebearer. Many varieties have been developed, with the primary difference between them being the shapes of their fins and tails. Guppies like plenty of food and will eat different foods. Most producers feed a dried commercial food. Guppies may reach lengths of 2.5 inches (6.35 cm). Females may give birth to 200 baby guppies at a time, though the average is about 50 babies. They prefer water that is 68 to 75°F (20 to 24°C). Guppies reproduce profusely, but the adults may eat the young.

Swordtails—Similar to guppies, a major distinction of the swordtail (*Xiphophorus helleri*) is the long, sword-like appearance of the caudal fin. Many different colorings are found on swordtails. They may reach lengths of 3 to 4.75 inches (8 to 12 cm) and are easily kept with other livebearers. Swordtails often eat their young when they are born, so producers use special traps through which the young can escape. Swordtails, live guppies, eat live and dried food. Their environmental requirements are similar to the guppy.

18-8. Attractive fresh-
water aquarium.

Mollies—Most mollies are of two species: small fin (*Poecilia sphenops*) and large fin (*Poecilia latipinna*). With some mollies, the fins get so large that swimming is nearly impossible. Most mollies are black. Other colors are increasingly in demand, such as the white molly. Mollies prefer water that is 72 to 82°F (22 to 28° C). Mollies like to eat vegetable foods, such as algae and cooked spinach. They like to form large schools in tanks with other mollies. It is best to keep several mollies together in an aquarium.

Platys—Platys are popular in aquaria and have been bred to achieve certain desired characteristics. The variatus platy (*Xiphophororus variatus hybrid*) is especially popular. It has a yellow color with a distinctive orange-red caudal fin and dark-lined vertical markings. Some platys may be brownish-yellow in color. They grow to about 2.5 inches long (6 cm) and prefer water that is 68 to 77°F (20 to 25°C). They are hearty eaters, being especially fond of live and dried food. Platys are easy to keep with other livebearers.

Saltwater Ornamental Fish

Saltwater ornamental fish are those that will survive and grow in saltwater. They are species that naturally live in seawater, which is commonly called saltwater. The water is a complex mixture of various minerals, with salt (sodium chloride) being one of the substances that is found in the largest amount. Other minerals in seawater include magnesium chloride, magnesium sulfate, and calcium carbonate. Very small amounts of several other minerals are found in seawater, such as zinc and molybdenum. Most saltwater ornamental fish require natural or synthetic seawater. A saltwater aquarium is a **marine aquarium.** All of the species included here are egg-layers.

18-9. Using a hydrometer to check saltwater. (Water in a marine aquarium should have a specific gravity of 1.020 to 1.025.)

Angelfish—There are many different species of angelfish; some able to live in freshwater. Most angelfish appear delicate,

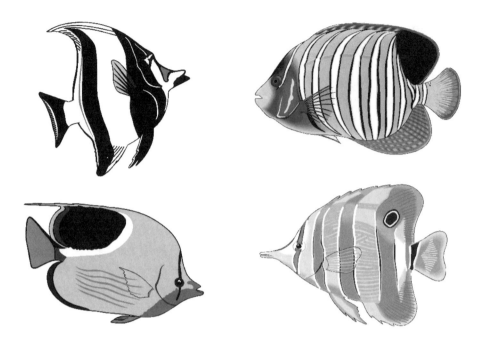

18-10. Four saltwater fish are moorish idol (top left), regal angelfish (top right), saddleback butterfly, (bottom left), and longnose butterfly (bottom right).

but they are hardy and are capable of living a long time in a well-managed aquarium. A popular angelfish is the French Angelfish (*Pomacanthus paru*), which has a black body and vertical yellow stripes. The coral beauty angelfish (*Centropyge bispinosus*) has a wide range of colors, including orange and black. Most angelfish prefer a water temperature of 77 to 86°F (25 to 30°C). Angelfish may be 6 inches (15 cm) long and 10 inches (26 cm) high (tip of fin to tip of fin). Angelfish eat a wide range of live food in natural environments, but will eat commercial food in captivity. Their eggs hatch in about 36 hours. As part of the incubation process, the eggs are carried in their parents' mouths and placed on the leaves of water plants and later in the sand.

Butterfly fish—Butterfly fish are some of the most beautiful and popular marine aquarium fish. Several different species are popular. Most of the butterfly fish need to have ample space in an aquarium. The long-nosed butterfly fish (*Forcipiger flavissimus*) has beautiful bright yellow and black colors, with the snout and throat being green. This fish will grow to lengths of 6 inches (16 cm). Butterfly fish prefer temperatures of 75 to 82°F (24 to 28°C). Butterfly fish are territorial and are preferably isolated from other butterfly species. Some people keep only one butterfly fish in a tank.

Basslets—Basslets are small colorful fish that are popular in marine aquaria. The royal gramma (*Gramma loreto*) is among the best known basslets. The royal gramma is a hardy fish and particularly good for the person who is just beginning a marine aquarium. Royal gramma are very aggressive toward others of their species. They are, therefore, usually isolated from other royal gramma. They can be mixed with other species of fish. Another popular basslet is the bicolor basslet (*Pseudochromis paccagnellae*), which has a yellow and violet color. The bicolor basslet is also an aggressive fish. Basslets prefer tiny, living food, such as brine shrimp. They like water that is 79 to 82°F (26 to 28°C).

MAINTAINING THE ENVIRONMENT: EQUIPMENT AND FACILITIES

Ornamental fish must have a good environment in order to survive and grow. This environment is artificially created by the equipment and facilities that are used. To some extent, what is needed varies with the species that are being grown. For example, goldfish can exist in simple systems, but other fish, such as angelfish, may require more elaborate equipment.

The basic equipment is a water container. Attached to the water container are other items intended to make the environment better for the fish. Water containers are often decorated with colorful gravel, aquatic plants, and other items.

WATER CONTAINERS

Several different kinds of water containers are used. Some are used to display fish; others are used to reproduce the fish and keep them in good health.

Aquaria

An ***aquarium*** is a container used to hold water. Sometimes the containers are known as "fish tanks" or "tanks." An aquarium should be water-tight. Most people begin with small aquaria and expand into larger

18-11. Setting up a new 29-gallon aquarium.

sizes as they develop skills in using them. Several smaller aquaria also have advantages over one larger aquarium, especially in disease control and water management.

A simple aquarium is the goldfish bowl. This is a rounded glass bowl with a water capacity of about 1 gallon (4 liters), but it may be smaller or larger. It is used with species that are hardy and can survive in water that may get fouled and is not regularly aerated.

The most common aquaria are rectangular, though some are square or spherical. Specially-built aquaria are available, but are often expensive. Aquaria range in size from 10 gallons (42 l) to 30 gallons (127 l), 50 gallons (212 l), 100 gallons (424 l), or larger.

Aquaria of all-glass materials are preferred, especially with saltwater. Aquaria that have metal corners or other parts will corrode and are more likely to leak. All-glass aquaria are easier to clean. The glass in an aquarium should be at least $^1/_4$ inch (64 mm) thick for small aquaria and $^3/_8$ inch (96 mm) thick for larger aquaria. A disadvantage of all-glass construction is that the larger aquaria are very heavy because thicker glass is used in making them. Only high-strength glass should be used in building an aquarium.

Aquaria need to be set up on stands that will support the weight of the tanks. The stands should be level and located where they will not need to be moved often. Moving an aquarium is a big job because of its weight and because the fish must be kept in good condition during the process.

Pools and Fountains

Decorative pools and fountains may be used for goldfish, koi, and similar species that are hardy. These structures are made of plastic, fiberglass, concrete, or other materials. The construction should be water-tight and easy to clean. The arrangement of the facilities should make it easy to manage the water and empty the container, should the need arise.

Vats

Vats are large tanks made of concrete, fiberglass, or a similar material. Vats are in a fixed location and cannot be moved about. These structures are used to raise and reproduce fish and not usually for displaying fish. Vats are rather large and may hold hundreds or thousands of gallons of water. The structures should be designed for ease of management. Water circulation, aeration, drainage, and other features should be designed into vats. Vats should be water-tight and designed to meet the needs of the species which are to be produced in them.

18-12. Vat with screen cover used to raise koi.

Earthen Ponds

Adapted ornamental species may be produced in small ponds made of earth, known as **earthen ponds.** These are frequently used with koi, goldfish, sunfish, and other species that are adapted to the local climate. The pond's construction should prevent seepage or overflow from nearby streams. Select a location that keeps predators and undesirable species out of the ponds. Grassy areas are needed around the edges to keep the water from getting muddy and to reduce erosion.

WATER QUALITY EQUIPMENT

The water in a fish tank should be kept appropriate for the growth of the species. This will require attention to several areas in water management.

Oxygenation

Oxygenation is the process of keeping adequate dissolved oxygen (DO) in the water. All species require oxygen to live. Oxygen that has been dissolved in the water is removed by the gills of the fish and used for life

18-13. Aquarium kits often have complete systems for water management.

processes, known as respiration. Water that has little movement in a tank takes on little new oxygen. The fish can soon use all of the DO up and die. The more densely populated the water is with fish (higher biomass), the greater the need for oxygen. A lone goldfish can survive a long time in a bowl where the water is not moving. Add more fish, and they may all soon die.

Several methods are available to keep dissolved oxygen in the water. Some are much more practical than others. Two methods of adding DO to water are commonly used: mechanical and biological.

Mechanical Oxygenation. *Mechanical oxygenation* involves using devices to add oxygen. These devices may splash water into the air or inject air into the water.

Aeration is a physical process of bubbling air into water or splashing water in the air. Aeration promotes the gas transfer between the water and the air. The supply of oxygen in the water is replenished by aeration.

With aquaria, air or oxygen may be injected into the water. An air pump may be used to force air through a plastic tube into the water at the bottom of an aquarium. With vats, aerators may be suspended on the top surface of the water and use fans to splash the water in the air. Splashing increases the rate of diffusion of oxygen in the water.

Biological Oxygenation. *Biological oxygenation* involves using phytoplankton (tiny plants) and other aquatic plants to replenish oxygen. These plants carry on photosynthesis to produce food. In the process, oxygen is released into the water. Biological oxygenation is especially important with large outdoor earthen pools.

Filtration

Fish create solid wastes in the water. These wastes foul the water and make it unfit for fish. *Filtration* is the process of removing solid particles and gases from the water to keep a good environment for the fish. Three kinds of filtration are used: biological, mechanical, and chemical.

Biological Filtration. *Biological filtration* involves using bacteria and other living organisms to convert harmful materials into forms that are less harmful. This is the most important kind of filtration in an aquarium. The bacteria feed on the uneaten feed, feces, and gases, such as nitrogen, in the water. Snails are sometimes used in aquaria to remove organic matter from the water. Crawfish may be used to scavenge food particles at the bottom of the aquarium.

Bacteria are the most important because their action changes nitrogen compounds into less harmful forms and replenishes the DO in the water. Aquarium wastes may form ammonia, a compound containing nitrogen. This material undergoes denitrification in the nitrogen cycle. The bacteria are primarily found in the filter bed of the aquarium.

Mechanical Filtration. *Mechanical filtration* involves using various kinds of filtration devices to remove particles from the water and keep the water clear.

Mechanical filtration involves flowing water over and through filters made of gravel, charcoal, and fibrous materials (known as floss). Filtering materials must be cleaned often; otherwise, they will become clogged with wastes and will not work very well. Many mechanical filtration systems are tied in with oxygenation systems.

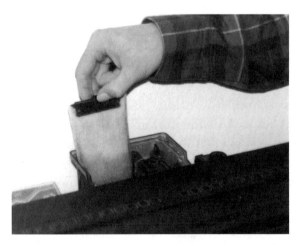

Chemical Filtration. *Chemical filtration* involves the use of chemical processes to filter water. Special kinds of chemical filters are used. In some cases, ozone and ultraviolet irradiation may be

18-14. Floss filters must be regularly changed or cleaned in a mechanical filtration system such as this one on a small aquarium. (Some of the filters contain charcoal.)

used, but they often are not essential. Chemical filtration often uses activated charcoal to help keep the water clear and prevent water yellowing.

18-15. Canister filter for a medium-size saltwater aquarium.

Combinations of Filtration Methods. Aquaria often use combinations of biological, mechanical, and chemical filtration. Devices are added to a fish tank so that filtration can take place. The major kinds of filters are undergravel filters, canister filters, outside power filters, and nitrifier/denitrifier filters.

Undergravel filters are placed at the bottom of the fish tank. The water is pulled through the bed of gravel. The gravel bed is both a mechanical and biological filtration system. The gravel particles screen solid materials out of the water and provide a place for beneficial bacteria to grow.

Canister filters are filtration systems outside the aquarium. Pumps and a system of tubes move the water through several layers of filtering material, such as activated charcoal and filter floss. The material removes solid wastes and provides a location for the bacteria to act. The material must be cleaned or replaced every few weeks, depending on the extent of fouling of the water.

Outside power filters hang on the back of the aquarium. Water is moved through the filtration material much as a canister filter. These are common on small aquaria.

Nitrifier/denitrifier filters supplement the action of bacteria in the nitrogen cycle. These filtration devices are expensive and used only in special situations in aquaria.

Thermometers and Heaters

Thermometers measure the temperature of the water. They often have thermostats attached to them. A thermostat is a control device that turns on the heater when the water temperature gets below a certain level.

Heaters help keep the water warm enough for fish. Most heaters are glass-enclosed electric heating elements that extend into the water. The

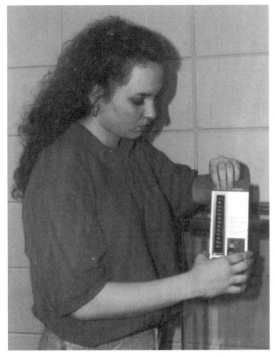

18-16. Preparing to install a thermometer on an aquarium.

18-17. Heater for a small aquarium.

thermostat in the heater turns the heating element on when the water is cool and off when the water has been warmed sufficiently.

Thermometers may be independent of thermostats and heaters. Each species of fish has its own temperature requirements. Most of the ornamental fish in aquaria prefer temperatures in the 70 to 80°F (21 to 27°C) range.

COVERS AND LIGHTING EQUIPMENT

Most fish tanks should be kept covered and properly lighted. Covers keep the fish from jumping out of the container and prevent predators or objects from falling into the fish tank. Predators, such as cormorants (fish-eating birds), can destroy large numbers of fish in outside tanks that are not covered. Cats or other animals can get into tanks and capture the fish. Covers may be made of meshed wire or solid sheets of metal, plastic, or other material. Covers are designed for easy removal or opening to feed the fish and remove or add fish.

18-18. Cover and lighting system for an aquarium.

Lighting equipment may be a built-in part of the cover. Lighting is needed so that people can see the fish as well as to encourage biological processes in the water. Water plants, depending on the species, need light to live and grow. Algae and other organisms may need light in order to grow. Lights often have reflective hoods that allow better lighting. Caution: Be very careful when using electricity around the water in a fish tank.

OTHER EQUIPMENT

A wide range of other equipment may be useful. Examples include dip nets, transfer containers, containers for aging water, decorative items for a fish tank, and hydrometers to determine the salt content of the water in a marine aquarium.

THE WATER ENVIRONMENT

Successfully keeping ornamental fish requires the right kind of water and keeping it in good condition. Freshwater and saltwater species usually cannot be mixed. Some species also prefer brackish water, which is a mixture of freshwater and saltwater. Managing the systems varies with freshwater and saltwater; however, many of the principles are the same.

ARTIFICIAL HABITAT

Establishing an aquarium involves establishing a certain kind of habitat. A *habitat* is a place where a plant or animal naturally lives. The aquarium is set up to duplicate the environment that nature provides. Knowing what to include requires careful study of natural habitats. In some cases, aquaria are designed for a particular climate, such as the Amazon River.

Various plants, animals, minerals, rocks, and other materials may be used in trying to achieve a balance in the artificial habitat. The items are arranged to provide a pleasing appearance as well as a good place for the fish to live and grow.

WATER SOURCE

A supply of quality water is needed. The water must be appropriate for the species that are being grown.

Freshwater

Freshwater can vary considerably in quality. Water quality depends on its source and the substances that it contains that make it impure. The dissolved chemicals in water can make it unfit for fish.

Water for fish tanks comes from three major sources: wells, tapwater, and rainwater.

Individuals who farm fish may drill wells into a water aquifer. The water is pumped from the earth and may or many not go into storage tanks. The water should be tested to determine its mineral content and other substances. In some cases, water may be unfit for fish tanks. Gases in water can be released by storing the water in an open container for a few days.

Tapwater is the water at the faucets in homes, offices, and businesses. Tapwater is common and readily available; however, it usually cannot be used as it comes from the tap. It comes from wells or is treated water from streams and lakes. Municipal water systems often add chlorine or other chemicals to water. These substances keep the water safe for humans to use, but make it dangerous to fish. Tapwater is often unfit to use if it is run directly into a fish tank, but it can often be readied by "aging." Aging involves collecting the water in an open container and allowing it to stand for several days. Chlorine and other substances are released into the atmosphere. Of course, a large aquar-

18-19. Collecting tapwater for aging.

ium will require a lot of "aged" water. Containers in which water is collected and stored must be clean. Dirty containers pollute the water. Tablets that prepare water for use can be purchased at a pet supply store and put into the tapwater.

Rainwater can be used in aquaria, but must be collected so that it is not polluted. Acid rain and other substances in rainwater may make it unfit for fish tanks.

Saltwater

Saltwater may be obtained as seawater or made using a synthetic salt mix. Natural seawater is often used near oceans or lakes with saltwater. Seawater is a complex solution of many chemical compounds. Natural seawater must be collected away from sources of pollution. This may involve going out into the ocean, away from in-flowing streams, and collecting the seawater in a large tank that is brought back to shore. The seawater must be handled properly to maintain its quality. Seawater can bring diseases to an aquarium if not properly conditioned by a process known as dark storage. Dark storage involves storing seawater for several weeks in covered plastic containers.

Synthetic seawater is made by mixing freshwater with a saltwater mix. These mixes can be bought at pet supply or aquaculture stores. Follow the instructions that are included with the mix for preparing the seawater. In most cases, it is easier and safer to make synthetic seawater than to use natural seawater.

WATER QUALITY

Water quality is the suitability of the water for use in a fish tank. Water has to be prepared for placing fish in it. The previous section of the chapter dealt with getting the water ready. This section is about maintaining water quality.

Temperature

Fish thrive best if the temperature of the water is within an appropriate range. The requirements for each species should be determined. Good references can be found in libraries or pet supply stores.

In addition to getting water ready for use, it must be at the right temperature to move fish into it. In some cases, the water will need to be warmed. The water into which fish are moved should vary only a few degrees from the water in which they have been living. Sudden changes in water temperature cause stress. Fish are susceptible to disease and other problems when stressed.

Most ornamental species prefer water that is 70 to 80°F (21 to 26°C). Saltwater species often prefer water that is slightly warmer than the freshwater species. As previously mentioned, heaters may be needed to help keep the water at a relatively uniform temperature.

pH

pH refers to the acidity or alkalinity of the water. It is measured on a scale of 0 to 14, with 7.0 being neutral. pH readings below 7.0 are increasingly acidic, the lower the number. pH readings are increasingly alkaline as the number goes up above 7.0. In nature, most water has a pH of about 8.0; therefore, ornamental species prefer water in fish tanks with a pH of 7.8 to 8.3.

Changes in pH result from chemical reactions in the water. Many things in an aquarium influence pH. For example, the materials used in a filter are continually being dissolved into the water. Those made from calcium-based materials can result in higher pH readings. The nitrification process and carbon dioxide in the water can lower the pH. Buffering solutions and powders can be bought at pet supply and aquaculture stores to help maintain water within the right pH range.

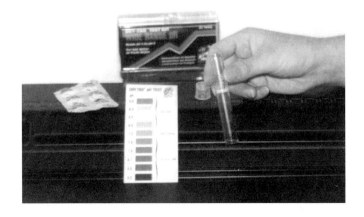

18-20. Using a kit to test the pH of water.

Nitrogen Cycle

The **nitrogen cycle** is the process of converting animal wastes into ammonia, ammonia into nitrites, and nitrites into nitrates. It is a natural process that occurs in water.

Ammonia is toxic to fish. Ammonia can sometimes be noticed as an odor above the water. It is more likely to be a problem when fish are fed too much or when solid wastes are not filtered out of the water. Commercial test kits can be used to test for ammonia. They must be used quickly and properly because, as a gas, ammonia can quickly go into the air. Over feeding and over stocking the fish tank contribute to ammonia problems.

Nitrites are formed when nitrifying bacteria act on ammonia. Most of this occurs in the filtration material, primarily small gravel, at the bottom of the fish tank. Nitrite is toxic to fish, but not nearly to the same extent as ammonia.

Nitrites are converted to nitrates by a chemical process. Plants in an aquarium also help to convert the nitrogen. Nitrite problems are more likely in newly-set-up fish tanks. The problems can be adjusted as the tank settles into supporting fish.

Dissolved Oxygen

Dissolved oxygen (DO) is gaseous oxygen that is in the water. The fish and other aquatic life remove the DO from the water. Most fish need at least 5.0 ppm of DO. Below 3.0 ppm DO level, some species show stress and die. Most fish die in a few minutes when DO is as low as 1.0 ppm. As described earlier, DO is maintained by using oxygenation equipment. Oxygen meters and test kits can be used to measure DO.

Salinity

Salinity is the amount of salt in the water. Freshwater species do not want and cannot survive in water with much salt. Saltwater species require sufficient salt in the water. Salinity is measured as specific gravity using a hydrometer. Water has a specific gravity of 1.0. Saltwater has a slightly higher specific gravity. Most saltwater species want water with a specific gravity of 1.020 to 1.024 at a water temperature of 77 to 80°F (25 to 26°C). Salinity can be increased by using commercially-prepared sea salts.

CHANGING THE WATER

The water in fish tanks should be changed periodically. Since this is a big job with large tanks, careful management should be followed to prevent water fouling. Water should be obtained and aged as with a new aquarium before it is used. Fish will need to be temporarily moved to another water tank. Care should be used to avoid stressing the fish at this time.

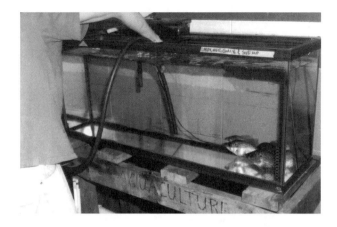

18-21. Using a hose to clean the solid materials from the bottom of an aquarium by siphoning. (This method of cleaning causes little stress to fish. Some replacement water will be needed.)

CARING FOR ORNAMENTAL FISH

Ornamental fish have needs that must be met. How long fish live and their rate of reproduction is often related to the kind of care they receive.

FEEDING

Feeding is providing the nutrients that ornamental fish need to live and grow. Nutrients are the substances in feed that provide nourishment. Health and longevity are closely related to the diet of a fish.

The amount of food a fish is fed each day is known as a ration. The ration should provide a diet that meets the nutrient needs of the fish. The amount of feed that fish eat is related to the temperature of the water as well as the kind of fish and their stage of life. As the water temperature rises, the use of food by the fish increases; therefore, they should be fed more.

The amount to feed is no more than the fish will eat in a few minutes. This involves careful observation of the behavior of the fish. Over feeding fouls the water.

Most ornamental fish growers feed a commercial food. These feeds are specially formulated for different fish. Some include animal protein, such as shrimp. Others include grain and related ingredients. All feed containers

18-22. Many specialized kinds of commercial food are available for ornamental fish.

should have labels that list the ingredients and nutrients. Most feeds are recommended for certain species.

HEALTH

Fish can get sick much like other animals. A good environment and proper feeding help to keep fish healthy. Minimizing stress also reduces the chance of a disease outbreak. Stress occurs when fish are handled or when there is a sudden change in their environment. Regularly and carefully observing fish helps people know the behavior of healthy fish. A disease should be suspected when one or more of the following are observed:

- Fish scratching on objects.
- Fish failing to eat.
- Increased respiration (gill movement) and gasping at the surface of the water.
- Clamped or folded fins.
- Cloudy eyes.
- Bloody spots or fuzzy patches on the body.
- White spots on the body.
- Frayed fins.
- Color changes.

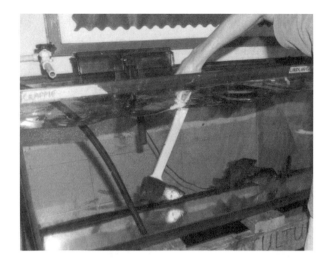

18-23. Scrubbing the sides of an aquarium to remove solid materials and algae helps keep the aquarium attractive.

- Protruding eyes or abdomen.

- Material hanging from the anus of the fish (could be a worm parasite).

Sick fish should be removed from the tank and kept in a separate tank. Disease treatments can be given to the diseased fish as well as the fish in the healthy tank. This involves putting an approved medication in the water. All medications should be used according to instructions.

When selecting new fish, observe the fish for any of the above signs. If they exist on a fish or in a tank, do not get the fish. You will likely be bringing disease to your aquarium! The reputation of the source of fish is very important. Get fish only from reputable sources.

MAINTENANCE SCHEDULE

Following a regular maintenance schedule can be valuable in keeping an aquarium in good condition. An *aquarium maintenance schedule (AMS)* is a listing of important activities and the frequency with which they should be performed in keeping a good aquarium environment. Some activities must be done each day, while others can be done weekly, monthly, quarterly, or annually.

- Daily—provide feed; check the heater, temperature, aeration, and filtration; remove any dead fish; and observe the fish for unusual behavior.

- Weekly—check the water level and pH; add water and chemicals as appropriate.

- Monthly—change and add water; siphon off dead material on the bottom; remove algae; and tend to any plants that may be in the aquarium.

- Quarterly—clean the filter; check electrical connections; check hoses and pump.

- Annually—clean the tank thoroughly; rinse the bottom gravel; and replace light bulbs/tubes.

REVIEWING

MAIN IDEAS

Ornamental fish are kept for beauty and personal appeal to people. Some ornamental fish are captured from the wild; others are produced in fish tanks. Many different species of ornamental fish are produced. Some are egg-layers; others are livebearers. Some live in freshwater; others live in saltwater.

Producing ornamental fish involves establishing and maintaining a good environment. The needed facilities include water containers, water quality equipment, and tank covers and lighting.

Water source is important to the success of ornamental fish production. Freshwater can be obtained from drilled wells, municipal water systems, and rain. Saltwater can be obtained from the sea or synthetically made using commercially available salts. In nearly all cases, the water must be conditioned or aged before it is used.

Water quality—the suitability of water for fish—must be maintained. It includes temperature, pH, nitrogen cycle, dissolved oxygen, and salinity. The solid materials must also be filtered out of the water.

Care of ornamental fish includes feeding, controlling disease, and performing maintenance activities on daily, weekly, monthly, quarterly, and annual bases.

QUESTIONS

Answer the following questions using correct spelling and complete sentences.

1. What are the sources of ornamental fish?

2. What are examples of ornamental fish? Name two examples each of fresh-water egg-laying fish, freshwater livebearing fish, and saltwater fish. Briefly describe each species.

3. What equipment is needed to produce ornamental fish?

4. What is oxygenation? Why is it important?

5. What kinds of filtration are used? Distinguish between the three common kinds.

6. What should be considered in selecting a source of water?

7. What are the important areas of water quality with ornamental fish?

8. What is the purpose of feeding? What guidelines should be followed in feeding?

9. Name three fish behaviors that indicate a possible disease problem.

10. Describe a maintenance schedule for ornamental fish production.

EVALUATING

CHAPTER SELF-CHECK

Match the terms with the correct definitions. Write the letter by the term in the space that is provided.

a. egg-laying fish	e. nitrogen cycle	i. salinity
b. aquarium	f. shoal	j. dissolved oxygen
c. oxygenation	g. guppy	
d. filtration	h. habitat	

1. _____ a container used to hold water for ornamental fish.

2. _____ fish that reproduce by the female laying eggs.

3. _____ the process of removing solids and gases from water.

4. _____ gaseous oxygen in water.

5. _____ salt content of water.

6. _____ the process of keeping adequate DO in the water.

7. _____ the process of converting wastes into ammonia, nitrites, and nitrates.

8. _____ place where a plant or animal naturally lives.

9. _____ group of fish living together.

10. _____ popular ornamental livebearing fish.

EXPLORING

1. Set up two fish tanks in the classroom. Establish a saltwater environment in one tank and freshwater in the other. Select the appropriate equipment and install the oxygenation, filtration, and lighting systems. Select the appropriate species and establish the tank. Establish an artificial habitat that is pleasing to view as well as appropriate for the species selected. Set up a routine aquarium maintenance schedule (AMS) for the tanks. (Additional resources may be needed. A visit to a local pet store will provide access to materials and information on the species to use.)

2. Take a tour of a large aquarium and study the different aquatic environments that have been installed. Ask the manager to let you observe the water filtration systems and other aspects of managing the aquarium.

Chapter 19

WILDLIFE ANIMALS

Animals in nature have always been a wonder and an inspiration to humans. Our ancestors had to depend on animals in the wild for food before livestock species of today were domesticated. We still use animals from the wild as food, but we also have learned to enjoy the aesthetic contribution of wild animals to our lives.

Wildlife are enjoyed in many ways. Watching animals is a favorite pastime with some people. Others enjoy hunting. Many animals are killed by sporting enthusiasts each year. People differ in how they view the use of wildlife. Some feel that legal hunting is a good sport and helps keep wildlife populations controlled; others disagree. People enjoy wildlife in many ways. What is your favorite?

As we have grown to appreciate and enjoy wild animals, we have also come to realize that a healthy population of wildlife demands planned management. As human populations grow and the amount of land available to wildlife decreases, the management task is becoming more difficult.

19-1. The Bald Eagle is a favorite for bird watchers. Once endangered, it is now increasing in numbers in Alaska and the Pacific Northwest.

OBJECTIVES

This chapter provides an introduction to animal wildlife and management. It has the following objectives:

1. Explain the importance of wildlife
2. Describe game fish and animals (major kinds)
3. Explain important practices in wildlife management
4. List important considerations for sports enthusiasts

TERMS

animal wildlife	habitat	pelt
biodiverse population	hunting	territory
carrying capacity	limiting factor	trapping
endangered species	niche	urbanization
game animal	non-game animal	

THE IMPORTANCE OF WILDLIFE

Animal wildlife are all animals that have not been domesticated. This includes insects, birds, fish, rodents, and many others including deer, elk, and bear. These animals benefit humans in many ways. Those hunted for food and other uses are known as *game animals*.

Why should we be concerned with wildlife populations? Our society has evolved to a point where we can go to the grocery store and buy any kind of meat or fish we could possibly want. There is no longer a need to hunt or fish for our food.

This argument would be true if the only reason we wanted to have fish and wildlife was to provide food. There are many other reasons to be concerned with wildlife and game fish populations. Several in this chapter include economic, beauty, recreation, and biodiversity.

ECONOMIC IMPORTANCE OF WILDLIFE

Large sums of money are spent on wildlife and game fish in North America each year. Millions of dollars are made every year in department stores that sell hunting and fishing supplies. Owners of wilderness hunting and fishing lodges have thousands of visitors every year that generate income. State departments of wildlife employ thousands of people with money made from the sale of hunting and fishing licenses. The communities around popular hunting, fishing, and wildlife areas and parks depend on the income from nature lovers, hunters, and fishers to operate the services of the community.

Hunting is killing animals with guns or other weapons. It is often equated with taking wildlife as a sport. The animals are used for meat, hides, or other products and to make trophies for home or office decorations.

Trapping is capturing animals for their products. The animals are usually caught in traps and remain alive until someone checks the traps. The animals are killed after being caught. Most of the animals are valued for their hides. Some trapped animals have other products, such as some local areas use the meat of the raccoon for food. People who trap animals are trappers.

A *pelt* is the whole hide with fur attached. Pelts are used to make valuable fur coats and other products. Some animals used for pelts include fox, rabbit, squirrel, raccoon, and mink.

Many people make a living by harvesting wildlife and fish. Commercial fishers, trappers, hunting and fishing guides, and outdoor outfitters are a few examples.

19-2. These pelts in a Sitka, Alaska, store were taken from wild animals by trappers.

19-3. Tourists go to Yellowstone National Park to see wildlife, such as this bull elk.

BEAUTY

Most people would stop to look at a deer grazing in a field or a young family of mallard ducks swimming on a pond. We enjoy the show that wildlife animals provide. Many people want to vacation in the country or at a park in hopes of seeing wildlife animals. The beauty of wildlife captures our attention and imagination.

Chances are that you have visited a zoo or been a part of a wildlife officer's program that let you get close to a wild animal. You may even be in a club formed to enjoy wildlife. Bird watching clubs, conservation clubs, and organizations at local parks have become a popular way to enjoy wildlife.

RECREATION

Every year hundreds of thousands of people enjoy fish and wildlife through hunting, fishing, photography, scouting, bird watching, and other recreational activities. Adults and children can enjoy nature through recreational activities.

Although hunting for food is no longer required for us to survive, it gives hunters a chance to relive what it must have been like for our ancestors. Most hunters are true sports-people who eat what they kill. Hunting is one of the oldest sports in the world.

Fishing with hook and line has been called relaxing, tense, laid back, and exciting—all are true! Many people who fish today still fish for food. Many others fish for sport, releasing most of their catch. We are fortunate to live in a country where we are never far from a fishing hole.

Photography, scouting for wildlife, bird watching, and other recreation activities are enjoyed by people who live in the country, suburbs, and in the city.

BIODIVERSITY

North America has a variety of wildlife animals. From the desert southwest to the maple-aspen forests of the northeast, many unique environments and many unique species of wildlife are found.

19-4. Black-tailed Jackrabbit in the Arizona desert.

Having a variety of wildlife species adds to the *biodiversity* of our planet. Biodiverse populations are less likely to be harmed by disease or illness. One concern is the growing number of wildlife animals that are *endangered species.* Endangered species are animals that are close to becoming extinct. Every time an animal reaches extinction, the biodiversity of our planet decreases.

MAJOR TYPES OF GAME AND NON-GAME ANIMALS, FISH, AND BIRDS

It is impossible to list all types of wildlife in North America. A few of the major types of fish and wildlife animals are listed here. The main categories of animals include: game animals, non-game animals, game fish, non-game fish, game birds, non-game birds, birds of prey, and endangered species.

GAME ANIMALS

Animals hunted for sport, meat, and other products are game animals. The kinds of animals used for game varies with where you live. Game animals can be large, like the white-tailed deer and the antelope. They can also be small animals, like the rabbit or squirrel.

In most places, game animals can be harvested by hunters during a regulated hunting or trapping season. Personnel in state departments of game and fisheries decide when the season occurs and how many animals can be harvested.

Popular species of large game animals include the white-tailed deer, mule deer, elk, antelope, bighorn sheep, and alligator. Small game species include the cottontail rabbit, squirrel, fox, mink, raccoon, and muskrat.

19-5. Squirrels are important to some hunters.

19-6. Alligators are hunted for skin and meat in the southeastern United States.

NON-GAME ANIMALS

There are many more non-game animal species than game animal species. These animals can be found deep in the forest or in urban areas, or just about any place in between! The diversity of non-game species is amazing.

Non-game animals do not provide us with a consumable product like meat or furs. They do give us beauty and add to the bio-diversity of our planet. Some non-game ani-

19-7. Jaguar is a non-game wild animal.

mals serve as a food source for predators. Others are scavengers and help nature recycle by eating dead animals. Non-game animals are an important part of nature.

Most people do not think of wildlife in their neighborhood. Many non-game animals can be found in urban and rural areas across the country. Mice, rats, opossum, and others are very comfortable in an urban setting.

GAME FISH

Game fish are sought to provide food or sport to the fisher. Game fish exist in both fresh and salt water and come in a variety of sizes. Most game fish species are caught with a hook and line.

Freshwater game fish can be caught in every state. Again, the types of game fish available from state to state will differ. Major types of freshwater game fish include black bass (largemouth, smallmouth, and spotted), sunfish, walleye, trout, pike, muskellunge, salmon, striped bass, crappie, and others. Popular saltwater species include groupers, sea trout, cobia, snappers, flounder, redfish, mackerel, tuna, sailfish, dolphin, and many more.

NON-GAME FISH

Just as non-game animals outnumber game animals, non-game fish outnumber game fish. Non-game species can be found in small freshwater creeks, ponds, lakes, and in saltwater habitats. Many species of non-game fish are too small to be caught on a hook and line, but some non-game

19-8. Bluebanded goby fish in the Pacific Ocean off the California coast.

19-9. The Bobwhite Quail is a highly desired bird by some hunters. (Courtesy, Brad Phares, Gainesville, Florida)

species are very large. For example, the sturgeon can grow to over 2,000 pounds!

GAME BIRDS

Game birds are hunted for food and sport. They can be divided into three basic categories that include migratory (non-waterfowl), waterfowl, and upland game birds.

The woodcock and the crow are examples of non-waterfowl game birds that are migratory. Game waterfowl include many species of ducks and geese. Upland game birds include the pheasant, grouse, and quail.

Harvest of game birds is regulated by state and federal law. Since migratory birds can travel to other countries, laws governing their harvest often have international input.

BIRDS OF PREY

Eagles, osprey, owls, and hawks are all examples of birds of prey. Most birds of prey are protected by laws that make shooting them or attempting to catch or harm them illegal. These birds are hunters. They eat small rodents, fish, snakes, reptiles, amphibians, and small mammals.

NON-GAME BIRDS

Non-game birds include all other birds. The sparrow, blue jay, robin, and pelican are all examples of non-game birds. Non-game birds live in a variety of environments. You can find these animals in wetlands, plains, uplands, and anywhere in between.

ENDANGERED SPECIES

Unfortunately, people have had a negative impact on many wild animals. Through **urbanization** (building cities, suburbs, and all of the support services needed by people in these areas), agricultural growth, pollution, and neglect, many species of wildlife are extinct or endangered.

One tragic example of an animal on the brink of extinction is the Florida panther. This large cat once ruled over the swamps of central and southern Florida. They need large areas to roam, but the urbanization of Florida is threatening their existence. There are an estimated 20 to 30 of these animals left in Florida today.

With work, the fate of some endangered species can be changed. The American alligator was also close to extinction. Today, the alligator is doing very well. In fact, the population is in such good shape that alligators can be legally harvested in some places in the southeastern United States!

WILDLIFE MANAGEMENT PRACTICES

Think for a moment about where you live. How many people live in your home? Does everyone have a bed and a chair at the table? People need space to live in. You would not be very comfortable if 10 people shared your bedroom, or if 50 of your classmates came to your house for dinner. There are limits to how many people your home can accommodate. Wildlife species are limited by the same factors.

HABITAT

Habitat is the place where animal wildlife live. It has four components: food, space, water, and cover. Think of your home as a habitat. It is a place where your basic survival needs are met. You have cover in a house or apartment. You also have food, water, and space to live.

Wildlife must have the proper habitat to meet their needs. All four components must be in place to support a wildlife species. If one component is missing or lacking, it is called a **limiting factor.** A deer needs to have small limbs to eat. If there are no small limbs, the deer cannot survive in that habitat—food is the limiting factor. Pheasants need tall grass to nest. If there is no tall grass in the habitat, there will be no pheasants—cover is the limiting factor.

19-10. Deer are found throughout North America. The species may vary with the location. (Courtesy, Brad Phares, Gainesville, Florida)

Even if a habitat is perfect for a particular species, there are limits to how many animals the habitat can sustain. The habitat's *carrying capacity* is the number of animals it can support. Carrying capacity is usually low in the winter and high in the summer. The population in a given habitat is directly related to the carrying capacity.

INCREASING WILDLIFE POPULATIONS

The best way to increase wildlife populations is to meet their basic needs in the habitat. While this is true, there are other considerations.

Niches

Every animal has a *niche* or a special place in a habitat where it belongs. The animal fills an important role in consuming and producing for the overall habitat. For example, white-tail deer and squirrels both live in the same habitat. The deer eats small branches and likes to bed in fields of high grass—this is its niche. The squirrel lives in the trees and eats nuts, berries, etc. The squirrel fills a different niche. Some animals may share similar niches, but they are seldom the same.

19-11. An egret has a special niche in a wetlands area.

Interspersion

When things are interspersed, they are mixed. For example, in an eight-acre plot of land, two acres could be dedicated to food, two acres to water, and the remaining four acres to cover. If these areas are arranged in large blocks, there would be very few places where animals could have easy access to all three areas. Placing these resources in smaller blocks would allow quicker access to all three and increase the number of animals in the habitat.

Territory

Some animals claim an area, or *territory,* which they protect and keep other animals from entering. Bears will mark their territory to warn others from invading. Space can quickly become a limiting factor for territorial animals.

19-12. Grizzly bears have territories and do not want others to invade their space.

CONSIDERATIONS FOR WILDLIFE SPORTS ENTHUSIASTS

Wildlife sports are important traditions in some communities. Sports include traditional hunting and trapping, as well as watching (observation), attracting wildlife, and photography. Sports enthusiasts need to keep some simple rules in mind when enjoying wildlife.

EXERCISE CAUTION

Hunting and trapping are sports that involve deadly force. Guns can kill or injure people as well as animals. Hunter and trapper safety courses are required in many states before a license can be obtained. These courses will teach you how to use traps and firearms and to respect their power. Many deaths and injuries occur each year at the hands of careless sports enthusiasts.

Another potential area of danger involves human/animal interaction. Wild animals are wild! They are not pets or zoo animals. They are driven by instinct and will try to get away from you and out of danger any way they can. Often, this includes biting, scratching, goring (puncturing with horns), or attacking, using any means available to them. When hunting, trapping, observing, or photographing, remember to stay at a safe distance.

TAKE ONLY WHAT YOU NEED

Wild animals are a renewable resource. Game fish and animal harvests are closely regulated by state and federal agencies. These agencies set limits as to the number of animals that can be taken by individual hunters, fishers, and trappers. To maintain healthy populations of wildlife, it is necessary for sports enthusiasts to obey those regulations.

19-13. The American Bison population was nearly destroyed by hunters killing animals they did not need.

There are times when a fisher or hunter may be having an excellent season and can harvest limits of fish or game on a daily basis. At times like these, the sports enthusiast needs to ask "how much will I use?" If the harvest will be wasted, should you take the animal even though you are within the legal limits? A conscientious sportsperson would answer "no."

RESPECT THE HOME OF ANIMALS

How would you like to have visitors in your home leave their trash on the floor? Would you be upset if someone came to your house and broke all of the furniture? You probably would not be happy with either situation.

People are often in the homes of animals. In their habitat, we need to respect where they live. Leaving paper, cans, and bottles, or damaging trees and animal homes, is not the way a responsible sports enthusiast uses the resource.

REVIEWING

MAIN IDEAS

Animal wildlife exists in large numbers across the country. These animals are important to us economically. They add beauty to our world. They provide recreation. In addition, they add to the bio-diversity of our planet. Major types of wildlife species include game and non-game animals, game and non-game fish, game and non-game birds, birds of prey, and endangered species.

Managers of wildlife must be concerned with providing food, water, cover, and space for the animals. Wildlife populations can be increased by providing niches for animals, practicing interspersion, and meeting territorial needs of animals.

Sports enthusiasts should exercise caution when dealing with wild animals. Those that are conscientious take wildlife that they will use. All people who use wildlife need to respect the animal's habitat and home.

QUESTIONS

Answer the following questions using correct spelling and complete sentences.

1. What are the four reasons that animal wildlife is important? Briefly explain each reason.

2. Distinguish between animal wildlife and game animals.

3. What is an endangered species? Give an example of an animal that was once endangered, but is no longer.

4. What is habitat? What are the four components of habitat?

5. What can be done to increase wildlife populations?

6. How should sports enthusiasts respond to animal wildlife?

EVALUATING

CHAPTER SELF-CHECK

Match the terms with the correct definitions. Write the letter by the term in the blank that is provided.

a. habitat   d. game animal   g. carrying capacity
b. niche   e. animal wildlife   h. pelt
c. endangered species   f. territory

    1. ____ the natural skin covering an animal used in making clothing and other products

    2. ____ animal close to becoming extinct

    3. ____ the number of animals a habitat can support

    4. ____ the place and role of an animal in its habitat

    5. ____ the place where animal wildlife naturally lives

    6. ____ an animal harvested for food or other products

    7. ____ the area an animal will defend as its own

    8. ____ animals that have not been domesticated

EXPLORING

1. Develop a personal library of information on game animals in your area. Collect brochures and other materials in your library. These are available from a local conservation officer, game and fish commission, or office of the Cooperative Extension Service.

2. Tour a wildlife refuge. Learn the kinds of animals in the refuge and the practices followed in improving habitat for the animals. Prepare a written or oral report on your findings.

3. Select a species of wildlife found in your area. Prepare a report that describes the animal, its habitat requirements, and other conditions that provide for its well-being.

Chapter 20

SERVICE AND SAFETY ANIMALS

Animals help people in so many ways! Think about the people you know who have animals. How do these animals help them? Not all animals are helpful. Sometimes people do not treat the animal with respect nor allow it to be helpful. What a waste!

Some animals are used in ways that do not provide food or pleasure. These animals are used to help people live better lives. In some cases, they help people live independently. Close your eyes for a moment and think about what it would be like to be unable to see. Can you find your way to school with your eyes closed? A trained leader dog for the blind helps visually-impaired people live independently.

Many different animal species are used for service and safety. Training is usually required so the animals will know what to do and how to do it. The training usually includes close bonding between the animal and its human owner.

20-1. Leader dogs are trained to be the "eyes" of visually-impaired people. (Courtesy, Beth Morgan, College of Veterinary Medicine, Texas A&M University)

445

OBJECTIVES

This chapter provides basic information about using animals for service and safety. The following objectives are included:

1. Explain service and safety animals and give examples of each
2. Describe management practices with service and safety animals
3. List the nutritional requirements of service and safety animals
4. Explain important health practices with service and safety animals
5. Describe facility and equipment needs with service and safety animals

TERMS

cria	K-9	service animal
handler	multipurpose dog	single-purpose dog
jack	predator	
jenny	safety animal	

KINDS AND USES OF SERVICE AND SAFETY ANIMALS

Service and safety animals help people in many ways. A ***service animal*** is an animal that gives assistance in some way. These animals often provide an important service. A ***safety animal*** is an animal that helps protect people and property. Service and safety are not widely thought of as uses for animals, but they are very important uses.

There are many kinds and possibilities when it comes to service and safety animals. Animals can do various tasks if they have the correct instincts and are properly trained. Just a few possibilities of service and safety animals are dogs, geese, llamas, and donkeys.

Service and safety animals help and protect. Dogs and geese in a person's yard warn of intruders. Llamas, donkeys, and dogs put out to pasture with sheep and goats protect them from ***predators.*** Predators are animals that live off of, kill, maul, or prey on other animals. Some common predatory animals are coyotes, wolves, bears, mountain lions, and domestic dogs. The predator animals in the past have been hunted, poisoned, trapped, shot, sterilized, and had their dens gassed. Today, many producers are using other animals to control the aggressive predators and protect their animals.

DOGS

One of the most common service and safety animals is the dog. Dogs are used in many ways, including as pets. Many homes have dogs. Dogs are used not only as companions, but to warn the occupants of the home when someone arrives.

Dogs have been called a human's best friend. One reason is that dogs are diverse animals. Not only can dogs perform several tasks, such as fetching things for their owners, but also protect the home and livestock while no one is there. The diversity of breeds of dogs is as vast as the number of things you can do with your dog.

Dogs have special uses. A good example is the well-trained police ***K-9,*** which is how the dog or canine is referred to by the police department. The K-9 is a highly versatile, dedicated member of the police department. When properly trained, the police K-9 can perform a variety of services

20-2. A K-9 unit German Shepherd in action.

for the department and community, as well as be a positive force for the department in community relations.

K-9 unit dogs are most commonly the German Shepherd breed because they are easily trained, strong, aggressive, and good tempered. It is important to have an aggressive dog for police work and also a good tempered dog to meet and greet the public. German Shepherds are known by many as "police dogs." However, some police departments use Standard Poodles and Rottweilers. There are many things to be considered before a police department decides upon the breed of dog to use.

Farmers and ranchers use trained guard dogs to stay with a flock of sheep and protect them. Guard dog breeds originated in Europe and Asia where they have been used for centuries to protect sheep from wolves and bears. The breeds most common for guard dogs in the United States are large dogs weighing 80 to 120 pounds (36 to 54 kg). They stand at least 25 inches (63 cm) at the shoulder and have flat ears covered with hair, short and blunt muzzles, and long tails. They are usually fawn, gray, or all white with dark muzzles. The most common breeds used are Great Pyrenees from France, Komondor and Kuvasz from Hungary, Maremma from Italy, Castro Laboreiro from Portugal, Shar Planinetz from Yugoslavia, Tibetan Mastiff from Tibet, and Akbash and Anatolian Shepherd, both from Turkey. The most successful breeds are the Akbash and Pyrenees.

20-3. Eight-month old Maremma male dog herding goats in Idaho. (Courtesy, Luck-E-G Nubians, Blanchard, Idaho)

DONKEYS

Some farms and ranches use donkeys as protection for sheep and goats. Donkeys were first used in the ranches of Montana and Utah. Word of their effectiveness has been spreading and they are now being tried across the country.

Donkeys are classified as either a *jenny,* which is a female, a *jack,* which is a breeding male, or a gelding, which is a castrated male.

20-4. Donkeys in a research program at Virginia Tech.

When selecting a donkey to use with sheep, it is important to know that jacks and geldings may be too aggressive with the sheep and kick or bite them. However, geldings are less likely to be aggressive and they are selected over jennies because they are stronger. Jennies work well with the sheep because they are less aggressive, unless they have newborn babies of their own. This may be remedied by removing the jenny and her newborn for a week or ten days, then return the two to the sheep flock. This usually works best.

Donkeys have a natural hate for all canine animals, which are members of the dog family. Donkeys do not kill predators, but they do take after them with their teeth bared, ears back, and striking at their feet. Guarding is a natural part of the donkey's temperament. Donkeys are replacing dogs in some places because they are cheaper to purchase and feed than dogs.

LLAMAS

Llamas, like donkeys, are used on some farms as protection for other animals, as opposed to protection for the homeowners themselves. Llamas were first brought to the United States as novelty animals. They have been used as pack animals. Initially, llamas were put with sheep because of convenience. Then producers realized they were losing fewer sheep to predators with the llamas running with the sheep.

Llamas that encounter predators get very aggressive and protective. They are very proud animals that remain calm and do not get scared and run from intruders. They take an aggressive role. When they encounter

20-5. A llama can become aggressive when danger approaches.

coyotes, dogs, foxes, or bears, they give an alarm call, walk or run toward the intruder, chase it, kick it, paw it, and sometimes kill it.

Unlike the donkey, llamas quickly bond to sheep. There is something mysterious about the relationship and the two animals get along well. The llama is also very instinctive about what to do with the sheep. There is no training involved for the llama, they take over and protect their new found sheep companions.

In some states, llama breeding is a government-sanctioned agricultural enterprise, which carries a helpful agricultural tax break. This may make it worth investigating if a producer is considering raising the animals.

The female lays down to be bred, and then 350 days after she has been bred she gives birth unassisted. The young are called *cria* and are usually born in the morning. The female, if approached by the male after she has been bred, will spit at him.

GEESE

Another type of safety animal, used for a different purpose is the goose. Geese are used to protect homes and their owners from intruders. A goose can be a very mean and ruthless animal. When an intruder arrives at the home, the goose begins honking and then chases the intruder. If the intruder is caught, the goose will hold onto the person with its mouth and beat them repeatedly with its strong wings. This may not seem too threatening, but wait until a goose comes to attack you!

Young female geese will begin laying eggs at approximately one year of age. They will lay more during their second and third years of

20-6. A goose can become aggressive toward strangers.

age. A female will lay between 10 and 60 eggs per season, which in the northern United States is late March to June. The laying period is determined by sunlight. The males, or the ganders, will be more active and better breeders after one year. It is important to note that geese are very selective breeders. Since they are so selective when selecting a mate, large flocks use one gander for three to five females.

MANAGEMENT AND PRODUCTION PRACTICES

Each species of animal used for service and safety has different requirements.

DOGS

K-9 units require special management practices. First, a **handler** is selected. This person works solely with the dog. The dogs are trained at one to three years of age. They go through training with their handlers for 12 weeks to enable them to go out on patrol. The dogs are trained to respond to commands and to bite and hold rather than to maul. Dogs trained in specialty areas, such as explosives and narcotics, require at least another 8 to 10 weeks of training. After that, the dog still goes through two days of training per month. K-9 units may be used for searching buildings for burglary suspects, tracking suspects of the crimes, locating lost persons, locating explosive or incendiary devices, searching out hidden narcotics, and protecting their officer partner.

20-7. A K-9 unit dog in training.

A dog is used on the police force until it is eight or nine years old. This depends largely on the area the dog works in. The harder the job, the earlier the dog is retired. The handler usually has the opportunity to adopt the dog and keep it. This usually happens since the dog has become such a vital part of the family.

The management of the dog is left up to the police department and depends upon the funds available to the department. There are *single-purpose dogs,* which are used for just one activity, such as patrols, tracking, narcotics, or explosives. There are also *multipurpose dogs,* which are used for more than one activity. This depends upon the philosophy and budget of the department. Departments' philosophies vary about the idea of the dog being trained for more than one activity. Budget restraints also dictate the involvement of the dog in more than one activity.

Guard dog management is vital. Guard dogs are not like donkeys or llamas who need very little, if any, training for their jobs. The training for a guard dog is extensive and takes a long time. First, the dog must be acquired when it is young, about six to eight weeks old. It should be intelligent, alert, and confident. There is no difference between the performance of male and female pups. However, if a female pup is selected, it is recommended that she be neutered at six months. This will eliminate male dogs coming to find the female when she is in heat.

20-8. A Border Collie pup lives next to a pen of lambs as part of his training.

The puppies that will work sheep must live and be reared with the sheep with little human contact from six to eight weeks of age. The pup must be monitored and bad behaviors must be corrected immediately. The training for the pup will take up to a year. Patience is essential. Give the pup the opportunity to learn with the flock. Start the pup out in small areas and keep working with it.

It takes pups up to two years or more to mature, so there will be expected puppy foolishness and guard dogs' genetics make them independent. Also, keep the goal in mind. This pup is not a herding dog, but a guard dog, so its job is to stay with or near the sheep and protect them from predators. Herding dogs, such as Border

Collies, Australian Shepherds, and Kelpies are to move sheep from area to area by biting, chasing, and barking at the sheep.

One-fourth of all dogs trained to be guard dogs will not make it for reasons such as genetics, starting the pup when it was too old, or improper rearing by the owner. Some dogs also will harass, injure, or kill stock, and roam away from the flock and even proper rearing cannot stop or correct the problem. Some dogs are more protective than others, so it is important to post a warning that there is a guard dog in the field so no humans get injured.

The number of guard dogs needed varies. It depends on how old the dog is and how much area the dog can properly handle. The sheep may graze close together or scattered; some dogs can cover an area more effectively than others. The rule is to start with one dog and let it gain knowledge and get established before adding others. A first-time guard dog user should start with one pup.

There are three basic behaviors vital to an effective guard dog: the dog must be trustworthy to not injure the stock; it must be attentive and stay close to the flock; and it must be protective whenever a predator shows up. A guard dog that is not with the sheep is not where it is supposed to be.

DONKEYS

The key to managing donkeys as security animals with your sheep is to make sure they bond with the sheep. Donkeys are gregarious, so they want to be around other animals. If there are no donkeys to be around, they will assume the role of a sheep and fit in with the crowd. This bond does take time. The donkey needs to get familiar with the flock. One donkey is needed for every 150 sheep.

20-9. This donkey has bonded with the herd of sheep.

Donkeys instinctively hate animals from the canine family, so they make good guards with no training at all. However, not every donkey is going to bond with sheep. It is important that a donkey is tested with sheep before it is released to graze with them. Often, the best bet for the producer is to buy a young female donkey raised on a sheep farm. If more than one donkey is out with the sheep, they will bond and spend time together, rather than mingling and bonding with the sheep.

Another vital aspect of using donkeys is to realize their limitations. Donkeys are very spunky and can kill dogs, coyotes, and foxes when they are aroused. However, a donkey can be killed by bears, mountain lions, and wolves. Donkeys know this and are afraid of these animals. It is important to know what your sheep need protection from to ensure that donkeys are the correct animals for the farming operation.

Besides knowing the limitations of the donkey, it is important to understand the techniques the donkey uses to protect the sheep. The donkey will bray loudly and stamp its feet at night if it hears or sees something unusual. They will chase dogs and coyotes that get too near.

GEESE

Geese used as protection should be obtained as goslings or baby geese. They should arrive in the late spring or early summer. Caring for the birds is easier when the weather conditions are more favorable. (More information on geese is in Chapter 13.)

20-10. Maremma puppies being fed highly-nutritious goat's milk. (Courtesy, Luck-E-G Nubians, Blanchard, Idaho)

NUTRITIONAL REQUIREMENTS

Service and safety animals must receive proper food if they are to perform effectively. Each species varies somewhat in requirements. Information on the feeding of K-9 unit dogs and guard dogs can be found in Chapter 17. Donkeys and llamas work well as guard animals for sheep because they eat the same food as sheep and sleep with them at night. This helps them know when threats occur to the flock's safety. Donkeys and llamas have the same lifestyle as sheep, so they fit naturally alongside them.

There are some additional dietary needs with the donkey. However, the donkey should not be fed in a separate area. This will lessen the bond between the donkey and the sheep. The donkey may even drive the sheep away so it can eat, especially early in the relationship. For a strong bond to exist, the animals must graze and sleep together and establish the same routine. (For more specific information about donkey nutritional requirements you may want to consult Chapter 15 on horses.)

Llamas may be fed with the sheep because they will stand aside and let the sheep eat while they look on. They eat the same foods as sheep and are very satisfied with a bale of hay and water.

Goslings should be fed all they can eat and drink. In other words, their dish should be full of food and water at all times. A 20 to 24 percent protein feed should be available at the elevator or feed store. If not, a chick starter or a duck or water fowl formula will be fine if the goslings are allowed outside to forage for food on their own. Goslings grow much faster than chicks, up to ten pounds in ten weeks. They need food in addition to the starter. Goslings will eat grass and bugs to get the necessary food to balance their ration. As long as they are allowed to forage, they will stay healthy because they find all of the necessary nutrients. Goslings also need water at all times. If weather permits, the goslings should be outside during the day and inside where it is safe at night.

Adult geese can be fed whole corn for feed and then left to forage on their own. Many times, geese are used to weed crops, such as strawberries, asparagus, sugar beets, and orchards. The geese will eat the weeds and not harm the plants. However, the geese should be removed from the strawberries when they start to ripen so they do not eat or ruin the berries. (Additional information on feeding is in Chapter 13, Poultry Production.)

HEALTH PRACTICES

K-9 unit dogs and guard dogs' health practices can be found in Chapter 17. Guard dogs do not usually live to a very old age. Many times, it is easy for the producer to forget to worm and vaccinate the dog since it is out with the sheep. Besides disease, some dogs just disappear, others are poisoned, shot, or hit by vehicles. Guard dogs may only live to be about three years old, so it is important to start training a new pup. Ideally, the training will occur while the older dog is working well with the sheep.

Donkeys do not require very much veterinary care because they are hardy animals. However, the requirements they do have are the same as those for their relatives in the equine family. See Chapter 15 for more specific information.

Llamas, much like donkeys, do not have many health threats. Also, they may last as long as 20 years as a guard animal with the sheep. In the end, they cost less money than a guard dog, which initially costs less, but requires more in care. A dog lasts only three years, of which, the first is only training. The llama is automatically trained because it is all instinct.

Goslings and geese are healthy animals. They need ample water to drink and to use to stay cool in the warm weather. Goslings, when they are very young, should not be outside in the rain or they become water-logged—their down soaks up all the moisture it will hold. Goslings have no feathers to repel the water. Their tiny legs cannot support the weight of the water. Geese are known for being disease-resistant, so little has to be done to maintain their health.

20-11. This K-9 unit dog has bonded with the handler.

FACILITY AND EQUIPMENT NEEDS

The decision of a police department to add a K-9 unit is a 24-hour-per-day, 365-day-per-year decision. The K-9 unit dogs have varying facilities. Some departments mandate the dog go home with its handler. The dog is with the handler 24 hours per day. The dog rides with the handler, does duty with the handler, and lives in the home with the handler and his or her family. Some officers say they spend more time with their dog than their spouse or children. Other K-9 unit dogs are kept either in kennels at the police department or in a kennel at the handler's home.

Guard dogs spend all of their time with the sheep, so they do not require special facilities separate from the sheep, except, possibly, in the winter if the sheep are penned. However, when the pup is weaned from the litter and the socialization with the sheep begins, it should be placed in an area not much larger that 150 square feet (14 m²) with three to six lambs. Orphan lambs work well because all of the animals are lonely and they are more likely to establish bonds. The pup should have an area it can escape to that the lambs cannot get to, which should contain the pup's food. The water dish for the pup and lambs should be shared to force bonding between the two. The pen should be checked several times the first few days to make sure all animals are healthy and adjusting to each other.

Some gentle play between the pup and the sheep is fine, but beware if the play gets too rough. If a lamb is dominating the pup, it should be removed and another lamb put in its place. The reason to put the pup alone with the lambs is to build confidence and to bond with the sheep. The pup may be petted by the owner each day when it is checked, but excessive handling must be avoided or the pup will not bond with the lambs.

The animals may have to be moved to bigger pens periodically if the pup grows too big for the pen. After the pup is 16 weeks of age, it and the group it was raised with can all be put out with the flock. The pup should not be put out into an area where there are predators until it is ready to take on the guarding role. The age varies with dogs, but it is sometimes as young as 4¹/₂ months. The dog will begin patrolling and taking on other activities of guarding. It should be large enough to defend itself from predators.

Donkeys should be kept with the sheep. They should sleep with the sheep since this is when the predators attack. If the donkey is not there

20-12. Border collie dogs can work in cold, snowy conditions if properly fed. (Courtesy, Red Bluff Ranch, Thermopolis, Wyoming)

during the hours it is needed, it cannot protect the sheep. It is important that the donkey is kept with the sheep so it bonds with the sheep. Donkeys can survive in the same climate as sheep, so no special housing is required.

Llamas can die of heat stress and require shade in hot weather. However, they do not have other special requirements. They can live happily on an eighth of an acre with only a three-sided windbreak for shelter. The "toilet habits" of the llama are clean and disease-free because they select one corner of their pen for defecation and do not deviate from that area.

Goslings require a warm, dry place for the first four weeks of their lives. If, in late May or June, the weather is still cold the goslings must be kept dry, warm, and away from drafts. A 250-watt heat lamp may be installed 15 to 18 inches above the floor to keep the goslings warm. If the goslings are piling up on each other, it is too cold and the weaker birds may get smothered. If the goslings are comfortable, they will sit evenly distributed on the floor under the heat lamp.

Goslings require litter on the floor of their pen. Litter, such as straw, sawdust, shavings, or crushed corn cobs will work. The material must be dry and absorbent since goslings are very messy.

REVIEWING

MAIN IDEAS

Several species of animals are used for service and safety. Dogs, donkeys, llamas, and geese are three examples.

The K-9 unit has become important in law enforcement. When properly used, the trained police dog is an asset that cannot be overlooked. It increases the chances of capture and the safety of the officer.

Guard dogs are used effectively as protectors for sheep. They require more work than either llamas or donkeys. The guard dog is a highly trained animal that works effectively with the sheep if it is reared correctly.

Donkeys make good protection providers for sheep. Not all donkeys perform as they should. A donkey should be tested for how well it bonds with sheep.

Llamas are proving to be successful protection animals because of their instinct to bond with and protect sheep. The viciousness of a llama can be provoked by a predator. Their popularity as companion animals is also growing.

Geese are animals that may not be used by everyone, but for those that choose the birds, they work out well. The birds are vocal and very aggressive, but also easy to care for and disease-resistant.

When selecting and considering types of animal protection, it is important to note the time commitment to train the animal, the efficiency of the animal, the length of time the animal can perform the job, the type of protection needed, and which animals are best suited for that type of protection.

QUESTIONS

Answer the following questions using correct spelling and complete sentences.

1. What animals are commonly used for service and safety?

2. What are the advantages and disadvantages of dogs, donkeys, llamas, and geese in service and safety?

3. Why is a good relationship between a K-9 and the handler important?

4. Why are donkeys and llamas particularly suited to guard sheep?

5. Why are donkeys and dogs not used together?

EVALUATING

CHAPTER SELF-CHECK

Match the terms with the correct definitions. Write the letter by the term in the blank that is provided.

a. handler   d. K-9   g. jack
b. service animal   e. jenny   h. single-purpose dog
c. safety animal   f. cria

1. ____ male donkey

2. ____ baby llama

3. ____ female donkey

4. ____ used for one safety or service activity

5. ____ person who directs a K-9

6. ____ animals that give assistance

7. ____ a trained police dog

8. ____ animals used to protect people and property

EXPLORING

1. Contact local law enforcement officials and learn the use they make of K-9 dogs. Interview the handler and determine how the dog is trained, what it does, and how it is managed. Prepare a written or oral report on your findings.

2. Prepare a report on using llamas, geese, or donkeys as service and safety animals. Interview owners of such animals and use library resources for information. Give an oral report in class on your findings.

Chapter 21

SCIENTIFIC AND LABORATORY ANIMALS

Animals have made many contributions to help people live better and longer lives. Scientific and laboratory animals have been used to find ways to improve the quality of life for humans. They have been used in research that led to cures for disease. Animal research has also improved food products, provided insight into human behavior, and given countless other services and benefits.

Did you know that laboratory animal research made it possible for you to avoid becoming a victim of polio? Polio was once a dreaded disease. A simple vaccine now keeps people from getting it. Wonderful!

Nearly every advancement in medical science in the last century can be attributed to animal research. Animals have played a major role in the development of antibiotics, insulin for diabetics, anesthetics, radiation and chemotherapy for cancer, open heart surgery, joint replacements, and hundreds of other medical advancements. People need to understand just how important laboratory animals are in scientific advancements.

21-1. Cultured enzymes from the pituitary glands of cattle are being studied in the sterile environment of a laminar flow hood. The application is to find a way to relieve human pain. (Courtesy, Corinne Mounier-Lee, Georgia Tech)

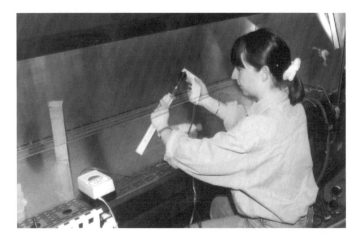

461

OBJECTIVES

The purpose of this chapter is to provide basic information on using laboratory research animals. The following objectives are covered:

1. Describe types of human and animal research

2. Identify the kinds and importance of animals in laboratory research

3. Discuss animal models used in modern research

4. Explain management practices used in animal research

TERMS

animal by-product tissues
animal tissue cultures
applied research
basic research
cavy

cheek pouch
clinical research
laboratory animal
living research animal
mathematical and computer
    model

models for animal
    research
non-human primate
non-living research
    system
primate

LABORATORY RESEARCH METHODS WITH ANIMALS

Scientists use research to find answers to questions. The research involves using procedures that are carefully planned and designed to provide accurate findings. Animals are sometimes used in the research.

ANIMALS IN RESEARCH

Answers to some questions can only be found by using animals. Agricultural researchers have long used animals in research. How would researchers know how animals respond to new feeds or methods of care without studying the animals?

Laboratory animals are animals raised and/or used in research laboratories. They are carefully tended to assure their well-being. Laboratory animals are often kept in controlled environments to assure accurate research. Only one organ or tissue may be the focus of study, or the entire animal may be involved in the research. Every effort is made to consider the well-being of research animals.

Animal by-product tissues are also used in research. These are tissues taken from animals when they are used for other purposes. The tissues, organs, or other parts of animals slaughtered for human food may be used in research. These materials are collected in slaughter plants and carefully preserved.

Animal products are used in research and to produce medicines and similar products. A good example is the chicken egg. Eggs contain albumin which is rich in nutrition and an excellent medium for the production of many vaccines.

TYPES OF RESEARCH

Animals are used in three types of research: basic, applied, and clinical. All three types help find answers to important questions.

21-2. Cattle by-product pituitary glands will be used to synthesize and artificially culture enzymes in a laboratory at Georgia Tech. This small package contains pituitary glands from 100 grown cows. (Courtesy, Corinne Mounier-Lee, Georgia Tech)

21-3. Observing animals in their natural environment is one form of basic research. Chimpanzees, higher-order primates, are often used in trying to better understand human behavior.

Basic research is carried out in a laboratory, with a computer, or in nature. According to the American Medical Association, the goal of basic research is to increase the knowledge and understanding of life processes and disease. Basic research does not have a predetermined focus. It uses observing, describing, measuring, and experimenting to provide data that is often used in the other types of research.

Scientists may have specific purposes in mind for their research. A good example is developing a new vaccine or a medical procedure that builds upon existing knowledge. This is known as *applied research.* Animals are used in applied research, but this type of research could involve computers, non-animal substitutes, or even people.

Research that takes place in a medical facility or clinical setting is called *clinical research.* Clinical research focuses on a specific human or animal problem. It often involves the animal species that the research is intended to help. Clinical research must be supported by knowledge gained in basic and applied research.

Scientific and laboratory animals are used in basic and applied research for human- and animal-related research objectives. They are used in clinical research when their species is the subject being studied. For example, research on chickens involves using chickens in carefully controlled settings. Most basic and applied research would be impossible to conduct without the use of scientific and laboratory animals.

TYPES OF SCIENTIFIC AND LABORATORY ANIMALS

Although many species of animals may be used in research, scientists have found that some animals are better than others for certain kinds of research. A few selected species of animals are used for basic and applied research. Clinical research animals involve any species that is being studied.

Most basic and applied animal research takes place in a laboratory setting. Scientists need animals with certain characteristics that make a particular species attractive for research purposes. Most research facilities have a limited amount of space; therefore, animals that require small amounts of space are most desirable. Larger animals also eat more food and create more waste. It is not surprising to learn that small animals are the animals of choice! In fact, over 90 percent of the animals used in biomedical research are mice, rats, and other rodents.

Another consideration is the similarity between the animal and the benefactor of the research. For example, a strain of mice has been developed that have immune systems which closely resemble humans. If the purpose of the study is to look at HIV and AIDS in humans it would make sense to use this strain of mice. Many scientific and laboratory animals have common traits with humans and other animals. Researchers must consider these similarities when selecting a research animal.

MICE

Mice (*Mus musculus*) are the most popular animals for scientific and laboratory use. They are small, relatively inexpensive animals requiring little space and food. Mice have a mild temperament and are easy to handle and work with.

Because they are efficient breeders, they are especially suited to research that requires large numbers of animals to conduct. Mice are most often used in genetic, immunology, and infectious disease research.

There are over 100 documented bloodlines of mice available to researchers. Some of the most common strains include C3H, C57BL/6, CBA, DBA/2, A, CF1, ICR, S, and SW.

Mice, like most other rodents, are nocturnal, which means that they are most active at night. They constantly groom themselves, in fact, an

21-4. Many mice can be raised in a small space. Note the clean bedding and feed containers.

21-5. The correct way to pick up a white rat in a laboratory is by its tail. Rats and mice have strong tails. Picking them up this way takes their well-being into account.

early sign of illness or disease is an ungroomed appearance. Like all rodents, mice's teeth grow constantly. This is not usually a problem in laboratory settings, as the food provided is hard and keeps the teeth trimmed.

If mice are kept together, the males may fight. Females are generally not aggressive.

RATS

A larger cousin of the mouse is the laboratory rat (*Rattus norvegicus*). It was developed from the wild brown Norwegian rat. Rats are commonly used in behavioral and nutritional research.

Five strains of rats account for most of the animals used in modern research. These strains include Wistar (WI), Sprague-Dawley (SD), Long Evans (LE), F344, and LEW.

Rats are also nocturnal rodents. They are usually docile and easily handled. They are larger and stronger than mice and greater effort is required to restrain them. Since they seldom fight, male rats can be kept in the same cage together.

HAMSTERS AND GUINEA PIGS

Although mice and rats are the most common animals involved in basic and applied research, other rodents play an important role.

Hamsters

The golden Syrian hamster (*Mesocricetus auratus*) is the most commonly used hamster for research. It is characterized by a stout body, short tail, and reddish-gold fir on the head, back, and sides, with grey-white fir on the underside. All hamsters have ***cheek pouches*** which they use to transport and temporarily store food.

Hamsters are aggressive and will bite when startled. They should be handled carefully and gently. When they are handled properly, they will become easier to work with. Females usually dominate the males and larger animals dominate smaller ones. Animals kept in the same cage develop a "pecking order" and are able to live peacefully.

Guinea Pigs

A Guinea pig (*Cavia porcellus*) is also known as a **cavy.** Guinea pigs differ socially and anatomically in several ways from other rodents. Important research strains include strain 13, strain 2, Dunkin Harlet (DH), and Shorthair (SH).

The cavy has a stocky body and short legs. It cannot climb as well as most other rodents and lacks agility. These characteristics, along

21-6. Guinea pig.

with their docile nature, make cavies easier to handle than most laboratory rodents.

One nutritional difference between guinea pigs and other rodents is their inability to produce vitamin C. Like humans, guinea pigs must supplement their diet with this vitamin. Vitamin C deficiency causes a fatal disease known as scurvy.

RABBITS

Rabbits (*Oryctolagus cuniculus*) were first domesticated in the 16th century. They serve as models for human and animal diseases, produce serum antibodies, and are used for drug testing. The New Zealand white rabbit is the most commonly used research animal.

Rabbits are used in several specialty areas of human re-

21-7. Rabbits in a clean cage.

search where rabbits and humans suffer from similar diseases and conditions. For example, rabbits can suffer from atherosclerosis (hardening of the arteries), emphysema, spina bifida, and cleft pallet. One focus of scientific and laboratory rabbit research is to better understand these conditions in humans.

Rabbits are curious animals that enjoy exploring their surroundings. They are gentle, usually docile, and easy to handle if not excited. Rabbits can be frightened by loud or sudden noise and often show fear by stomping their hind feet. A frightened rabbit may try to bite or scratch.

CATS

The cat (*Felis catus*) has been living with humans for over 5,000 years. The domestic cat is of particular importance in brain research. The cat brain seems to be at a development stage between primates and lower mammals. A great deal of the present human and animal brain research has grown from basic research with cats. The American shorthair is the dominant research breed.

Healthy cats are curious and aware of their surroundings. Their ears are erect and eyes are bright and alert. Most cats enjoy being petted and purr when handled.

21-8. A cat in a cage with food, water, and a litter pan.

DOGS

Dogs (*Canis familiaris*) have been used for experimentation for over 300 years. The most common research with dogs includes physiology, pharmacology, and surgical studies. Beagles are the breed of choice for most non-surgical research.

Daily contact improves the dogs physical and psychological well-being. A healthy dog greets its handler with bright eyes, alert ears, and a wagging tail.

Good facilities for dogs include a large cage with waterer, feeder, and rest area. Indoor and outdoor runs are used to ensure that the animal is able to exercise.

NON-HUMAN PRIMATES

A *primate* is an animal with an opposable thumb used for grasping things. The human being is a primate. Look at your thumb. How important is it to you?

Several species of *non-human primates* are used in a broad range of scientific and laboratory settings be-

21-9. Beagle.

cause of their close resemblance to humans. We frequently think of these animals as monkeys. Monkeys can be divided into two basic categories: old world and new world.

Old world monkeys come primarily from Africa and Asia. They have close-set nostrils that open downward and cheek pouches capable of carrying food. Old world monkeys also have pads on their buttocks. New world monkeys have nostrils that open to the side, and do not have buttocks pads or cheek pouches. New world monkeys also have long tails used for grasping and climbing. A unique characteristic of non-human primates that separates them from other scientific and laboratory animals is their social behavior. Communication through verbal and non-verbal means is of special interest to researchers.

Most research is conducted with conditioned animals that are taken from the wild. Very few non-human primates are bred domestically for scientific and laboratory use.

ANIMAL MODELS FOR SCIENTIFIC AND LABORATORY INQUIRY

It is important to understand that the use of animals in research is strictly controlled and monitored. Scientists have developed a set of acceptable *models for animal research.* These ensure that scientific and laboratory animals will contribute to knowledge and not be misused.

Four basic models exist for the use of animals in research. They include:

1. Living animals

2. Living animal systems (tissue culture)

3. Non-living systems

4. Mathematical or computer models

Using these models, scientists can better understand the human body, human behavior, diseases, treatments, and effects of drugs on the body.

Living research animals (including humans) provide the best vehicles for psychological and human research. Living animals are responsive to treatments, can react to their environment, and are readily examined. However, the trend in modern research is to move away from using as many living animals and move toward utilizing other models.

Animal tissue cultures can tell scientists a great deal about how body systems will react to stimuli. The tissues are taken from animals, including humans, and grown in a laboratory. A great deal of the research in cancer treatment through radiation and chemotherapy is done using tissue cultures.

Non-living systems research is using mechanical models developed by scientists that mirror animal activity. They are used in research that is related to movement or effects of injury on living animal systems. An example related to humans is the development of the artificial hip. Many models had to be developed in the laboratory before a working design could be developed.

Mathematical and computer models are developed only after many hours of basic research. The models do not involve animal subjects in research. These models are designed to mimic animal behaviors or systems. Scientists can manipulate these models on the computer and see results of their manipulation in seconds, whereas conducting the experiment with live animals in the laboratory would take much longer. Computer and mathematical models are wonderful tools; but they are limited to what we know about a particular animal or situation and, therefore, are not perfect.

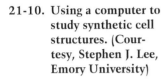

21-10. Using a computer to study synthetic cell structures. (Courtesy, Stephen J. Lee, Emory University)

MANAGEMENT PRACTICES USED IN ANIMAL RESEARCH

Researchers who use animals in their laboratories must meet a wide range of standards that allow them to work with living beings. They must prove their ability to provide basic care for the animals, as well as conduct appropriate research. The welfare of the animals is a top priority for researchers and agencies overseeing research with scientific and laboratory animals.

REGULATION OF ANIMAL RESEARCH

Animal research is heavily regulated by the researchers themselves. Researchers are concerned with keeping animals healthy both physically and psychologically. Animal research is also regulated from the outside.

Universities utilizing animal research must follow strict regulations. Research institutions are members of the Institutional Animal Care and Use Committee (IACUC). All animal research conducted must be approved by this committee. This committee looks at the research proposal to make sure that protocols of animal treatment are followed. For example, the committee will look at medical care provided for the animals, the qualifications of the personnel conducting the research, and the use of anesthetics. The committee also examines the institutions' overall use of animals in research and determines if animals are being utilized correctly.

External review of the use of animals in research is based on the *Animal Welfare Act* of 1985. This federal legislation was developed to

21-11. Large laboratory animals often have space much as they would in a natural environment. This llama is being used in veterinary medical clinical research. Except for the shaved areas on its neck, the llama appears as any other llama. Its well-being is important to the researchers.

21-12. Watering and feeding equipment are built into the design of this pen.

ensure the proper use of animals in research activities. The U.S. Department of Agriculture is the federal agency responsible for ensuring animal welfare in research. This agency regulates and inspects animal research facilities through the Animal and Plant Health Inspection Service (APHIS).

In addition, the Public Health Service (PHS), a division of the U.S. Department of Health and Human Services, the Environmental Protection Agency, and the Food and Drug Administration have animal welfare regulations. Most states also have regulations that are intended to protect the welfare of animals.

SCIENTIFIC AND LABORATORY ANIMAL NUTRITION

It is important that the animals used in research are fed a nutritious, balanced diet. If an animal is malnourished, the results of the research may be clouded by the effect of nutrition and not by the effect of the experimental treatment! Care is needed on the part of researchers and laboratory technicians to give animals involved in research studies a proper diet.

Animal diets will vary greatly depending on the type of animal, the animal's size, maturity level, and activity level, and the environment. The animal's diet includes water, protein, carbohydrates, fats, vitamins, and minerals. These elements provide energy and nutrients that help the animal grow and maintain itself.

SCIENTIFIC AND LABORATORY ANIMAL ENVIRONMENT

How would you feel if you were reading this book in a room with poor lighting, loud noise in the background, and a temperature below freezing? You probably would not be concentrating on what is written here! Just as humans have comfort levels in their environment, animals also have ranges in which they are comfortable. The only life experience most scientific and research animals have is in the laboratory environment. Making their existence comfortable is a major concern of animal researchers.

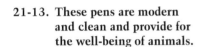

21-13. These pens are modern and clean and provide for the well-being of animals.

Probably the biggest concern for animal welfare in scientific and laboratory settings is the animal's living space. An animal must be given enough space to eat, sleep, and live out its life in relative comfort. Space requirements will vary with the type and size of the animal. Suggested minimums range from 6 square inches (15 cm²) for a small mouse to 144 square feet (4.4 m²) for horses. Space requirements are available from the National Institutes of Health in publication 86-23.

Animals are usually most comfortable at a temperature between 65 and 85°F (16.6° to 29.4° C). Relative humidity of the room should fall in the 30 to 70 percent range for most animals.

Lighting in the facility should be uniformly distributed and be bright enough for researchers to see and effectively work with the animals. Animals require a daily period of darkness to help them rest. The light and dark periods should remain on a set schedule to avoid stressing the animals.

Fresh air provided through adequate ventilation eliminates odors and reduces the number of airborne disease. Fresh air is a necessity to maintaining a healthy environment.

Although it is impossible to eliminate noise from the environment, the animals should not be exposed to loud noises or areas of high activity (loading docks, building entrances). Noise can cause stress in animals and even lead to health and behavioral problems.

SCIENTIFIC AND LABORATORY ANIMAL HYGIENE

Hygiene is a primary concern for researchers in a laboratory setting. Many illnesses and diseases are the result of poor sanitation and can ruin experiments by causing animal sickness and even death.

Sterilization of equipment and materials that come in contact with animals is a fact of life in the laboratory. Bedding that is used to line the cages, instruments used for examination, even the cages used to house the animals must be sterilized to avoid spreading disease.

Disinfectants are used to kill or inhibit the growth of bacteria and viruses. These chemicals are usually too strong to use directly on animals and are applied to cages and equipment used by animals.

Sanitation is the practice of controlling bacteria and other organisms by removing objects that they live on which come in contact with the animals. For example, removing animal waste, washing the cages, and cleaning examination tables are common sanitation practices. Sanitation would include changing the animals bedding or cleaning the cage floor.

21-14. Stray animals are often captured and used for research and education. This animal control officer has a snare designed to capture a dangerous animal with a minimum of injury to the animal.

REVIEWING

MAIN IDEAS

Animals contribute in many ways to the welfare of people and to the welfare of other animals. Scientific and laboratory animals are used in basic and applied research and to a lesser extent in clinical research.

There are many types of scientific and laboratory animals, but the most popular animals are the rodents. Over 90 percent of the animals in biomedical research are rodents. Other popular research animals include rabbits, cats, dogs, and non-human primates.

The advancements made in animal research have allowed scientists to develop four basic models that utilize animals in research. Only one model, the living animal model, actually conducts research with live animals.

Management in the laboratory or research facility is a top priority for scientists. Animal welfare is the goal of animal research scientists. Scientific and laboratory animal research is strictly regulated, internally and externally.

The living environment is regulated to provide a healthy, comfortable life for the animals. Space requirements, temperature, humidity, noise, and lighting are all components of the animals' environment. Keeping the animals healthy through hygiene is accomplished through sterilization, disinfection, and sanitation.

QUESTIONS

Answer the following questions using correct spelling and complete sentences.

1. Distinguish between using animals and using animal by-products in research.

2. What are three kinds of research? Briefly define each.

3. What animals are used in laboratory research? Which is most popular?

4. What animal models are used in scientific and laboratory research?

5. How is laboratory animal research regulated?

6. How is the well-being of animals provided for in a laboratory environment?

EVALUATING

CHAPTER SELF-CHECK

Match the terms with the correct definitions. Write the letter by the term in the blank that is provided.

a. cavy   d. animal by-product tissue   g. Animal Welfare Act
b. laboratory animal   e. non-human primate   h. clinical research
c. basic research   f. living animal model

1. ____ a federal law that assures proper use of animals in research

2. ____ a guinea pig

3. ____ research animals that are kept alive for study

4. ____ animals similar to humans

5. ____ research in a medical facility

6. ____ an animal raised for and/or used in laboratory research

7. ____ tissues from an animal used primarily for purposes other than research

8. ____ research that increases knowledge to help understand animals

EXPLORING

1. Prepare a report on the use of laboratory animals to alleviate a human disease. Be sure to investigate the disease and its effect on humans. Do you feel that animals should have been used for this research?

2. Organize a class debate or discussion to explore the use of animals in research. What is the conclusion drawn from the information that is shared? Do you agree with the positions taken by different individuals in the debate or discussion? Why?

Chapter 22

EXOTIC ANIMALS

Animals are special in many ways! What would a circus be without elephants, or a zoo, without a zebra? We look forward to the entertainment animals can provide. Just watching them go about life is fascinating.

People enjoy working with animals. Some people work in caring for them; others own and raise them. New and interesting kinds of animals are occasionally grown. The people who grow them are pioneers. They learn by trial and error. They do not have the benefit of research and written information to help them.

Special animals often need little help from humans to survive. The animals may need to be cared for in protected areas or receive a little extra attention. There may be at least one of these that is highly interesting to you.

22-1. Elephants are favorite animals.

OBJECTIVES

This chapter provides background information on exotic animals. It has the following objectives:

1. Describe and identify major species of exotic animals
2. Explain how to assess opportunities in exotic animal production
3. Explain important management practices with exotic animals

TERMS

animal refuge	exotic animal	performing animal
biologist	exotic animal dealer	zoo
circus	exotic animal investment species	zoo curator

USES OF EXOTIC ANIMALS

An *exotic animal* is an animal that is not native to where it is being raised. Exotic animals are often new or are used in new ways. In some cases, they need extra care to survive. Exotic animals are used for a wide range of purposes. Often, formal marketing structures are not in place for them or their products.

Exotic animals are relatively newly domesticated animals or animals that are beginning to be used for food and fiber production. They may also be used as investment animals and zoo and performing animals.

EXOTIC FOOD AND FIBER ANIMALS

Some exotic animals are used for human food and other products. In a broad sense, all current domesticated livestock started as an exotic species. Exotic animals used for food are new agricultural animals or animals being used for new purposes. A few species in the United States that are now considered exotic are the ostrich and emu, llama, alligator, eel, and bison. Many other animal species are raised to produce agricultural commodities. Some exotic animals are harvested from the wild.

Ostrich and Emu

Not native to the United States, the ostrich and emu are large birds that are beginning to be used for their meat and hide. These birds cannot fly, but are extremely fast runners. Their heavily-muscled legs are the source of most of the bird's meat. The meat is tender and dark in color. Health-conscious consumers are very interested in the "red meat" texture with very little fat content. Besides being very good to eat, ostrich and

22-2. Young emu. (Courtesy, Texas Department of Agriculture)

22-3. The head of a llama is distinct and alert.

emu eggs are large enough to feed most families! (These animals were also included in Chapter 13, Poultry Production.)

Llama

Llamas are mammals that have a potential of being raised commercially for their coat. Llama meat has a low fat content and is delicious. At this time, few llamas are eaten. Breeding animals can sell for more than $10,000 and to eat one would make for a very expensive meal. Some llamas are used as guard animals with sheep and goats. In remote parts of the world, llamas are used as pack animals.

Alligator

The American alligator, once on the brink of extinction, is now considered an exotic food animal. Alligators are raised on farms in many southern states. Their meat is light in color, has a firm texture, and a distinct flavor. Their skin is used for making expensive leather products. Alligators grow and reproduce well in captivity.

Bison

The American bison (sometimes called buffalo) is being used to produce an alternative to beef. Although few purebred bison are grown for meat, the genetics of the bison are used in cross-breeding programs with domesticated beef animals. This cross, often called a beef-alo, yields meat similar to beef with a lower fat content and less cholesterol.

22-4. A bison calf in a herd of bison. (Courtesy, American Bison Association)

Other Exotic Food and Fiber Animals

Several species of animals are raised for food, fur, and fiber that are considered exotic. Eel, elk, turtles, mink, and upland game birds, deer, and game animals are a few examples.

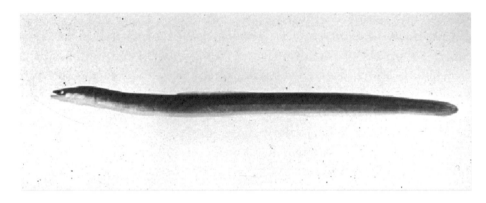

22-5. Eel culture has emerged in selected locations in North America. Eel meat is used for food and the skin for fine leather. This shows the American Eel *(Anguilla rostrata)*. (Courtesy, American Fisheries Society)

22-6. Elk are being domesticated and grown as exotic animals for meat and hides. (Courtesy, North American Elk Breeders Association)

Exotic Species Harvested in the Wild

Some businesses have started to market products from exotic wild animals. Meat from the antelope, deer, elk, and small game is sold throughout the country. Other wild products include snake meat, shark steaks, and skins, feathers, hides, and hair from a variety of birds and animals.

EXOTIC INVESTMENT ANIMALS

Since exotic animals are new animals, or animals being used for a new purpose, most people are afraid to invest the money needed to start an exotic animal business. Fortunately, there are always some entrepreneurs who are willing to take the risk. Investing in exotic animals can pay off well or it could leave you broke.

An *exotic animal investment species* is an animal raised in hopes of making a large sum of money. If successful, they are more profitable than cattle. There is a high degree of risk and many people fail to earn big profits.

Early investors in the pot-bellied pig market paid thousands of dollars for the first breeding animals that came to the United States. They hoped to make their money back selling these animals as pets. Although there was interest in the animals, the American public did not buy many pot-bellied pigs. Some investors were stuck with expensive pets.

22-7. Pot-Bellied Pig.

Other investors in exotic animals have fared better. A trend for sports enthusiasts has been hunting in animal preserves. With the areas available to hunters decreasing, animal preserves have become a good alternative. Investors are realizing a boom in business, while sports enthusiasts are provided with a place to practice their sport.

Investing in Agriculture

One area of investment is in agriculture. Several species are available for an entrepreneur to choose to invest in today. Good business people keep their eyes open for new investment opportunities. No one knows for sure what the next big exotic species opportunity will be!

Investing in Pets

There will always be people who want to own unique pets. New companion animals could be a good investment for the investor. Roadblocks for the new companion animals include state and federal laws restricting new animals, international law, and the compatibility of the animal with the environment and other animals.

Investing in Research

Many animals, especially strains of mice and rats, are used in research. A potential investor could fund development of a new strain or improvement of an existing strain. New research animals could also be added to assist with research.

EXOTIC ZOO AND PERFORMING ANIMALS

Zoo and performing animals are raised in captivity or captured from the wild. Training often begins at a young age to prepare animals that will be performers.

Zoo

A *zoo* is a zoological garden. It is a parklike area where animals are kept for viewing. Many of the species are exotic animals that are not native to the area where the zoo is located. Sometimes, the animals will breed in captivity. Environments are constructed that resemble the wild environments for the animals. Probably the best known zoo in North

22-8. These flamingos are kept in an environment similar to their native habitat.

22-9. Four animals commonly found in a zoo are: sable antelope (top left), tiger (top right), zebra (bottom left), and grizzly bear (bottom right).

America is the San Diego Zoo in California, which has a total of 3,200 animals representing about 800 species. The Smithsonian Institute Zoo in Washington, D.C., is well known for its Panda bears. Many cities have one or more zoological parks.

Zoos have four roles: education, entertainment, research, and animal conservation. Many people with special training in zoology work in a zoo. A *zoo curator* is a person trained in exotic animal zoology who looks after the animals.

Zoos get animals in several ways. Some get animals by breeding and raising them. This works only with the species that are on display. If a zoo does not have a suitable male and female pair, this is not possible. In some cases, zoos exchange animals. For example, one that has raised a young giraffe might exchange it with another zoo for an animal of equal value. Many zoos get their animals from animal dealers. An *exotic animal dealer* buys and sells exotic animals. They help meet the needs of zoos for certain species of animals.

Animal well-being is a high priority. Each animal is fed according to the findings of research and careful observation. Health care is provided to keep the animals disease-free. Occasionally, the animals must be protected from people who are diseased. Some species get diseases from humans. An example is the closure of a popular and attractive monkey exhibit at a major zoo because the animals contracted tuberculosis. They got the disease from eating food scraps thrown into their area that had been infected by humans!

Some zoos have trained animals and provide performances. Others limit their work to caring for the well-being of the animals.

Performing Animals

A *performing animal* is one taught to go through a routine of unusual activity. Performing animals may be domesticated or exotic species. Dogs and horses are often taught to perform tricks. Exotic animals, such as elephants, tigers, and monkeys, are also taught similar activities.

Many circuses have trained animals as part of the show. A *circus* is a variety show with animal and human performances. Humans are closely involved with the animals.

Who do you think breeds and raises circus animals? The people who do so have skills in agriscience! It is true that some animals come from the wild, but many are bred and raised in captivity.

ANIMAL REFUGES

An *animal refuge* is a protected place where animals are cared for to assure their well-being. Some animal refuges are for wildlife; others include exotic animals. A bird sanctuary is a kind of refuge.

There are about 475 National Wildlife Refuges in the United States. These range from small areas to protect one species of bird to large land areas for the protection of several species. Private individuals also have refuges that cater to exotic animals.

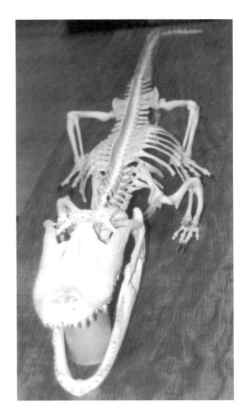

22-10. Veterinary medical specialists use this alligator skeleton to learn more about how the animals move about.

Animal refuges study animals and develop new ways of meeting their needs. They often try to maintain and reestablish a threatened species. Biologists often work with the animals. A **biologist** is a person trained in special areas about living organisms. Wildlife biologists have specific training in the biology of wildlife.

OPPORTUNITIES IN EXOTIC ANIMAL PRODUCTION

Many opportunities exist for the entrepreneur to get into the exotic animal business. But, before you get started, you should examine the requirements to start an exotic animal production business.

PROFITABILITY

Making a profit is the goal of all businesses. The exotic animal business is no different. If you decide to enter this part of livestock and poultry production, you need to know the costs, advantages, and regulations that will affect your business.

Start-up Costs

Start-up cost depends greatly on the species you choose to raise. As mentioned earlier, llama production would be a very expensive enterprise. Investing in a species of tropical fish, on the other hand, would be much less expensive. One must consider start-up costs of breeding stock, facilities, and equipment.

22-11. Tropical fish starter kit.

Operating Costs

Operating costs include feed, medicine, labor, marketing, utilities, permits, insurance, and any other cost of being in business. You can get information to estimate these costs from extension agents, feed salespeople, or others in the same business.

CHOOSING A SPECIES TO RAISE

Suppose you have decided to raise an exotic animal. How will you decide which species to raise? Here are a few questions you should ask.

Location Limitations

Many exotic species are environment-specific. For example, alligators would not grow well in Ohio. They need to be in a subtropical region to grow and reproduce. Fortunately, there are other species that are not as environmentally sensitive. Also, some species (tropical fish for example) can be kept indoors. You need to be aware of limitations you have because of your location.

Comparative Advantages

You should also compare advantages and disadvantages between species. If you are considering several alternatives, look at ease of production, costs, marketing, and ease of leaving the industry.

22-12. Raising crickets on a small scale for fish bait does not require a large investment.

Regulations

Ask the proper agencies in your state and community about regulations for exotic animals. You must find out if you are allowed to possess the species or if you need a license to produce the animals.

Regulations on exotic animals are strict. States like Florida, Louisiana, and Mississippi have experienced many problems with exotic animals invading the wild. One example is the nutria. This large rodent is not native to the United States. It was accidentally released and has now become a major problem. The animal is a vegetarian with a big appetite! These animals are destroying thousands of acres of wetland vegetation every year.

Regulations are also in place to prevent the spread of disease, protect people and property, and ensure that animals are treated humanely. Housing, zoning, shipping, and purchasing regulations must be explored before starting your business.

Market

You could be the best nutria producer in the world and be out of business, if you do not have a market for your product. Make sure you answer this question before investing your first penny.

You should decide whether you want to sell grade or purebred animals. Decide if you are going to sell directly to consumers or sell to a dealer or processor. You may even need to work on developing your market before you start.

22-13. Some exotic animals can be sold through pet stores.

EXOTIC ANIMAL
MANAGEMENT PRACTICES

Although there are many similarities between exotic and other agricultural animal species, there are important differences. To raise an animal is to be responsible for its well-being. Responsible producers will know the needs of their animals.

SIMILARITIES BETWEEN EXOTIC AND OTHER AGRICULTURAL ANIMALS

The basic rules for animal production apply to producing exotic animals. They must have a balanced ration. They must be given adequate space to live. Their living area should be kept sanitary and safe. The environment must be comfortable. They must receive preventive care and timely treatment for illness and injury. In these ways, raising exotic animals is very much like raising hogs or sheep.

DIFFERENCES BETWEEN EXOTIC AND OTHER AGRICULTURAL ANIMALS

While there are many similarities, the differences are key to being a successful exotic animal producer. Each species has its own characteristics that make it a unique animal. Major areas of difference between traditional domestic livestock and exotic species are nutritional requirements, health needs, housing and living space, and reproduction.

Nutritional Requirements

Although every animal needs a balanced ration, every animal differs in its nutritional requirements. Domestic livestock rations have been carefully studied. We know with a great degree of certainty what they require in their diet. Information on most exotic animals is sketchy at best.

What would you feed alligators? How can you find out what they need? Fortunately, alligator diets have been researched by universities and alligator producers. This information would be easy to find from these sources.

Most other exotic animals are the subject of research and you can learn about their requirements from universities or the cooperative extension service.

22-14. Facilities used for exotic animals must be kept clean. This shows an area used for exotic waterfowl being hosed down.

Health Needs

Veterinarians and other animal health care professionals are familiar with the health needs of domestic animals. Unfortunately, your local veterinarian may not be familiar with the needs of exotic animals. The veterinarians trained to work with exotic, zoo, circus, and performing animals are usually at universities or work for agencies that have exotic animals.

Housing and Living Space

Housing needs will vary with the species. You need to find out whether animals can be kept together, how much space they need, what temperature they prefer, and the type of housing they need.

The exotic animal owner will also need to know how to restrain their animals. For example, keeping an ostrich in confinement requires strong fences!

Reproduction

Animal production requires reproduction. Some species will reproduce well in captivity, while others need to be given great care to encourage reproduction.

REVIEWING

MAIN IDEAS

An exotic animal is any new species of animal that is being used for agricultural purposes, or an existing animal that is being used for a new purpose. The major types of exotic animals include food and fiber animals, investment animals, and zoo and performing animals.

Opportunities exist for entrepreneurs in the exotic animal industry. One must consider the profitability of the animal, the costs of production, and choose a species to raise.

Exotic animals and domestic animals have many characteristics in common. They all need proper nutrition, health care, sanitary living conditions, and space to live. They also have differences that must be attended to. To find out about these differences, you should seek help from universities, the cooperative extension service, or veterinarians that specialize in exotic animal care.

QUESTIONS

Answer the following questions using correct spelling and complete sentences.

1. What is an exotic animal? How are they different from other domesticated animals?

2. What are three species of exotic animals used for food and fiber?

3. Why did investors often lose money with the pot-bellied pig?

4. How does a zoo get its animals?

5. Give an example where zoo animals contracted disease from humans.

6. What are the major questions to answer in trying to determine if an exotic animal business will be profitable?

7. What should you consider in deciding on the species of exotic animal to raise?

8. What are the major practices in managing exotic animals?

EVALUATING

CHAPTER SELF-CHECK

Match the terms with the correct definitions. Write the letter by the term in the blank that is provided.

a. zoo curator   d. animal refuge   g. biologist
b. performing animal   e. exotic animal   h. exotic animal invest-
c. zoo   f. exotic animal dealer   ment species

1. ____ an animal raised with hopes of having a large profit

2. ____ person who buys and sells exotic animals

3. ____ animal that is not native to the area where it is raised and is being used for new purposes

4. ____ parklike area where animals are kept for viewing

5. ____ animal taught to perform a routine of unusual activities

6. ____ person trained in understanding living organisms

7. ____ person trained in exotic animal zoology

8. ____ place where animals are protected to assure their well-being

EXPLORING

1. Visit a zoo or animal refuge. Develop a list of the animals that you see. Study the kind of environment provided. Interview a zoo curator about feeding and other management practices with the animals. Prepare a report on what you observed.

2. Select an exotic animal. Use various reference materials and prepare a report on the origin of the animal and the kind of environment it needs to survive. Report in class.

Scientific Classifications of Agricultural Animals

DAIRY AND BEEF

a. **Kingdom Animalia**—animals collectively; the animal kingdom

b. **Phylum Chordata**—one of the approximately 21 phyla of the animal kingdom, in which there is either a backbone (in the vertebrates) or the rudiment of a backbone, the chorda

c. **Class Mammalia**—mammals or warm-blooded, hairy animals that produce their young alive and suckle them for a variable period on a secretion from the mammary glands

d. **Order Artiodactyla**—even-toed, hoofed mammals

e. **Family Bovidae**—ruminants having polycotyledonary placenta; hollow, nondeciduous, up-branched horns; and nearly universal presence of a gall bladder

f. **Genus *Bos***—ruminants quadruped, including wild and domestic cattle, distinguished by a stout body and hollow, curved horns standing out laterally from the skull

g. **Species *Bos taurus* and *Bos indicus*—*Bos taurus* includes ancestors of European cattle and of the majority of cattle found in the United States; *Bos indicus* is represented by the humped cattle (Zebu) of India and Africa and the Brahman breed of America**

SHEEP

a. **Kingdom Animalia**

b. **Phylum Chordata**

c. **Class Mammalia**

d. **Order Artiodactyla**

e. **Family Bovidae**

f. **Genus *Ovis*—**the genus consisting of the domestic sheep and the majority of wild sheep; horns form a lateral spiral

g. **Species *Ovis aries*—**domesticated sheep

SWINE

a. **Kingdom Animalia**

b. **Phylum Chordata**

c. **Class Mammalia**

d. **Order Artiodactyla**

e. **Family Suidae—**the family of nonruminant, ariodactyl ungulates, consisting of wild and domestic swine

f. **Genus *Sus*—**the typical genus of swine, restricted to the European wild boar and its allies, with the domestic breed derived from them

g. **Species *Sus scrofa* and *Sus vittatus*—***Sus scrofa* **is a wild hog of continental Europe from which most domestic swine have been derived;** *Sus vittatus* **was the chief, if not the only, race or species of East Indian pig that contributed to present day domestic swine**

HORSE

a. **Kingdom Animalia**

b. **Phylum Chordata**

c. **Class Mammalia**

d. **Order Perissodactyla—**nonruminant, hoofed mammals, usually with an odd number of toes, the third digit the largest, and in line with the axis of the limb; includes the horse, tapir, and rhinoceros

e. **Family Equidae—**members of the horse family may be distinguished from the other existing perissodactyla (rhinoceros and tapir) by their comparatively more slender and agile build

f. Genus *Equus*—includes horses, asses, and zebras

g. Species *Equus caballus*—the horse is distinguished from asses and zebras by the longer hair on the mane and tail; the presence of the "chestnut" on the inside of the hindleg; and by other less-constant characteristics, such as larger size, larger hoofs, more arched neck, smaller head, and shorter ears

POULTRY

a. Kingdom Animalia

b. Phylum Chordata

c. Class Aves—those animals having feathers

d. Subclass Neornithes—modern birds (as opposed to fossil birds)

e. Suborder Carinatae—those birds with keel-like breast bones; this superorder has 2,810 genera and 8,616 species; in addition, there are over 25,000 subspecies; the genus-species of domesticated birds are:

1. chicken—*Gallus domesticus*

2. duck—*Anas domestica*

3. goose—*Anser domestica*

4. guinea fowl—*Numida meleagris*

5. pigeon—*Columba domestica*

6. turkey—*Mealeagris gallapavo*

Common Terms or Names of Animals

SEX CLASSIFICATION NAMES

Animal	Male of Breeding Age	Mature Female	Young Male	Young Female
cat	tomcat (tom)	queen (or pussy)	—	—
catte	bull	cow	bull	heifer
dog	dog	bitch	puppy dog	puppy bitch
goat	buck	doe	buck kid	doe kid
horse	stallion	mare	colt	filly
chicken	cock	hen	chick	chick
sheep	ram	ewe	ram lamb	ewe lamb
swine	boar	sow	shoat	gilt

NEWBORN, UNSEXED, AND GROUP NAMES

Animal	Newborn	Unsexed Male	Unsexed Female	Group
cat	kitten	gib	neuter	bevy
cattle	calf	steer	spayed	herd
dog	pup	castrate	spayed	pack
goat	kid	wether	spayed	band
horse	foal	gelding	spayed	herd
chicken	chick	capon	—	flock
sheep	lamb	wether	spayed	flock
swine	pig (or piglet)	barrow	spayed	drove

Glossary

Accelerated lambing—improving reproductive efficiency with sheep by using out-of-season lambing techniques

Aeration—physical process of bubbling air into water or splashing the water

Agriscience—the use of science in producing food, fiber, and shelter

Albumen—the white part of a poultry egg that surrounds the yolk

Allele—different forms of genes

Alveoli—tiny structures in a mammary gland that remove nutrients from the blood and convert the nutrients into milk

Amino acids—the building blocks of protein

Anaplasmosis—a parasitic disease, primarily of cattle, caused by a protozoan that attacks the red blood corpuscles

Anatomy—study of the form, shape, and appearance of animals

Anestrous—when a female does not cycle through estrous; nonbreeding season

Animal by-product tissues—tissues taken from animals when they are used for other purposes, such as from slaughtered cattle

Animal domestication—taking animals from nature and raising them in a controlled environment

Animal health—condition in which an animal is free of disease

Animal industry—all of the activities in raising animals and meeting the needs of people for animal products

Animal marketing—movement of animals from the farm or ranch to the consumer; includes preparing the products for consumption

Animal model—the genetic evaluation for dairy cattle production using pedigree data and production related factors

Animal producer—a person who raises livestock, poultry, or other animals

Animal production—raising animals for food and other uses

Animal refuge—protected place where animals are cared for to assure their well being

Animal services—assistance provided in animal production from off-farm sources

Animal supplies—materials needed in animal production that are from off-farm sources

Animal tissue cultures—tissues taken from living animals for laboratory study

Animal well being—caring for animals so that their needs are met and they do not suffer

Animal wildlife—all animals that haven't been domesticated

Anthelmintic—chemical compounds used for deworming cattle

Anthrax—an acute infectious disease that can affect most endothermic animals

Antibody—immune substance produced in the body of an animal

Applied research—research carried out to achieve a specific purpose

Aquacrop—commercially-produced aquatic plants, animals, and other species

Aquaculture—the production of aquatic plants, animals, and other species

Aquarium maintenance schedule—a listing of important activities in keeping a good aquarium

Aquarium—container used to hold water for ornamental fish

Arthropoda phylum—animals with exoskeletons, segmented bodies, and jointed legs

Artificial insemination—when semen collected from a male animal is placed in the reproductive tract of a female of the same species

Atom—smallest unit of an element

Automated feeding—mechanical animal feeding

Backgrounding systems—production system that takes calf from the time of weaning to the feedlot phase; weight is added before going to a feedlot

Bacteria—one-celled organisms that may cause disease; disease examples include tuberculosis, brucellosis, and mastitis

Balanced ration—a ration which contains all the nutrients that the animal needs in the correct proportions

Barrow—a male hog castrated before sexual maturity

Basic research—research carried out to increase the knowledge and understanding of an area

Beef—the meat of cattle

Behavior—reaction of an organism to a stimulus; how an organism responds to its environment

Billy—male goat; also known as buck

Biodiverse population—having a wide variety of members in a population, especially a wildlife population

Biological filtration—using bacteria and other organisms to convert harmful materials in aquaculture water into forms that are less harmful

Biological oxygenation—using phytoplankton (tiny plants) to replenish oxygen in water

Biologist—a person with specialized training in living organisms

Biotechnology—the management of biological systems for the benefit of humans; biology is used to produce new products

Bitch—female dog

Blackleg—an acute, highly infectious disease that primarily affects young cattle; usually results in death

Boar—sexually mature male hog

Bos indicus—cattle of Indian origin; descendants of the zebu and have humps on their necks, large droopy ears, and loose skin; include cattle with brahma blood

Bos taurus—cattle of European origin; includes common breeds such as Hereford and Angus

Brackish water—water that is a mixture of saltwater and freshwater

Brand—a permanent method of cattle identification involving burning a scar on the hide of cattle

Breed—a group of animals of the same species that share common traits

Breeding period—the time of the year when the female has estrus periods and estrous cycle

Breeding—helping animals reproduce and grow so that the desired offspring result

Broiler—young meat-type chicken between 5 and 12 weeks of age

Broodfish—sexually mature fish kept for reproduction

Browse—woody and broad leaf plants; forage for goats, deer, and certain other animals

Brucellosis—a disease of the reproductive tract of cattle, sheep, goats, and hogs that causes abortion; also known as bang's disease

Buck—male goat; also known as billy

By-product feeds—feeds made from by-products of manufacturing foods, such as pulp from citrus or beets

Cage—a structure for confining aquacrops in water; cages float on the water

Candling—shining light through an egg to see its features

Capon—castrated male chicken

Carbohydrate—food component that provides energy; includes sugars, starches, and cellulose

Career—general direction that a person takes with work

Carnivore—flesh-eating animal

Carrying capacity—with wildlife, the number of animals a given habitat area can support

Castration—removing or destroying the testicles of a male so that it does not breed

Cavy—a guinea pig

Cell—basic unit of life; building blocks for organisms

Chammy—soft, pliable leather made from the skin of sheep and goats

Cheek pouch—pouches in the cheeks of some rodents, such as the hamster, where food can be stored

Chemical filtration—using chemical processes to filter water; activated charcoal is often used for chemical filtration in an aquarium

Chordata phylum—animals that have spinal cords

Chromosome—thread-like parts of a cell nucleus containing the genetic material

Circulatory system—body system that moves nutrients, oxygen, and metabolic wastes

Circus—variety show with animal and human performers

Clinical research—research that takes place in a medical clinic or other laboratory

Cloning—the production of one or more exact genetic copies of an animal

Cloven hoof—divided hoofs; examples are cattle, sheep, and goats

Coccidiosis—a parasitic disease affecting poultry; prevented and treated with anticoccidials in feed and water

Cock—male chicken over one year of age

Cockerel—male chicken less than one year old

Cold housing—using unheated buildings to house animals

Colostrum—the first milk given by an animal after parturition; baby animals need colostrum to help develop disease immunity

Companion animal—a domesticated animal kept by humans for enjoyment in a long-term relationship

Companion fish—ornamental fish kept for human companionship or as pets

Concentrate—a feed high in energy, such as corn, oats, and wheat

Conception—the time when an egg is fertilized to create new life

Confinement method—raising sheep, hogs, and other animals in confined areas; complete environmental control may be used

Conformation—the type, shape, and form of an animal

Connective tissue—tissue that holds and supports body parts

Consumer—people who eat, wear, and otherwise use goods and services

Consumption—the use of products and services that have value

Contagious disease—a disease spread by direct or indirect contact

Contract production—when an agreement is made between a producer and buyer before the hogs or other animals are raised

Cow-calf system—a cattle production system that involves raising calves for sale to other growers

Creep feeding—feeding calves supplementally so they grow faster while still nursing

Cria—young llama

Crossbreeding—when animals of the same species but of a different breed are mated

Crustacean—animal with an exoskeleton, such as shrimp and lobsters

Curry comb—oval shaped plastic or metal device use to loosen sweat, manure, and other foreign materials in animal hair

Cutability—the amount of saleable retail cuts that may be obtained from a carcass

Dairy cattle—cattle raised and kept for milk production

Dairy product—milk and any products in which milk is the major ingredient

Dairy Herd Improvement Program—a national dairy production testing and record-keeping program

Debeaking—removing the tips of beaks so birds cannot peck each other

Declawing—cutting the claws from the feet of a cat

Dehorning—used with horned breeds of cattle to remove horns

Deoxyribionucleic acid (DNA)—protein-like nucleic acid on genes that controls inheritance

Digestive system—system that breaks food into smaller parts for use by the body

Disease—disturbance of body functions or structures

Disinfectant—a cleaning substance that destroys the microbial causes of disease

Disposal pit—a pit in the earth for disposing of dead birds; also known as a compost pit; microorganisms decompose the dead birds

Dissolved oxygen—gaseous oxygen that is suspended in water; used by aquatic organisms; measured as DO

Diurnal—animals that sleep during the night and awake during the day

Docking—cutting of all or part of the tail; keeps manure from soiling the wool

Doe—female goat; also known as a nanny

Dominant—traits that cover up or mask the alleles for recessive traits

Down—soft feathery covering on young poultry

Draft—pulling of an animal to move an object

Draft animal—an animal which has been trained and is used for pulling heavy loads

Draft horse—a horse standing 14.2 to 17.2 hands high and weighing over 1,400 pounds (635 kg)

Drake—mature male duck

Drenching—orally giving liquid medications

Dry cow—a cow that has stopped producing milk; most cows produce milk 305 days in a herd for each lactation

Dual-purpose breed—breed of cattle raised for both meat and milk

Dubbing—removing the comb and wattles on day-old chicks

Duckling—young duck that hasn't grown feathers

Earthen pond—water impoundments dug into the earth; small ponds are used for ornamental fish

Ectotherm—an animal that adjusts its body temperature to its environment

Egg injection—a method of vaccinating chicken embryos in the egg on the eighteenth day of incubation

Egg—sex cell of the female animal

Egg-laying fish—fish that reproduce by laying eggs

Embryo—a young animal developing in the reproductive tract of a female from fertilization to the completion of differentiation.

Embryo transfer—moving an embryo from the reproductive tract of one female animal to that of another female animal of the same species

Employment—the primary way people use time and earn a living

Endangered species—animals that are close to becoming extinct

Endocrine system—body system that secretes hormones that regulate metabolism, growth, and reproduction

Endotherm—an animal that maintains a constant body temperature; sometimes called warm-blooded

Entrepreneur—person who takes risk by trying to meet the demands of people for a product

Epithelial tissue—tissue that covers body surface

Equine sleeping sickness—a viral disease of horses, mules, and wild rodents; transmitted by biting insects

Equitation—riding and managing horses

Estrous cycle—time between periods of estrus

Estrous synchronization—bringing a group of female animals into heat at the same time

Estrus—time when the female is receptive to the male and will stand for mating; also known as heat

Ewe—female sheep

Excretory system—system that rids the body of wastes

Exotic animal—an animal that is not native or is being given a new use; exotic animals often are not domesticated

Exotic animal dealer—buyer and seller of exotic animals

Exotic animal investment species—exotic animal that raised in hopes of making a lot of money; giving a good return on investment

Expressing—manually squeezing the eggs from a female fish; sperm may be squeezed from a male using a similar procedure

External parasite—a parasite that lives on the external parts of its host

Farm flock method—a flock of sheep kept on a farm to produce market lambs

Farrier—a person who cares for the feet of horses, especially the design and fitting of shoes

Farrowing—the process of a sow giving birth to pigs

Fat soluble vitamin—kinds of vitamins that are stored in body fat and released by the fat as the body needs them

Fats—food component that provides energy; contains 2.25 times more energy than carbohydrates; form in which excess energy is stored by an animal

Feather—long hair on legs of horses

Feed—what an animal eats to get nutrients

Feed additive—a substance added to a feed to meet a particular need

Feeder pig—a pig that weighs 30–60 pounds and is sold to another farm to feed to market weight

Feedstuff—ingredients used in the feeds for animals

Fertilization—union of sperm and egg

Fetus—a young animal developing in the reproductive tract of a female after the organs have begun to form; the fetus matures for birth

Fiber—the material left after food has been digested

Filly—female horse under three years of age, except for Thoroughbreds which are under four years of age

Filtration—process of removing solid materials from water

Fin—structures on a fish that help with locomotion and balance in the water

Fingerling—an immature fish larger than a fry

Finishing system—beef cattle production system that feeds growing calves for slaughter

Fish—vertebrate aquatic animals

Floating—using a file to remove the sharp edges on the teeth of horses

Foal—young horse of either sex that hasn't been weaned

Foaling—process of a female horse giving birth

Food fish production—raising fish for human use as food

Foot and mouth disease—a viral disease affecting animals with cloven hoofs; no known treatment

Forage—feeds made mostly of leaves and tender stems of plants

Forager—cattle that eat grasses and plants and make good use of these foods in meeting their nutritional needs

Free access—when animals eat all the feed they want; feed is always available

Freshwater—water with little or no salt content

Freshwater ornamental fish—ornamental fish that live in freshwater

Fry—young, newly hatched fish

Functional type—the traits of a cow that are good enough to allow her complete a useful life in a dairy herd

Fungi—one of five kingdoms of living organisms; fungi are unicellular (one-celled) organisms; some fungi may cause disease

Gaggle—a group of geese

Gainful employment—employment that provides benefits, such as salary or wages

Gait—the way a horse moves (walks, runs, etc.)

Game animal—animals hunted for sport and food

Gamete—sex cell that can unite to form other cells

Gander—mature male goose

Gelding—male horse castrated before reaching sexual maturity

Gene—segments of chromosomes that contain hereditary traits of organisms

Gene mapping—locating the genes on a chromosome

Gene transfer—moving a gene from one organism to another

Genetic code—sequence of nitrogen bases in a DNA molecule

Genetic engineering—a form of biotechnology in which genetic information is changed to make a new product

Genetics—laws and processes of inheritance by offspring from parents

Genome—the genetic material for an organism

Genotype—the genetic makeup of an organism

Gestation—the period of pregnancy

Gilt—a young female hog that has not farrowed

Gizzard—muscular organ inside a bird that grinds food

Glucose—a simple sugar in food that is the source of energy for most cells

Goal—level of achievement that people set for themselves

Goal setting—establishing goals to be achieved and ways and means of reaching them

Gosling—baby goose of either sex

Grade cattle—cattle that are not registered; may be purebred or of mixed breed.

Gregarious behavior—instinct of animals to form groups, such as flocks or herds

Growth—increase in size of muscle, bones, and organs of the body of animals

Grubs—internal parasites caused by heel flies; move through the body forming bumps to form under the skin on the backs of cattle; also known as warbles

Habitat—place where animal and plant wildlife live

Hairball—wad of hair that collects in the digestive tracts of cats; hair is from licking their coats, known as grooming

Hand—a measurement for horses equal to 4 inches (10.2 cm)

Handler—a person who trains and works with K-9 dogs or other animals

Harness—attachments placed on an animal to provide control and allow it to pull

Harvesting—capturing an aquacrop so that it can be marketed

Hatchery—place where eggs are incubated for hatching

Health—the condition of an animal; how well the functions of life are performed

Herding dog—dog trained to assist in herding sheep and other animals

Heredity—the passing of traits from parents to offspring

Heterozygous—an organism having different alleles for a particular trait

Hierarch of classification—a way of showing relationships and differences among animals

Hog cholera—a viral disease of swine; nearly eradicated in the United States

Homeostasis—a characteristic of animals with organ systems in which a relatively constant internal environment is maintained by the animal

Homogenization—breaking fat droplets in milk into very small particles so that they stay in suspension

Homozygous—an organism having similar alleles for a trait

Hound—dog that tracks; follow scent left by animals or people

Hunting—catching or killing wild animals for sport or food; guns or other weapons are often used

Hunting and jumping horse—a horse specially bred and trained for use in fox hunting

Immunity—resistance of an animal to disease

Immunoglobulin—antibodies passed from a cow to her calf in colostrum; provides passive immunity to disease

Incinerator—a device for burning dead chickens or other animals

Incubation—the time required for the embryo to develop in a fertile egg; with fish it is the time between spawning and hatching

Insemination—placing sperm in the reproductive tract of a female

Internal parasite—a parasite that lives inside a host

Invertebrate—animal that does not have a backbone

Isolation—separating diseased and non-diseased animals

Jack—a male donkey of breeding age

Jenny—female donkey

Job—specific work; work carried out at a site

Jog—a two-beat horse gait

K-9—trained police dog

Kid—young goats under a year of age

Kidding—process of a doe (nanny) giving birth

Kitten—baby cat

Laboratory animal—an animal raised and/or used in research laboratories

Lactation—secretion of milk by the mammary glands; the production of milk by a female mammal

Lactation ration—a ration high in nutrients needed by lactating animals

Lamb—a young sheep less than one year old; the meat from a young sheep

Lamb feeding—specialized lamb production where the lambs go to feedlots after weaning

Lambing—the process of a ewe giving birth

Leptospirosis—a bacterial disease of cattle, sheep, and other animals

Lice—species of external biting and sucking parasites of cattle, hogs, and other species

Life science—science dealing with living things

Light horse—a horse standing 14.2 to 17 hands and weighing 900–1,400 pounds (408–635 kg)

Limiting factor—a component of a wildlife habitat that is missing so that wildlife cannot survive

Linear evaluation—coding 15 primary traits of dairy cattle for use in corrective mating.

Litter—the floor covering in a poultry house; also applies to floor coverings for other animals; a group of young animals born at a single birth, such as pigs or kittens

Litter pan—container with absorbent material for pets to use in urinating and defecating

Livebearing fish—fish that reproduce by giving birth to live young

Livestock—animals produced on farms and ranches for food and other products

Living research animal—living animals used in research

Lope—three-beat horse gait

Lymphatic system—body system that circulates a fluid known as lymph; lymph protects the body from disease

Macromineral—minerals needed in larger amounts

Maintenance—no loss or gain in weight by the body

Mammal—a class of animals whose offspring are fed with milk secreted by the mammary glands of the female

Management intensive grazing—a pasture system that concentrates large numbers of dairy cows on small pasture areas to harvest forage by grazing plant leaves and stems; the cattle are moved to another pasture after 12–48 hours of grazing.

Marbled—intramuscular fat that makes beef more palatable to humans

Mare—mature female horse; over four years of age except for Thoroughbreds, which are over five years of age

Mariculture—producing aquacrops in saltwater

Marine aquaria—aquaria with saltwater for growing marine species

Market hog—young hog raised for meat and weighing about 220 pounds

Mastitis—a bacterial disease of the mammary glands; the udder of cattle may be warm and hard to the touch; causes lost dairy production

Mathematical and computer model—carrying out research related to animals without using animals; the models mimic animal behavior

Meat animal—animal raised for meat

Meat animal by-product—products made from the parts of animals not used as food

Meat-type hog—a hog whose carcass that gives the greatest amount of lean meat in the areas of high-value cuts

Mechanical filtration—using devices to remove particles from the water and keep the water clear

Mechanical oxygenation—using devices that bubble or splash water to add dissolved oxygen

Metabolic disorder—disease related to nutrition

Microinjection—process of injecting DNA into a cell using a fine diameter glass needle and microscope; used to produce transgenic animals

Micromineral—minerals needed in smaller amounts; also known as trace minerals

Milking parlor—a concrete platform raised above the floor to make it easier for people to attach milkers to dairy cows and perform other functions on the udder

Mineral—inorganic elements needed for a healthy body

Models for animal research—controls over the use of laboratory animals to assure that the use will contribute to knowledge and not abuse the animal

Mohair—a product from angora goats used in making clothing

Molecular biotechnology—changing the structure and parts of cells in order to change an organism

Molecule—smallest unit of a substance

Mollusca phylum—animals with hard shells

Mollusks—aquatic animals with thick, hard shells

Molting—process of poultry shedding and renewing feathers

Mule—hybrid offspring of a male ass (jack) and female horse (mare)

Multi-purpose dog—K-9 dog used for more than one activity

Muscular system—body system that creates bodily movements; acquires materials and energy

Muscular tissue—tissue that creates movement of body parts

Mutation—changes that naturally occur in the genetic material of an organism

Mutton—meat from a sheep more than one year old

Nanny—female goat; also known as doe

Natural insemination—when a male animal deposits semen in the reproductive tract of a female during copulation

Natural selection—when animals breed without control by humans

Needle teeth—eight sharp teeth in pigs at birth; usually clipped off to prevent injury to the sow

Nervous system—a body system that coordinates body functions and activities

Nervous tissue—tissue that responds to stimuli and transmits nerve impulses

Niche—the special place in a habitat where every animal fits

Nitrogen cycle—process of converting animal wastes into ammonia, ammonia in nitrites, and nitrites into nitrates; naturally occurs in water

Nocturnal—animals that sleep during the day and awake during the night

Non-game animal—animal that does not provide products such as meat and fur

Non-human primate—primates closely resembling humans; often used in research; examples are monkeys and chimpanzees

Non-living research system—using mechanical models that resemble animal activity

Non-sporting dog—dog developed for a specific purpose; primarily used as pets

Noncontagious disease—a disease that is not spread by animal contact; includes nutritional, physiological, and morphological diseases

Nutrient—substance that is necessary for an organism to live and grow

Nutrient dense—a food or feed with large amounts of nutrients relative to calories

Nutrition—food needs and how food is used by an animal; process by which animals eat and use food

Occupation—specific work that can be described and has similar duties in different locations

One-stage production—a swine production system that raises pigs to a weight of 40–60 pounds at which time they are sold to another producer to grow them out

Operculum—structure on a fish that covers the gills

Organ system—a system formed when two or more organs work together to perform an activity

Organelles—tiny organ-like structures within a cell, such as the nucleus and mitochondria

Organism—any living thing

Organismic biotechnology—improving intact or complete organisms without artificially changing the genetic makeup

Ornamental fish—fish kept for their appearance and personal appeal

Orphaned lamb—a lamb whose mother has died

Ovary—primary reproductive organ of the female

Ovulation—when a mature ovum is released by the ovary

Ox—a bovine draft animal

Oxygenation—process of keeping adequate dissolved oxygen in water

Palatability—how well an animal likes food; how food feels in the mouth and on the tongue

Parasite—multicellular animal organisms that live in or on another animal

Part-time animal producer—a person who raises animals on a part-time basis; the person may have income from other sources

Parturition—process of giving birth

Pasteurization—heating milk to destroy bacteria

Peacock—male peafowl

Pedigree—record of an individual's heredity

Pelt—the whole hide of an animal with the fur attached

Pen—a structure that confines aquacrops in water; pens are attached to the earth at the bottom of the water

Per capita consumption—the average amount consumed in a year by a person

Performing animal—an animal taught to go through a routine of unusual activity

Pet carrier—a carrying case to safely and securely carry a pet

Phenotype—an organisms physical or outward appearance

Physiology—study of the functions of cells, tissues, organs, and systems of an organism

Piglet—baby pigs

Plankton—tiny plants and animals that float in water; serve as food to some aquatic animals

Pleasure animal—animals kept as pets or to use in other ways for the enjoyment of people

Plug—a horse with poor conformation and common breeding

Polled—cattle naturally without horns

Polo mount—a horse especially bred for use with polo

Pond—water impoundment made by building earthen dams or levees

Pony—horse standing under 14.2 hands high and 500 to 900 pounds (227–408 kg); small horse

Porcine somatotropin (pST)—a growth hormone for swine

Porcine stress syndrome—a nonpathological disorder in heavily muscled hogs that results in sudden death

Poult—young turkey not grown to the point where its sex is readily identifiable

Poultry—fowl raised on farms for use as food

Poultry science—study and use of science in raising poultry

Power—the combination of pulling capacity and speed; measured in horsepower

Preconditioning—preparing animals for stress associated with hauling

Predator—animal that lives off of other animals; kills, mauls, or preys on them

Predicted transmitting ability—estimate of the traits an animal will transmit to its offspring

Pregnant—condition of a female carrying unborn young

Primate—animal with an opposable thumb used for grasping things

Probe—a tool for measuring backfat on a hog

Production cycle—the complete cycle in the production of a crop or growth of an animal

Production intensity—the number or weight of aquatic species in a volume of water; biomass

Progeny—offspring of animals

Prolific—producing large numbers of young or offspring

Protein—food substance important for animal growth, maintenance, reproduction, and other functions

Protozoa—unicellular organisms; simplest form of animal life; some protozoa may cause disease

Puberty—time when sexual maturity is reached; the animal is capable of reproduction

Pullet—young female chicken

Purebred—animal eligible for registry in a breed association; parents must meet standards of the association

Purebred flock—flock of sheep of purebred breeding

Queen—female cat; also known as pussy

Racehorse—running or harness horses used in racing

Raceway—water impoundment that uses flowing water

Ram—male sheep kept for breeding purposes

Range band method—production system used with sheep where a herder moves the flock over a large area of land

Ration—the food an animal consumes each day; the total amount of feed an animal is given in a 24-hour period

Recessive—genetic traits masked by dominant traits

Recombinant DNA—gene splicing; genes are cut and moved to a cell to be altered

Regurgitate—when ruminants return eaten but unchewed food to their mouths for chewing

Reproduction—process by which offspring are produced

Reproduction ration—a ration high in nutrients needed by breeding animals

Reproductive efficiency—timely and prolific replacement of a species

Reproductive system—system by which new individuals of the same species are created

Reptile—an ectothermic animal with dry scaly skin and lungs for breathing; examples are lizards and snakes

Respiratory system—body system that regulates gas exchange of an organism

Riding horse—a horse that is ridden for pleasure

Risk—the possibility of losing; in entrepreneurship, the possibility of losing financial investment

Roan—mixture of white and colored hairs on cattle and horses

Roaster—young chicken somewhat older and larger than a broiler

Roundworm—an internal parasite of animals

Ruminate—when a ruminant chews their cud

Sac fry—newly hatched fish with the yolk sac still attached

Safety animal—an animal that helps protect people and property

Salinity—the amount of salt in water; measured as ppt (parts per thousand)

Saltwater—water that has a high salt content; typically 33–37 ppt salt

Saltwater ornamental fish—ornamental fish that live and grow in saltwater

Scheduled feeding—providing feed at certain times of the day

Science—knowledge about the world we live in

Scientific classification system—a method of distinguishing animals, plants, and other organisms on the basis of physical characteristics

Self-feeding—when animals have feed available all of the time

Service animal—an animal that assists people in living and work

Sexual reproduction—union of sperm and egg to form a new individual

Shipping fever—an environmental disease of cattle and sheep caused by stress; more likely to affect underfed animals

Single-purpose dog—K-9 dog used for just one activity

Skeletal system—bony structure that gives a framework for the body

Smolt—immature salmon larger than a fry

Social ranking—the order that an animal falls within a group of animals of the same species

Sow—sexually mature female hog

Spawning—release of eggs by a female fish and subsequent fertilization by a male of the same species

Spawning nest—a place where fish spawn; artificial spawning nests are used in aquaculture

Spaying—surgical procedure that removes the ovaries and uterus of a bitch to prevent breeding; also used with other animals

Specific pathogen free—animals free of disease at birth and raised in an aseptic environment; most often associated with hogs but used with other species, including shrimp

Spent hen—hen no longer used for egg production; processed into cooked poultry products

Sperm—sex cell of the male animal

Sporting dog—dog trained to help hunters find and retrieve game

Spring water—water from natural openings in the earth; similar to well water

Stallion—mature male horse; over four years of age except for Thoroughbreds, which are over five years of age

Stock horse—popular mixed breed horses often used with cattle

Stud horse—male horse kept specifically for breeding

Superovulation—getting a female to release more than the usual number of eggs during a sing estrous cycle

Supplement—a feed material high in specific nutrients; used to assure an animal has sufficient nutrients

Surface runoff—excess water from precipitation

Synthetic biology—using chemical substances to create systems with some of the characteristics of living organisms

Synthetic seawater—made for saltwater aquaria by using a commercially available mix

Tail docking—clipping the tail from baby pigs to prevent tail biting

Tapeworm—an internal parasite in the intestines of animals; may grow several feet long

Tapwater—the water from faucets in homes and offices; usually must be treated before using with fish

Taxonomy—science of classifying organisms

Technology—the practical use of science

Territory—the area an animal defends and protects

Testicle—primary reproductive organ of the male

Three-stage production—swine production system in which pigs are farrowed in stalls and kept there until weaning and being moved to pens until they reach 100 pounds, where they are moved again to be fed to market weight

Tom—male turkey

Tomcat—male cat

Total mixed ration—a ration that provides all needed feed ingredients in each mouthful of feed a cow eats

Toy breed—small dogs used for companionship

Transgenic animal—an animal which has stably incorporated a foreign gene into its cells

Trapping—capturing animals for their products using traps that keep the animals alive until detected by the trapper

Tri-purpose animal—an animal used for work, milk, and meat; oxen is an example

Tropical fish—small brightly-colored fish that are popular in home and office aquaria

Two-stage production—swine production system in which pigs are farrowed, nursed, weaned, and raised until they weigh 60 pounds and moved to a finishing unit

Type production index—a system for ranking dairy cows based on overall performance

Udder—mammary glands, teats, and associated structures on female mammals

Ultrasonics—equipment that uses bursts of high frequency sound in assessing the conditions of animals, including amount of lean meat and reproductive condition

Urbanization—building cities, suburbs, and other developments for the convenience of people; often destroys wildlife habitat

Vat—large tank made of concrete, fiberglass, or similar material for growing fish

Vertebrate—animal with a backbone

Vertical integration—when an agribusiness is involved in more than one step in providing agricultural products; chicken industry is a good example

Virus—a tiny particle too small to be seen with an ordinary microscope that often causes disease

Vitamin—organic food substances needed in small amounts for good health; important in body functions

Walk—a four-best horse gait

Warm housing—heating buildings in which animals are kept in the winter

Water facility—structures in which aquacrops are grown

Water quality—suitability of water for a particular use

Water soluble vitamins—vitamins dissolved by water and need to be consumed each day

Well water—water pumped from aquifers deep in the earth

Wether—male sheep castrated before sexual maturity

Whelp—process of a bitch giving birth

Wool—soft coat on sheep used as a fiber for human clothing and other products

Work ethic—how people feel about work

Working dog—dog used by people to get work done

Yoke—wooden bar in a harness that holds two animals together so that they work together as a team; primarily used with oxen; also a device placed on the necks of animals to keep them in a pasture or pen

Yolk—the center, yellow part of a chicken egg; provides nourishment for a developing embryo

Zoo—zoological garden; parklike area where animals are kept for viewing

Zoo curator—person trained in exotic animal zoology who looks after the animals in a zoo

Zygote—a fertilized ovum

Bibliography

Ackefors, Hans, Jay V. Huner, and Mark Konikoff. *Introduction to the General Principles of Aquaculture*. Binghamton, New York: The Hawarth Press, Inc., 1994.

American Kennel Club, The. *American Kennel Club Dog Care and Training*. New York: MacMillan Publishing Company, 1991.

American Kennel Club, The. *The Complete Dog Book*. New York: MacMillan Publishing Company, 1992.

Austic, R. E. and M. C. Nesheim. *Poultry Production*. Philadelphia: Lea and Febiger, 1990.

Baker, MeeCee and Robert E. Mikesell. *Animal Science Biology & Technology*. Danville, Illinois: Interstate Publishers, Inc., 1996.

Bearden, H. J. and J. W. Fuquay. *Applied Animal Reproduction*. Englewood Cliffs, New Jersey: Prentice-Hall, Inc., 1992.

Blasiola, George C. *The New Saltwater Aquarium Handbook*. Hauppauge, New York: Barron's Educational Services, Inc., 1991.

Burton, DeVere. *Ecology of Fish and Wildlife*. Albany, New York: Delmar Publishers, Inc., 1995.

Cheeke, Peter R. *Impacts of Livestock Production*. Danville, Illinois: Interstate Publishers, Inc., 1993.

Coppinger, Raymond P. "A Valuable Asset to the Small Farm," *Journal of Range Management*, April-May, 1988.

Crunkilton, John R., Susan L. Osborne, Michael E. Newman, Edward W. Osborne, and Jasper S. Lee. *The Earth and AgriScience*. Danville, Illinois: Interstate Publishers, Inc., 1995.

511

Ensminger, M. E. *Beef Cattle Science*. Danville, Illinois: Interstate Publishers, Inc., 1987.

Ensminger, M. E. *Horses and Horsemanship*. Danville, Illinois: Interstate Publishers, Inc., 1990.

Ensminger, M. E. *Animal Science*. Danville, Illinois: Interstate Publishers, Inc., 1991.

Ensminger, M. E. and R. O. Parker. *Sheep and Goat Science*. Danville, Illinois: Interstate Publishers, Inc., 1986.

Ensminger, M. E. *Poultry Science*. Danville, Illinois: Interstate Publishers, Inc., 1992.

Ensminger, M. E. *Dairy Cattle Science*. Danville, Illinois: Interstate Publishers, Inc., 1993.

Ensminger, M. E., J. E. Oldfield, and W. W. Heinemann. *Feeds & Nutrition Digest*. Clovis, California: Ensminger Publishing Company, 1990.

Etgeu, W. M., R. E. James, and P. M. Reaves. *Dairy Cattle: Feeding and Management*. New York: John Wiley and Sons, 1987.

Gillespie, James R. *Modern Livestock and Poultry Production*. Albany, New York: Delmar Publishers, Inc., 1992.

Graves, Will. *Raising Poultry Successfully*. Charlotte, Vermont: Williamson Publishing, 1985.

Green, Jeffrey, Roger A. Woodruff, and Robinette Harman. "Livestock Guarding Dogs and Predator Control: A solution or Just Another Tool?" *Rangelands Magazine*. April, 1984.

Herman, H. A., Jere R. Mitchell, and Gordon A. Doak. *The Artificial Insemination and Embryo Transfer of Dairy and Beef Cattle*. Danville, Illinois: Interstate Publishers, Inc., 1994.

Hieronimus, Harro. *Guppies, Mollies, Platys and other Live-bearers*. Hauppauge, New York: Barron's Educational Series, Inc. 1991.

Holland, Barbara. "An Exotic Creature Makes a Useful Change in Careers," *Smithsonian*. August, 1994.

Langdon, John. *Horses, Oxen and Technological Innovation*. New York: Syndicate of the University of Cambridge, 1986.

Lee, Jasper S. and Diana L. Turner. *Introduction to World AgriScience and Technology*. Danville, Illinois: Interstate Publishers, Inc., 1994.

Lee, Jasper S. and Michael E. Newman. *Aquaculture: An Introduction*. Danville, Illinois: Interstate Publishers, Inc., 1992.

Romans, John R., William J. Costello, C. Wendell Carlson, Marion L. Greaser, and Kevin W. Jones. *The Meat We Eat*. Danville, Illinois: Interstate Publishers, Inc., 1994.

Schraer, William D. and Herbert J. Stoltze. *Biology: The Study of Life*. Englewood Cliffs, New Jersey: Prentice Hall, 1995.

Smith, Donald J. *Horses at Work*. Wellingborough, Northants, England: Patrick Stephens Limited, 1985.

Spotte, Stephen. *Marine Aquarium Keeping*. New York: John Wiley & Sons, Inc., 1993.

Taylor, Robert E. *Scientific Farm Animal Production*. Englewood Cliffs, New Jersey: Prentice Hall, 1995.

Telleen, Maurice. *The Draft Horse Primer*. Emmaus, Pennsylvania: Rodale Press, 1971.

Towle, Albert. *Modern Biology*. Austin, Texas: Holt, Rinehart and Winston, 1993.

Warren, Dean M. *Small Animal Care and Management*. Albany, New York: Delmar Publishers, 1995.

Watson, Peter R. *Farming with Draft Animals*. Washington: Transcentry Corporation, 1981.

____. *Anyone for a Few Geese?* East Lansing: Michigan State University, Cooperative Extension Service, Bulletin E-1049, 1987.

____. *The Care of Dogs and Puppies*. Chevy Chase, Maryland, National 4-H Council, n.d.

____. *New Perspectives in Our Lives with Companion Animals*. Philadelphi: University of Pennsylvania Press, 1983.

____. *Pocket Pets Series: Gerbils, Guinea Pigs, Hamsters, and Rats and Mice*. East Lansing: Michigan State University, 4-H Youth Programs, n.d.

____. *Raising and Training a Livestock-Guarding Dog.* Corvallis: Oregon State University, 1986.

____. *Your Cat and You: A Complete Guide to Cat Selection and Care.* Washington: U. S. Department of Agriculture, Central Region Extension Publication, 1991.

____. *Animal Damage Control.* Washington: U. S. Department of Agriculture, Animal Plant Health and Inspection Service, 1991.

____. *Livestock Guarding Dogs: Protecting Sheep from Predators.* Washington: U. S. Department of Agriculture, Animal Plant Health and Inspection Service, 1991.

____. *Michigan Agriscience and Natural Resources Advanced Specialized Curriculum, Equine Science.* East Lansing: State of Michigan, Administrative Board, 1991.

____. *Freshwater and Marine Aquarium Fishes.* New York: Simon and Schuster, Inc., 1977.

____. *Sire Summaries: 1995.* Battleboro, Vermont: Holstein Association, 1995.

Index

515